PLASTICS PROCESSING
DATA HANDBOOK

PLASTICS PROCESSING DATA HANDBOOK

Donald V. Rosato, Ph.D.

Polysar of Nova, Inc.

Dominick V. Rosato, P.E.

Rhode Island School of Design
and
Plastics FALLO
Chatham, Mass.

VNR Van Nostrand Reinhold
New York

Copyright © 1990 by Van Nostrand Reinhold
Library of Congress Catalog Card Number 89-8991
ISBN 0-442-31869-3

Printed in the United States of America

Van Nostrand Reinhold
115 Fifth Avenue
New York, New York 10003

Van Nostrand Reinhold International Company Limited
11 New Fetter Lane
London EC4P 4EE, England

Van Nostrand Reinhold
480 La Trobe Street
Melbourne, Victoria 3000, Australia

Nelson Canada
1120 Birchmount Road
Scarborough, Ontario M1K 5G4, Canada

16   15   14   13   12   10   9   8   7   6   5   4   3   2   1

Library of Congress Cataloging-in-Publication Data

Rosato, Donald V.
    Plastics processing data handbook / by Donald V. Rosato and
Dominick V. Rosato.
        p.   cm.
    Bibliography: p.
    Includes index.
    ISBN 0-442-31869-3
    1. Plastics—Handbooks, manuals, etc.   I. Rosato, Dominick V.
II. Title.
TP1130.R67   1989
668.4—dc20   89-8991
                                                                      CIP

PREFACE

This comprehensive book provides guidelines for maximizing plastics processing efficiency in the manufacture of all types of products, using all types of plastics. A practical approach is employed to present fundamental, yet comprehensive, coverage of processing concepts. The information and data presented by the many tables and figures interrelate the different variables that affect injection molding, extrusion, blow molding, thermoforming, compression molding, reinforced plastics molding, rotational molding, reaction injection molding, coining, casting, and other processes.

The text presents a great number of problems pertaining to different phases of processing. Solutions are provided that will meet product performance requirements at the lowest cost. Many of the processing variables and their behaviors in the different processes are the same, as they all involve basic conditions of temperature, time, and pressure. The book begins with information applicable to all processes, on topics such as melt softening flow and controls; all processes fit into an overall scheme that requires the interaction and proper control of systems. Individual processes are reviewed to show the effects of changing different variables to meet the goal of zero defects. The content is arranged to provide a natural progression from simple to complex situations, which range from control of a single manual machine to simulation of sophisticated computerized processes that interface with many different processing functions.

Many of the procedures used in an individual process are identical or similar to those of other processes; so the reader interested in a process can gain a different perspective by reviewing the other processes. When an obvious correlation exists, a cross reference is made to the other processes. Other pertinent information provided here includes processing guidelines for troubleshooting, auxiliary equipment, testing procedures, and quality control.

This book may be used and understood by people in production, design, engineering, marketing, quality control, R&D, and management. Sufficient information is presented to ensure that the reader has a sound understanding of the principles involved, and thus recognizes what problems could exist—or, more important, how to eliminate or compensate for potential problems. Knowledge of all these processing methods, including their capabilities and limitations, helps one to decide whether a given product can be fabricated and by which process.

Both practical and theoretical viewpoints are presented. Persons dealing with practicalities will find the theoretical explanations enlightening. In turn, theorists will gain insight into the practical limitations of equipment and plastics (which are not perfect). The various process limitations are easily understood if properly and simply introduced to processors and potential processors.

Information contained in this book may be covered by U.S. and worldwide patents. No authorization to utilize these patents is given or implied; they are discussed for information only. Any disclosures are not a license to operate, or a recommendation to infringe, any patent. With few exceptions, no attempt has been made to refer to patents by number, title, or ownership.

Special acknowledgment must be made to Dr. Nick R. Schott for his extensive assistance and encouragement in the preparation of this book. Dr. Schott is Professor and Head of the Plastics Engineering Department at the University of Lowell, whose undergraduate program in plastics engineering was the first to be accredited in the United States. This department also offers master's degrees and doctorates in plastics engineering.

D. V. ROSATO

CONTENTS

PLASTICS PROCESSING DATA HANDBOOK

Chapter 1

OVERVIEW
OF PROCESSING OPERATIONS

INTRODUCTION

The various processes reviewed in this book are used to fabricate all types and shapes of plastic products, ranging from household convenience packages to electronic devices and many others—including the strongest products in the world, used in space vehicles, aircraft, building structures, and so on. Proper process selection depends upon the nature and requirements of the plastic, the properties desired in the final product, the cost of the process, its speed, and product volume. (Note that a plastic also may be called a polymer or a resin.) Some materials can be used with many kinds of processes; others require a specific or specialized machine. Numerous fabrication process variables play an important role and can markedly influence a product's aesthetics, performance, and cost.

This book will provide information on the effects on performance and cost of changing individual variables during processing, including upstream and downstream auxiliary equipment. Many of these variables and their behaviors are the same in the different processes, as they all relate to temperature, time, and pressure. This chapter contains information applicable to all processes characterized by certain common variables or behaviors, such as plastic melt flow, heat controls, and so forth. It is essential to recognize that for any change in a processing operation, there can be advantages and/or disadvantages. The old rule still holds: for every action there is a reaction. A gain in one area must not be allowed to cause a loss in another; changes must be made that will not be damaging in any respect.

All processes fit into an overall scheme that requires interaction and proper control of different operations. An example is shown in Fig. 1-1, where a complete block diagram pertains to a process. This FALLO (*Follow ALL Opportunities*) approach can be used in any process by including those "blocks" that pertain to the fabricated product's requirements.

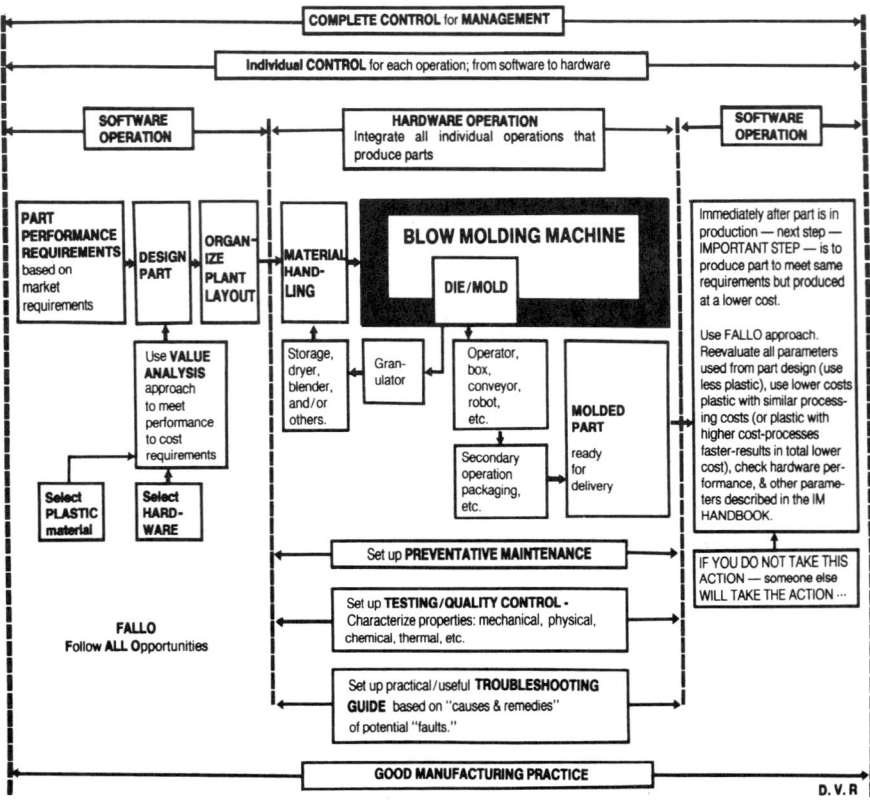

Fig. 1-1. Complete extrusion and injection blow molding operation; the FALLO approach.

The FALLO concept has been used by many manufacturers to produce acceptable products at the lowest cost. Computer programs featuring this type of layout are available. The FALLO approach makes one aware that many steps are involved in processing, and all must be coordinated. The specific process (injection, etc.) is an important part of the overall scheme and should not be problematic. The process depends on several interrelated factors, such as: (1) designing a part to meet performance and manufacturing requirements at the lowest cost; (2) specifying the plastic; (3) specifying the manufacturing process, which basically requires (a) designing a tool "around" the part, (b) putting the "proper performance" fabricating process around the tool, (c) setting up necessary auxiliary equipment to interface with the main processing machine, and (d) setting up "completely integrated" controls to meet the goal of zero defects; and (4) "properly" purchasing equipment and materials, and warehousing the materials (1-9).*

Worldwide plastics consumption is at least 125,000 million pounds (by weight). About 36 percent is processed by extruders, 32 percent by injection molding, 10 percent by blow molding, 6 percent by calenders, 5 percent in coatings, 3 percent in compression, 2 percent in powder form, and 6 percent using other processes. These percentages do not correlate with the number of machines used; for example, there are three times more injection machines than extruders.

Major advantages of using plastics include formability, consolidation of parts, and providing a low cost to performance ratio. For the majority of applications that require only minimum mechanical performance, the product shape can help to overcome the limitations of commodity resins such as low stiffness; here improved performance is easily incorporated in a process. However, where extremely high performance is required, reinforced plastics or composites are used (see Chapter 7).

Knowledge of all processing methods, including their capabilities and limitations, is useful to a processor in deciding whether a given part can be fabricated and by which process. Certain processes require placing high operating pressures on plastics, such as those used in injection molding, where pressures may range from 2,000 to 30,000 psi (13.8 to 206.9 Pa). Because of these pressures and the fact that three-dimensional parts are molded, injection molding is the most complex process; but it is easily controlled (see Chapter 2). Lower pressures are used in extrusion and compression, ranging from 200 to 10,000 psi (1.4 to 69 Pa); and some processes, such as thermoforming and casting, operate at relatively little pressure. Basically, with the higher pressures it is possible to develop tighter dimensional tolerances with higher mechanical performance; but there is also a tendency to develop undesirable stresses (orientations) if the processes are not properly understood or controlled. A major exception is reinforced plastics processing at low or contact pressures (see Chapter 7). Regardless of the process used, its proper control will maximize performance and minimize undesirable process characteristics (1-4, 7-13, 102-114).

COORDINATION OF PRODUCT REQUIREMENTS WITH MACHINE PERFORMANCE

Practically all processing machines can provide useful products with relative ease, and certain machines have the capability of manufacturing products

---

*Parenthetical numbers are used to cite references, which are gathered in a "References" section at the end of the book.

to very tight dimensions and performances. The coordination of plastic and machine facilitates these processes. This interfacing of product and process requires continual updating because of continuing new developments in manufacturing operations. The information presented in this book should make past, present, and future developments understandable in a wide range of applications.

Most products are designed to fit processes of proven reliability and consistent production. Various options may exist for processing different shapes, sizes, and weights (Table 1-1). Parameters that help one to select the right options are: (1) setting up specific performance requirements; (2) evaluating material requirements and their processing capabilities; (3) designing parts on the basis of material and processing characteristics, considering part complexity and size (Fig. 1-2) as well as a product and process cost comparison (Table 1-2); (4) designing and manufacturing tools (molds, dies, etc.) to permit ease of processing; (5) setting up the complete line, including auxiliary equipment (Fig. 1-1 and Chapter 9); (6) testing and providing quality control, from delivery of the plastics, through production, to the product (see Chapter 10); and (7) interfacing all these parameters by using logic and experience and/or obtaining a required update on technology.

PROCESSING FUNDAMENTALS

Polymers usually are obtained in the form of granules, powder, pellets, and liquids. Processing mostly involves their physical change (thermoplastics),

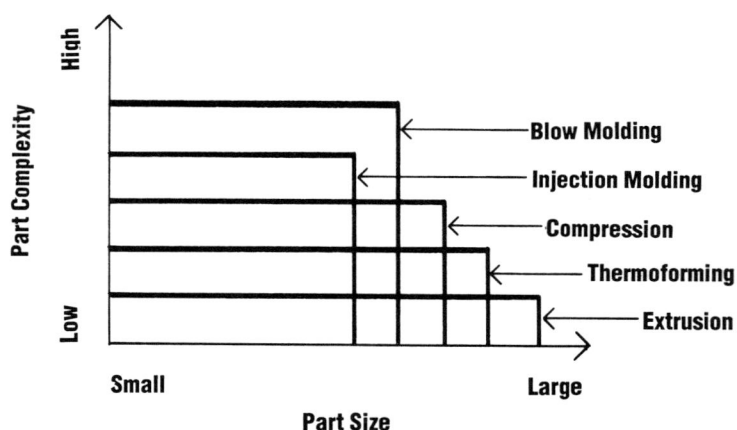

Fig. 1-2. Processing characteristics.

Table 1-1. Competitive Processes.

	INJECTION MOLDING	EXTRUSION	BLOW MOLDING	THERMO-FORMING	REACTION INJECTION MOLDING	ROTATIONAL MOLDING	COMPRESSION AND TRANSFER MOLDING	MATCHED MOLD, SPRAY UP
Bottles, necked containers, etc.	2, A		1	2, A		2		2
Cups, trays, open containers, etc.	1			1	1		1	2
Tanks, drums, large hollow shapes, etc.			1	2, A		1		2
Caps, covers, closures, etc.	1			2	2		1	1
Hoods, housings, auto parts, etc.	1		2	2	2		1	
Complex shapes, thickness changes, etc.	1						1	2
Linear shapes, pipe, profiles, etc.	2, B	1					2, B	
Sheets, panels, laminates, etc.		1, C					2	2

1. Prime process.
2. Secondary process.
A. Combine two or more parts with ultrasonics, adhesives, etc. (see Chapter 9).
B. Short sections can be molded.
C. Also calendering process.

Table 1–2. Cost Comparison of Plastic Products and Different Processes (Cost Factor × Material Cost = Purchased Cost of Product).

	COST FACTOR	
PROCESS	OVERALL	AVERAGE
Blow molding	$1\frac{1}{16}$ to 4	$1\frac{1}{8}$ to 2
Calendering	$1\frac{1}{2}$ to 5	$2\frac{1}{2}$ to $3\frac{1}{2}$
Casting	$1\frac{1}{2}$ to 3	2 to 3
Centrifugal casting	$1\frac{1}{2}$ to 4	2 to 4
Coating	$1\frac{1}{2}$ to 5	2 to 4
Cold pressure molding	$1\frac{1}{2}$ to 5	2 to 4
Compression molding	$1\frac{3}{8}$ to 10	$1\frac{1}{2}$ to 4
Encapsulation	2 to 8	3 to 4
Extrusion forming	$1\frac{1}{16}$ to 5	$1\frac{1}{8}$ to 2
Filament winding	5 to 10	6 to 8
Injection molding	$1\frac{1}{8}$ to 3	$1\frac{3}{16}$ to 2
Laminating	2 to 5	3 to 4
Match-die molding	2 to 5	3 to 4
Pultrusion	2 to 4	2 to $3\frac{1}{2}$
Rotational molding	$1\frac{1}{4}$ to 5	$1\frac{1}{2}$ to 3
Slush molding	$1\frac{1}{2}$ to 4	2 to 3
Thermoforming	2 to 10	3 to 5
Transfer molding	$1\frac{1}{2}$ to 5	$1\frac{3}{4}$ to 3
Wet lay-up	$1\frac{1}{2}$ to 6	2 to 4

though in some cases chemical reactions occur (thermosts). A variety of processes are used. One group consists of the extrusion processes (pipe, sheet, profiles, etc.). A second group takes extrusion and in certain cases injection molding through an additional processing stage (blow molding, blown film, quenched film, etc.). A third group consists of injection and compression molding (different shapes and sizes), and a fourth group includes various other processes (thermoforming, calendering, rotational molding, etc.).

The common features of these groups are: (1) mixing, melting, and plasticizing; (2) melt transporting and shaping; (3) drawing and blowing; and (4) finishing. Mixing, melting, and plasticizing produce a plasticized melt, usually made in a screw (extruder or injection). (See Chapter 2 on the heating/shearing action required to obtain the melt.) Melt transport and shaping involve applying pressure to the hot melt in order to move it through a die or into a mold. The drawing and blowing technique stretches the melt to produce orientation (see Chapter 3) of the different shapes (blow molding, forming, etc.). The final feature of processing, finishing, is the usual solidification of the melt.

The most common feature of all processes is deformation of the melt with its flow, which refers to rheology. Another feature is heat exchange, which involves the study of thermodynamics. Changes in a plastic's molecular structure are chemical. These properties are reviewed briefly in the following paragraphs, and will be discussed in detail throughout the book, with a focus on how they influence processes.

Rheology

The flow of plastics is compared to that of water in Fig. 1–3, to show their different behaviors. With plastics there are two types of deformation or flow: viscous, in which the energy causing the deformation is dissipated, and elastic, in which that energy is stored. This combination produces viscoelastic plastics.

Viscosity is a material's resistance to viscous deformation (flow). Its unit of measure is Pascals·second (Pa·s) or pounds·second/in^2 (lb·s/in^2). Plastic melt viscosities have a range from 2 to 3,000 Pa·s (glass 10^{20}, water 10^{-1}). The resistance to elastic deformation is the modulus of elasticity (E), which is measured in Pascals (Pa) or pounds per square inch (psi); its range for a plastic melt is 1,000 to 7,000 kPa (145 to 1,015 psi), which is called the rubbery range (Fig. 1–4).

Not only are there two classes of deformation; there are also two modes in which deformation can be produced: simple shear and simple tension. The actual action during melting, as in a screw plasticator, is extremely complex, with all types of shear–tension combinations. Together with engineering design, deformation determines the pumping efficiency of a screw

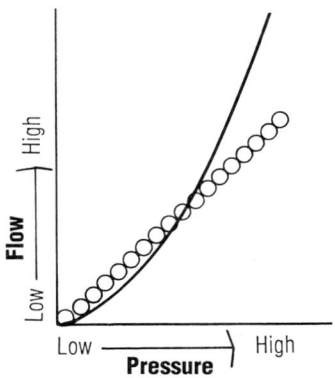

Fig. 1–3. Rheology/flow properties of plastics.

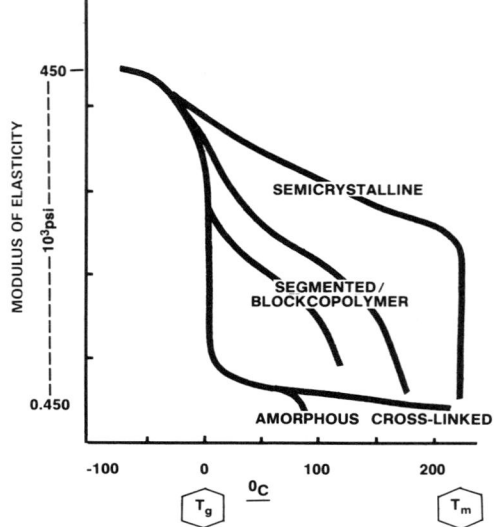

Fig. 1–4a. Example of the dynamic-mechanical properties of plastics.

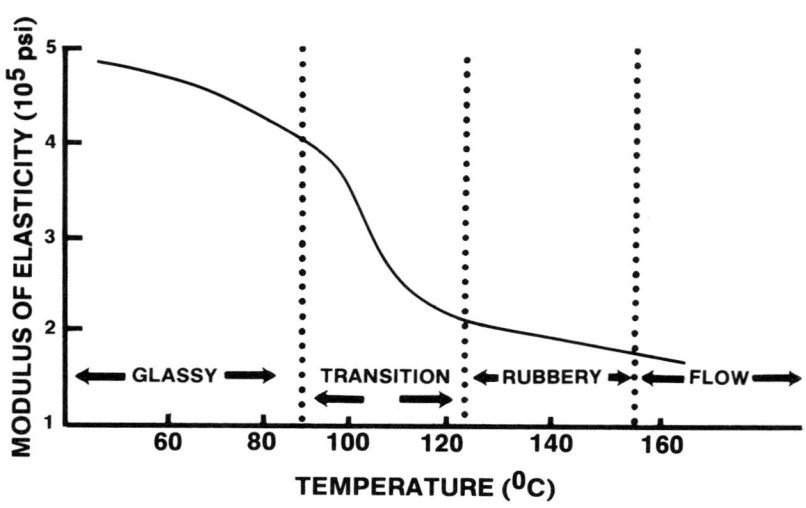

Fig. 1–4b. Example of the modulus of elasticity of plastics.

plasticator and controls the relationship between output rate and pressure drop through a die system or into a mold.

Shear Rate

When a melt moves in a direction parallel to a fixed surface, such as a screw barrel, mold runner, or die wall, it is subject to a shearing force. As the screw speed increases, so does the shear rate, with potential advantages and disadvantages. Advantages of an increased shear rate are a less viscous melt and easier flow. This shear-thinning action is required to "move" plastic. When water (a Newtonian liquid) is in an open-ended pipe, pressure can be applied to move it; and by doubling the water pressure, the flow rate of the water is doubled. Water does not have a shear-thinning action. However, in a similar situation but using a plastic melt (a non-Newtonian liquid), if the pressure is doubled, the melt flow may increase from 2 to 15 times, depending on the plastic used. As an example, linear low density polyethylene (LLDPE), with a low shear-thinning action, experiences a very low rate increase, which explains why it can cause more processing problems compared to other PEs in certain equipment. The higher-flow melts include polyvinyl chloride (PVC) and polystyrene (PS).

A disadvantage observed with the higher shear rates is that too high a heat increase may occur, potentially causing problems in cooling, as well as degradation and discoloration. A high shear rate can lead to a rough product surface (melt fracture, etc.). For each plastic and every processing condition, there is a maximum shear rate beyond which such problems can develop.

Shear in the channel of the screw is equal to $\pi DN/60h$ (where D = average barrel inside diameter, N = screw RPM, and h = average screw channel depth). This formula does not include the melt slippage between the barrel wall and screw surfaces, but the shear rate obtained is still useful for purposes of comparison. A $2\frac{1}{2}$ in. screw with a 0.140 in. channel rotating at 100 RPM results in a shear rate of 93.5 reciprocal seconds (rsec). This value is approximately the desired value in most extrusion processes, with 100 rsec generally the target.

The same formula can be used to determine the shear rate of the slippage between the barrel and screw. With a new barrel, which usually has a small clearance of 0.005 in., a very high shear rate of about 2,618 rsec can exist. With this small clearance only a small amount of melt is subject to the higher heat, so that any overheating is overcome by the mass of melt it encounters (mixes with). As the screw wears, more melt flows through enlarged clearances, but the shear rate is lower. The effect of wear on overheating is usually very small and is not the main reason why the complete melt overheats.

Shear rates also can be determined in melt flow through mold cavities and particularly in extrusion dies. The formulas applicable to the different-shaped dies usually do not account for slippage of melt on the die surfaces, but they can be used to compare the processability of melts and to control melt flow. The formula for a die extruding a rod is $4Q/\pi R^3$, for a long slit it is $6Q/wh^2$, and for an annulus die it is $6Q/\pi Rh^2$ (where Q = volumetric flow rate, R = radius, w = width, and h = die gap).

Molecular Weight Distribution

Plastics are made up of molecules arranged in long flexible chains. These chains become entangled with each other, and these entanglements are largely responsible for high viscosity in melts. Shear can be envisioned as sliding molecules in rotation; this rotation causes the chains to disentangle. At low shear, molecular chains become entangled; as the shear rate increases, they gradually disentangle, and the viscosity is reduced. The result, expressed as a so-called flow curve (Fig. 1-3), is related to the processability of the plastic material.

One method of defining plastics uses their molecular weight (MW), a reference to the plastic molecules' weight/size. Here MW refers to the average weight of a plastic, which is always composed of different-weight molecules. These differences are important to the processor, who uses molecular weight distribution (MWD) to evaluate materials. A narrow MWD enhances the performance of plastic products (to be discussed later). Melt flow rates are dependent on the MWD, as illustrated in Fig. 1-5.

Elasticity

As a melt is subjected to a fixed stress (or strain), the deformation vs. time curve will show an initial rapid deformation followed by a continuous flow (Fig. 1-6). The relative importance of elasticity (deformation) and viscosity (flow) depends on the time scale of the deformation. For a short time, elasticity dominates; over a long time, the flow becomes purely viscous. This behavior influences processes: when a part is annealed, it will change its shape; or, with post-extrusion (Chapter 5), swelling occurs. Deformation contributes significantly to process flow defects. Melts with small deformation have proportional stress–strain behavior. As the stress on a melt is increased, the recoverable strain tends to reach a limiting value. It is in the high-stress range, near the elastic limit, that processes operate.

Molecular weight, temperature, and pressure have little effect on elastic-

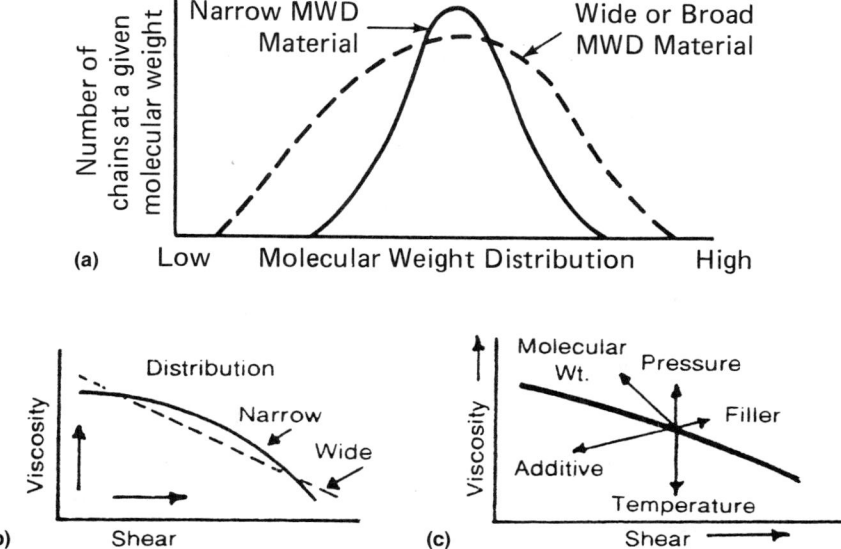

Fig. 1–5. (a) Molecular weight distribution (MWD) curves. (b) Viscosity vs. shear rate as related to MWD. (c) Factors influencing viscosity.

ity; the main controlling factor is MWD. Practical elasticity phenomena often exhibit little concern for the actual value of the modulus and the viscosity. Although the modulus is influenced only slightly by MW and temperature, these parameters have a great effect on viscosity and thus can alter the balance of a process.

Flow Performance

In any practical deformation there are local stress concentrations. Should the viscosity increase with stress, the deformation at the stress concentration will be less rapid than in the surrounding material; the stress concentration will be smooth, and the deformation will be stable. However, when the viscosity decreases with increased stress, any stress concentration will cause catastrophic failure.

Flow Defects

Flow defects, especially as they affect the appearance of a product, play an important role in many processes. Flow defects are not always undesirable,

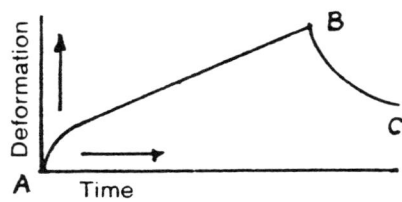

(a)

A-B: Viscoelasticity with slow deformation
B: Load removed
B-C: Viscoelastic recovery

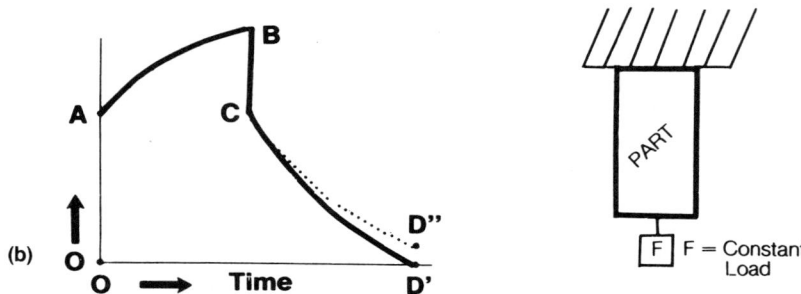

(b)

O-A: *Instantaneous loading* produces *immediate strain.*
A-B: *Viscoelastic deformation* (or creep) gradually occurs with sustained load.
B-C: Instantaneous *elastic recovery* occurs when load is removed.
C-D: Viscoelastic recovery gradually occurs; where no permanent deformation (D') o˙ ˙vith a permanent deformation (D''-D'). Any permanent deformation is related to type plastic, amount & rate of loading and fabricating procedure.

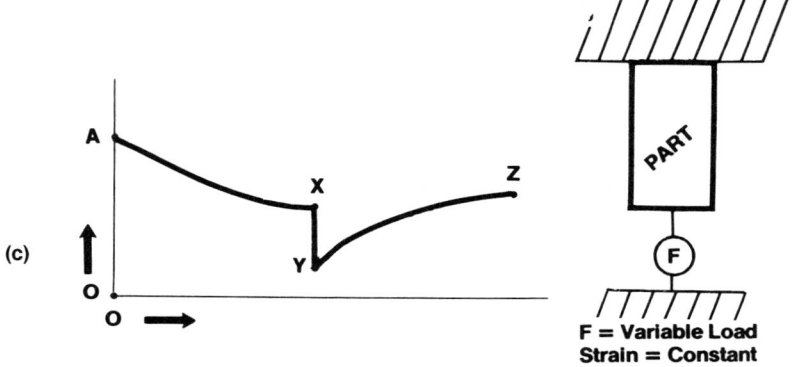

(c)

O-A: *Instantaneous loading* produces *immediate strain.*
A-X: With strain maintained gradual *elastic relaxation* occurs.
X-Y: *Instanteous deformation* occurs when load is removed.
Y-Z: *Viscoelastic deformation* gradually occurs as residual stresses are relieved. Any permanent deformation is related to type plastic, amount & rate of loading and fabricating procedure.

as, for example, in producing a matt finish. Six important types of defects can be identified, and are applied here to extrusion because of its relative simplicity. These flow analyses can be related to other processes; and even to the complex flow of injection molding.

Nonlaminar Flow. Ideally, a melt flows in a steady, streamlined pattern in and out of a die. Actually, the extrudate is distorted, causing defects called melt fracture or elastic turbulence. To reduce or eliminate this problem, the entry to the die is tapered or streamlined.

Sharkskin. During flow through a die, the melt next to the die tends not to move, whereas that in the center flows rapidly. When the melt leaves the die, its flow profile is abruptly changed to a uniform velocity. This change requires a rapid acceleration of the surface layer, resulting in a high local stress. If this stress exceeds some critical value, the surface breaks, giving a rough appearance (sharkskin). With the rapid acceleration, the deformation is primarily elastic. Thus the highest surface stress, and worst sharkskin, will occur in plastics with a high modulus and high viscosity, or in high-molecular-weight plastics of narrow MWD at low temperatures and high extrusion rates. The addition of die lip heating, locally reducing the viscosity, is effective in reducing sharkskin.

Nonplastication. This condition produces uneven stress distribution, with consequent lumpiness. The product could appear ugly or have a fine matt finish. With a wide MWD, there could be a lack of gloss.

Volatiles. Many plastics contain small quantities of material that boil at processing temperatures; or, they may be contaminated by water absorbed from the atmosphere. These volatiles may cause bubbles, a scarred surface, and other defects. See Chapters 2 and 9 for methods of removing volatiles (vented barrels and dryers).

Shrinkages. The transition from room temperature to a high processing temperature may decrease a plastic's density up to 25 percent. Cooling causes possible shrinkage (up to 3 percent) and may cause surface distortions or voiding with internal frozen strains. As reviewed in other chapters, this situation can be reduced or eliminated by special techniques, such as cooling under pressure in the injection molding process.

Melt Structures. High shear at a temperature not far above the melting point may cause a melt to take on too much molecular order. In turn, distortion could result. This subject is discussed further in this chapter in the section "Effect of Process on Material Properties."

Fig. 1-6. (a) Basic deformation vs. time curve. (b) Stress–strain deformation vs. time (creep effect). (c) Strain–stress deformation vs. time (stress-relaxation effect).

Thermodynamics

With the heat exchange that occurs during processing, thermodynamics becomes important. It is the high heat content of melts (about 100 cal/g) combined with the low rate of thermal diffusion (10^{-3} cm^2/s) that limits the cycle time of many processes. Also important are density changes, which, for crystalline plastics, may exceed 25 percent as melts cool. Melts are highly compressible; a 10 percent volume change for 10,000 psi force (700 kg/cm^2) is typical. Surface tension of about 20 g/cm may be typical for film and fiber processing when there is a large surface-to-volume ratio.

Chemical Changes

The chemical changes that can occur during processing include: (1) polymerization and cross-linking, which increases viscosity; (2) depolymerization or damaging of molecules, which reduces viscosity; and (3) complete changes in the chemical structure, which may cause color changes. Already degraded plastics may catalyze further degradation.

Trends

Because melts have many different properties and there are many ways to control processes, detailed and factual predictions of final output are difficult. Research and hands-on operation have been directed mainly at explaining the behavior of melts or plastics. Modern equipment and controls are overcoming some of the unpredictability. Ideally, processes and equipment should be designed to take advantage of the novel properties of plastics rather than to overcome them.

PROCESS CONTROL AND INSTRUMENTATION

Adequate process control and its associated instrumentation are essential for product quality control. The goal in some cases is precise adherence to a control point. In other cases, maintaining the temperature within a comparatively small range is all that is necessary. For effortless controller tuning and lowest initial cost, the processor should select the simplest controller (of temperature, time, pressure, melt flow rate, etc.) that will produce the desired results. This section primarily will review temperature control, which is applicable to all the processes, as it is usually the most critical parameter. The basic approach described here can be applied to the other required controls (2).

Temperature Control

Heat is usually applied in various amounts and in different locations, whether in a metal plasticating barrel (extrusion, injection molding, etc.) or in a metal mold/die (compression, injection, thermoforming, extrusion, etc.). With barrels a thermocouple is usually embedded in the metal to send a signal to a temperature controller. In turn, it controls the electric power output device regulating the power to the heater bands in different zones of the barrel. The placement of the thermocouple temperature sensor is extremely important. The heat flow in any medium sets up a temperature gradient in that medium, just as the flow of water in a pipe sets up a pressure drop, and the flow of electricity in a wire causes a voltage drop.

Barrels are made of steel, which is not a particularly good conductor of heat (being ten times worse than copper). Thus there is a gradient in the steel barrel from the outside of the barrel to the inside next to the plastic. In $3\frac{1}{2}$ in. (88.9 mm) and $4\frac{1}{2}$ in. (114.3 mm) extruder barrels, these gradients or differences in temperature can routinely be 75 to 100°F (23.9 to 32.8°C) or more, as the zone heaters pump in heat or zone coolers take excess heat out. Yet, for years users routinely accepted extruders with sensors mounted in very shallow wells, or, even worse, mounted in the heating/cooling jacket.

Consider a barrel with a shallow well for its sensor. Assume a perfect temperature controller set at 400°F (204°C). There is a 75°F gradient from the outside to the inside of the barrel; thus the actual temperature down near the plastic would be 325°F with the sensor set at 400°F. If the extruder started to generate too much heat, the temperature could reach 475°F before the sensor detected the increase. With this on–off control action, even with the controller set at 400°F the plastic temperature variation is 150°F. The result could be poor product performance and increased cost to process the plastic.

A deep well sensor will respond much more quickly than a shallow one to changes in the plastic's temperature. However, it responds slowly to changes, for example, in the heater line voltage or in the cooling water heat. The time constant for heat to propagate from the heater down to a deep well location is about 6 minutes in a $3\frac{1}{2}$ in. barrel. Thus an upset due to a cooling water temperature change might take 20 minutes or more to settle out. This system does respond to ambient conditions rapidly, but it retains part of the temperature error inherent in the use of shallow wells. In the example just given for a shallow well with 150°F variation, the variation would be only half as great, or 75°F, if two sensors were used, one deep and one shallow.

The DUO-Sense process (Holton/Harrel Inc., U.S. Patent 4,272,466 June 9, 1981) solved this problem, retaining the advantages of both deep and shallow wells by using a cascade control loop. The primary temperature loop is a shallow well, and a secondary loop senses the deep well temperature, using it to adjust the set point of the shallow well. This system offers such advantages as preventing the temperature of the heater from rising as high as it otherwise would, greatly extending the heater band life, and so on.

These on–off controllers are unsatisfactory for a loading having a long time constant, such as an extruder barrel, a die adapter, a die, and so forth. The temperature will oscillate violently at an amplitude that is set not by the characteristic of the controller, but by the delay in the load, as reviewed in Fig. 1–7. To reduce this variation, a proportional control was developed. It is similar to the on–off, but operates in between full on and off, with its output proportional to the deviation of temperature from the set point value (Fig. 1–8). Variations still exist with this system, but they are less than those of the on–off control.

This type of temperature controller has three characteristics: (1) the actual temperature of a single proportional controller will never be at the set point; (2) the error in temperature, or droop, of a proportional controller, will vary over a considerable portion of a proportional band as the process varies; and (3) in the case of a large time lag, the proportional band of a simple proportional controller will have to be quite large. A considerable portion of the proportional band will normally be used; so the temperature will vary considerably during normal operation of the extruder. Thus a sim-

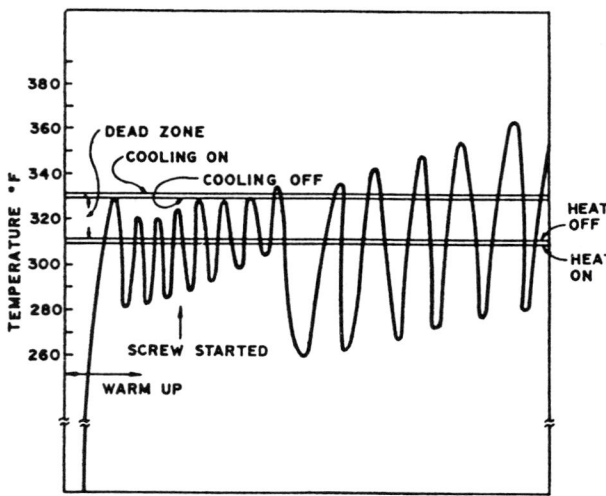

Fig. 1–7. Temperature variations with time in a typical extruder barrel using on–off controls.

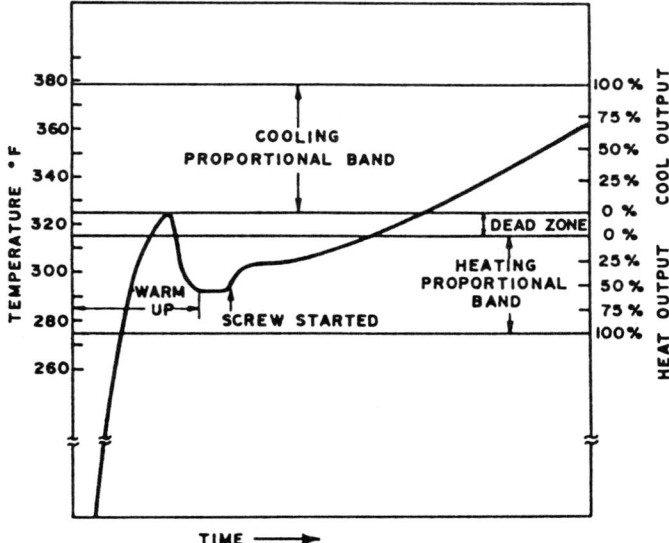

Fig. 1-8. Example of temperature variation with proportional control of an extruder barrel (no automatic reset).

ple proportional controller is better than an on–off control but does not do the best job of controlling temperature.

The introduction of automatic reset into controllers for the plastic processor made it possible to hold the temperature constant even in the presence of extremely long lags. Automatic reset is a characteristic added to a proportional controller that functions as an integrating, or averaging, system, looking at the droop, or temperature error, over a period of time and adjusting the output so that the droop goes to zero. As a result, the actual temperature goes to the set point (Fig. 1-9). Automatic reset is almost always used with an additional "rate" term, which adds an anticipatory characteristic that does not affect steady state performance but does speed up the response to changes in operating conditions. A modern proportional plus automatic reset plus rate—a three-mode controller—is capable of controlling within 1°F (0.6°C) of the set point all the way from full heating to full cooling, even when controlling from a deep well sensor.

Temperature Sensors

Two common type of sensors are used for temperatures in the ranges experienced in plastic processing equipment: the thermocouple (TC) and the

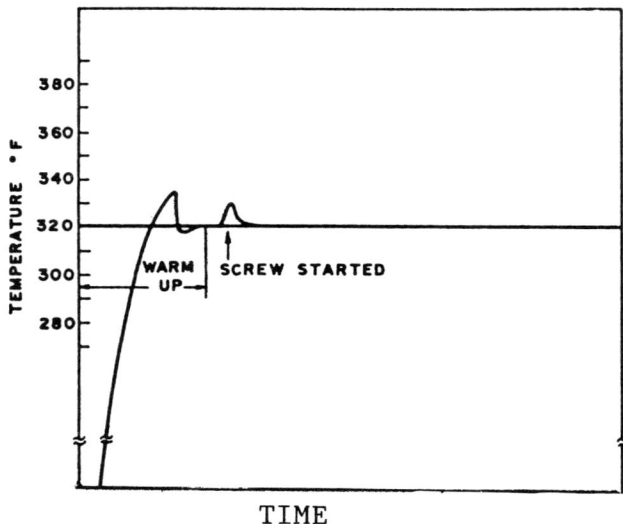

TIME

Fig. 1-9. Variation of temperature in an extruder barrel with time, using proportional plus automatic reset control.

resistance temperature detector (RTD). The TC is by far the more common type. It depends on the fact that every type of metallic conductor has a characteristic electrical barrier potential, and whenever two different metals are joined together, there will be a net electrical potential at the junction. This electrical potential changes with temperature.

The RTD sensor is based on the fact that the resistance of some metals changes markedly with temperature, whereas the resistance of platinum, the metal most commonly used in RTDs, is extremely stable. Its variation in temperature is both repeatable and predictable to a high degree of accuracy. In the past, TCs offered major cost advantages; but with the advent of low-cost solid state dc amplifiers, the use of RTDs has become more realistic.

RTDs have about 60 times higher sensitivity than TCs, their amplifiers are less expensive and much less sensitive to electrical noise disturbance, they offer better linearity (are twice as linear as TCs), and they are twice as interchangeable. The RTD does not have the TC's compensating cold junction; so only the desired temperature is involved. With TCs both ends of the wire are sensitive to temperature changes; there is no way of distinguishing between a change in the process and a change in the ambient temperature, so there is some residual drift. Although the RTD itself costs more than a TC, an RTD system that includes the sensor plus an amplifier is almost always less expensive for an equivalent quality level. Processors should be aware of the availability and superiority of RTDs (2).

Melt Temperature Profile

Usually the melt temperature only is taken or estimated from the inside of the barrel or the surface of the melt as it moves through the barrel. Various techniques can be used (such as IR sensors) that look at melt temperatures across the entire melt stream, as, for example, when it exits an extruder (or when it exits an injection molding nozzle into space, etc.). An automatic thermocouple system (patented by AutoProbe, Normag Corp., Hickory, NC) has a motor-driven, retractable melt thermocouple, which moves across the melt stream while simultaneously displaying temperature and temperature profile position (10). The system shows that temperature variations within the melt stream can be considerably wider than expected.

It had been generally accepted by most extrusion processors and suppliers that the melt temperature variance at the end of an extruder was negligible. Stationary thermocouples had been immersed in melts, but very limited useful data could be obtained, as probes tended to disturb the melt flow or be damaged. Obtaining the profile with a standard immersion thermocouple required that an operator position the probe manually, plot the position, and so on. Results were not repeatable, or were tentative at best.

Temperatures of the automatic retractable TC have ranged from 402°F on the melt channel wall to 325°F in the center of the melt stream. Melt flowed through a one-inch melt pipe processing LDPE with a melt index of 2. The flow rate was 1,000 lb/hr. The temperature profile was computer-generated with 20 separate readings across the melt stream.

Automatic Tuning

There is one major disadvantage to using an automatic reset barrel temperature controller: the coefficients of the proportional, the reset, and the rate terms all have to be adjusted properly to obtain desired performance. It is not difficult to do this, but it can be time-consuming. One must follow the manufacturer's instructions.

Timing and Sequencing

Most processes operate more efficiently when functions must occur in a desired time sequence or at prescribed intervals of time. In the past, mechanical timers and logic relays were used. Now electronic logic and timing devices predominate, based on the so-called programmable logic controller. These devices provide sophisticated sequencing and timing, and lend themselves easily to reprogramming if it becomes necessary. Different suppliers provide special consoles that can be plugged in, and logic sequences can be added by means of "ladder diagrams" representing the desired functions and/or timing.

Screw Speed Control

Many processes require speed controls. Performance and reliabilities of these controls are very similar to those of the temperature controls—you get what you purchase. Early speed controllers, like the temperature controllers, were mechanical. Speeds were held within 5 percent, resulting in poor plastic melt control. Where better speed control is desired, the solution is the same as in temperature control; only the equipment names are changed. A device is added to the motor, and an "integral" characteristic is provided, corresponding to the automatic reset with heat. It brings the speed closer to the set point. A "derivative" characteristic, corresponding to the rate in heat, ensures a prompt response to any upsets.

The arguments for the use of integral or derivative control of speed are the same as for temperature. Different systems are available, the latest being the all-digital speed control on plastic processing machines that require speed control. These speed controls permit accuracies of 0.5 percent or less. An all-digital phase locked loop system permits all motors in a machine and/or a processing line to be synchronized with each other exactly or in a desired speed ratio, just as if they were mechanically geared together.

Weight, Thickness, and Size Controls

By controlling important factors such as final product weight, thickness, and/or size, enhanced product performance and a definitely lower product cost can be achieved. Many different systems are used to ensure product performance (from different suppliers); so product controls will directly influence process control (1–4). In controlling the feeding in many processes, it is very important to accurately meter material by volume and/or by weight (see Chapter 9).

Microprocessor Control Systems

Many microprocessor systems are available to control all of the different sequences required, including those described in this book for all the different processes. Some are dedicated and have been designed for specific processes. All these systems consist of electronic devices that can receive and store all types of information (machine operation, melt process, materials handling, and other upstream activities, as well as downstream operations—takeoff, weighing, quality control, decorating, packaging, etc.—and others). Some microprocessors can be programmed to make decisions on inputs and/or machine operation responses. These devices have memories that can store and retrieve all the pertinent data.

Today's microprocessors react and scan inputs quickly (10 ms), in a time that includes mathematical analysis, and tomorrow's systems will do even more; so it is important to consider using microprocessors that can be easily updated and integrated with new developments that will improve performance and cost of products.

Problems in Process Control

Purchasing a sophisticated process control system is not a foolproof solution that will guarantee perfect parts. Solving problems requires a full understanding of their causes, which may not be as obvious as they first appear. Failure to identify contributing factors when problems arise can easily result in the microprocessor's not doing its job. The conventional place to start troubleshooting a problem is with the basics—temperature, time, and pressure requirement limits. Often, a problem may be very subtle, such as a faulty control device or an operator making random control adjustments. Process controls cannot usually compensate for such extraneous conditions; however, if desired, they may be included in a program that provides the capability to add functions as needed.

Most controls, particularly the older ones, are the open loop type. They merely set mechanical or electrical devices to some operating temperature, time, and pressure. If this is all that is required, the control may remain in operation. However, this setup is subject to a variety of hard-to-observe disturbances that are not compensated by open loop controls. Thus process control must close the loop to eliminate the effect of process disturbances.

There are two basic approaches to problem solving: (1) find and correct the problem, applying only the controls needed; or (2) overcome the problem with an appropriate process control strategy. The approach one selects depends on the nature of the processing problem, and whether enough time and money are available to correct it. Process controls may, in most cases, provide the most economical solution. To make the right decision, one must systematically measure the magnitude of the disturbances, relate them to product quality, and identify their cause so that proper control action can be taken.

Before investing in a more expensive system, the processor should methodically determine the exact nature of the problem, to decide whether or not a better control system is available and will solve it. For example, the temperature differential across a mold (for injection molding, etc.) can cause uneven thermal mold growth. The mold growth also can be influenced by uneven heat on tie-bars (uppers can be hotter, causing platens to bend—a change that could be reflected on the mold). Once the cause is determined, appropriate corrective action can be taken. (For example, if the mold heat has varied, perhaps all that is needed is to close a large garage door nearby, to eliminate the flow of air on the mold that has caused the problem; or it may be necessary just to change the direction of flow from an air conditioning duct.)

EFFECT OF PROCESS ON MATERIAL PROPERTIES

Plastics comprise a large and diverse class of materials that, like metals and other materials, possess a very broad range of properties and processing characteristics (Fig. 1-10). As is true of any materials, the advantages and disadvantages of plastics properties and processing should be understood if they are to be used intelligently. There are two forms of plastics: thermoplastics (TPs) and thermosets (TSs) (Fig. 1-11). Thermoplastics repeatedly soften when heated and harden when cooled. Based on their molecular structure, TPs melt, and their performance and behavior change (this behavior is reviewed later in this chapter). Thermosets go through a soft plastic stage only once, then harden irreversibly and cannot be softened again, as a chemical action occurs during the heat cycle.

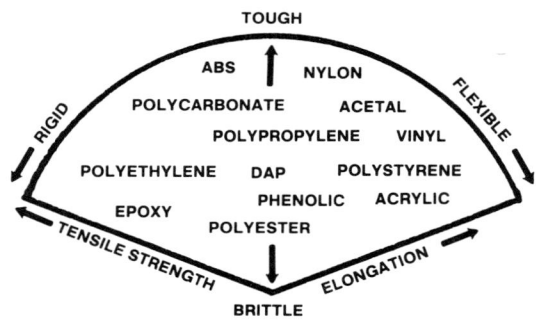

Note: With formulation changes (via additives, fillers, reinforcements, alloying, etc.) position of plastic can move practically any place in the "pie."

Fig. 1-10. Example of range of mechanical properties of plastics.

Thermoplastic:

These plastics become soft when exposed to sufficient heat and harden when cooled, no matter how often the process is repeated.

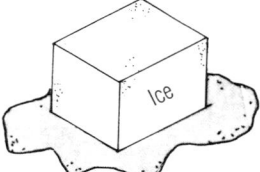

Thermosetting:

The plastics materials belonging to this group are set into permanent shape when heat and pressure are applied to them during forming. Reheating will not soften these materials.

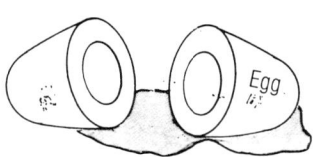

Fig. 1-11. Characteristics of thermoplastics (TPs) and thermosets (TSs).

The dividing line between TP and TS is not always distinct. Cross-linked TSs are thermoplastic during the initial heat cycle and prior to the chemical cross-linking action. Other plastics may behave like crossed-link high density polyethylene, which normally is a TP but can be cross-linked, by high energy radiation or chemically while being processed, to become a TS called XHDPE (1-3, 8, 9, 11).

By using different modifiers, additives, fillers, and reinforcements with the various plastics, more than 15,000 compounds are commercially available. They are classified as commodity resins (90 percent) and engineering resins. Commodities such as PE, PP, PS, and PVC account for two-thirds of plastics sales. Engineering resins are characterized by better heat resistance, higher impact strength, high stiffness, and/or many other "improved" properties, and thus bring a higher price than commodity resins. Among the more significant engineering resins are PA, PC, PS, PEEK, ABS, and so on. Some commodity resins contain certain reinforcements, and/or are alloyed with other resins, that put them into the engineering categories. Perhaps the major distinction between them is cost. Table 1–3 lists various plastics and their abbreviations.

For each family of plastics, such as polyethylene, there are many different grades; and each grade is tailored to provide certain performance and/or process characteristics. This diversity allows the processor and product designer considerable freedom in selecting a particular grade. Actually, only a few dozen important families of plastics are used and blended into different grades (Table 1–3); so rather than 15,000 grades, only a few hundred are the big sellers. The others usually meet speciality requirements, and may be of limited use. Plastic selection should be based on the performance required and process capability. Just as there are guidelines in this book for selecting the process, there are guidelines in the references for selecting the plastic (3, 12–21). Even though the selection can be complex because of the many variations that are available, a logical approach will provide the answer. However, compromises or tradeoffs are inevitable when one is dealing with the complex but controllable (within limits) operation of processing plastics.

Relationship of Processing to Molecular Weight of Plastics

As there are many different plastics, a number of techniques for defining and quantifying their characteristics exist. Important techniques that relate to processing will be reviewed in this book. Molecular weight (MW), mentioned earlier, relates to the size of the molecules that make up a resin. These molecules are not of the same length or weight, and MW has a significant effect on processability and performance. Resins with low MWs are easier to process but are weaker and more brittle than those with high MWs. The latter are tougher, more chemically resistant, and so on, and require tighter process controls. Generally processing the higher-MW plastics requires more energy in the form of temperature and pressure (1–3, 8–30).

The molecular weight distribution (MWD), also discussed earlier, is an indication of the relative proportions of molecules of different weights/ lengths. It shows the breadth of distribution, or the ratio of large, medium, and small molecular chains in the resin. If most of the molecules have about the same MW, the MWD is classified as "narrow." A "wide or broad" MWD implies a large variation in MW. Figure 1–5 compares wide and narrow MWDs. The MWD is independent of both density and melt index (MI) (see Chapter 10), and must be taken into account in considerations of both processing and product performance. A narrow MWD enables much better,

Table 1-3. Types of Plastics.

Acetals
Acrylics
  Polymethylmethacrylate (PMMA)
  Polyacrylonitrile (PAN)
Alkyds
Allyl diglycol carbonate (CR-39)
Allyls
  Diallyl phthalate (DAP)
  Diallyl isophthalate (DAIP)
Amino
  Urea formaldehyde (UF)
  Melamine formaldehyde (MF)
Cellulosics
  Cellulose acetate (CA)
  Cellulose acetate propionate (CAP)
  Cellulose acetate butyrate (CAB)
  Ethyl cellulose (EC)
Chlorinated polyethers
Epoxy (EP)
Ethylene vinyl acetate (EVA)
Ethylene vinyl alcohol (EVOH)
Fluorocarbons
  Polytetrafluoroethylene (PTFE)
  Fluorinated ethylene propylene (FEP)
  Polyvinylidene fluoride (PVDF)
  Polyvinyl fluoride (PVF)
Liquid crystal polymers (LCP)
  Aromatic copolyester TP
Polyaryl ether
Polyaryl sulfone (PAS)
Phenolic
  Phenol formaldehyde (PF)
Polyamide (nylon) (PA)
Polyamide-imide (PAI)
Polyaryl ether
Polyarylates (PAR)
Polybenzimidazole (PBI)
Polycarbonate (PC)
Polyesters
  Aromatic polyester (TS)
  Thermoplastic polyester
    Polybutylene terephthalate (PBT)
    Polyethylene terephthalate (PET)

  Crystallized PET (CPET)
Unsaturated polyesters (TS)
Polyetheretherketone (PEEK)
Polyetherimide (PEI)
Polyimides (PI)
  Thermoplastic PI
  Thermoset PI
Polymethyl pentene
Polyolefins (PO)
  Polyethylene (PE)
  Low density PE (LDPE)
  Linear LDPE (LLDPE)
  High density PE (HDPE)
  Ultra high molecular weight PE (UHMWPE)
  Polypropylene (PP)
  Ionomer
  Cross-linked PE (XLPE)
  Polybutylene (PB)
  Polyallomers
  Chlorinated PE (CPE)
Polyphenylene ether (PPE)
Polyphenylene oxide (PPO)
Polyphenylene sulfide (PPS)
Polyurethanes (PUR)
Silicones (SI)
Sulfones
  Polysulfone (PSU)
  Polyether sulfone (PES)
  Polyphenyl sulfone (PPS)
Styrenics
  Polystyrene (PS)
  High-impact PS (HIPS)
  Acrylic styrene acrylonitrile (ASA)
  Acrylonitrile butadiene styrene (ABS)
  Styrene acrylonitrile (SAN)
  Styrene butadiene (SB)
Vinyls
  Polyvinyl chloride (PVC)
  Polyvinyl acetate (PVAc)
  Polyvinylidene chloride (PVDC)
  Polyvinyl alcohol (PVA)
  Polyvinyl butyrate (PVB)
  Chlorinated PVC (CPVC)

and narrower, process control. Two plastics with the same MI and density will process very differently if their MWDs are dissimilar.

Plastic Structures/Morphology

In addition to the size of molecules and their distribution, the shapes or structures of individual molecules also play an important role in determining the processability and properties of a plastic. Some plastics are formed by the alignment of long chains of molecules; in others, molecules with side branches or lateral connections form complex structures. All the forms exist in two and three dimensions; and because of the geometry or morphology of such molecules, some can come closer together than others. Thus, plastics that can be packed closely can easily form crystalline structures in which the molecules are aligned in some orderly pattern; and during processing, they tend to develop higher strength in the direction of the molecules. These are the crystalline thermoplastics—although commercially perfect crystalline polymers are not produced, so technically they are semicrystalline TPs. Other TPs do not align themselves; they go in all directions and twist like cooked spaghetti. They are noncrystalline, and are better known as amorphous TPs. These normally transparent plastics undergo only small volume changes (compared to the cyrstalline plastics) when melting or solidifying during processing. Table 1–4 compares the basic performance of crystalline and amorphous plastics; however, exceptions do exist, particularly with the many blends and fillers/reinforcements that are compounded (1–4, 8, 9, 11–26).

Processing conditions influence these polymers. For example, heating a crystalline material above its melting point and then quenching it can produce a polymer that has a far more amorphous structure than its original

Table 1–4a. Morphology of Thermoplastics (TPs).

CRYSTALLINE		AMORPHOUS
No	TRANSPARENT	Yes
Excel	CHEM. RESIST	Poor
No	STRESS-CRAZE	Yes
High	SHRINKAGE	Low
High	STRENGTH	Low*
Low	VISCOSITY	High
Yes	MELT TEMP	No
Yes	CRITICAL T/T**	No

*Major exception is PC.
**T/T = Temperature/time.

Table 1–4b. General Properties of TPs During and After Processing.

PROPERTY	CRYSTALLINE	AMORPHOUS
Melting or softening	Fairly sharp melting point	Softens over a range of temperature
Density (for the same material)	Increases as crystallinity increases	Lower than for crystalline material
Heat content	Greater	Lower
Volume change on heating	Greater	Lower
After-molding shrinkage	Greater	Lower
Effect of orientation	Greater	Lower
Compressibility	Often greater	Sometimes lower

Typical crystalline plastics are: polyethylene, polypropylene, nylon, acetals, and thermoplastic polyesters.
Typical amorphous plastics are: polystyrene, acrylics, PVC, SAN, and ABS.

Table 1–4c. Influence of Reinforcements and Fillers on TPs.

RESIN	REINFORCEMENTS	FILLERS
Amorphous:	+ Can more than double tensile strength	− Lower tensile strength
ABS		+ Can more than double flexural modulus
SAN	+ Can increase flexural modulus fourfold	
Amorphous		+ Raise HDT slightly
nylon	+ Raise HDT slightly	− Embrittle resins
Polycarbonate	± Toughen brittle resins embrittle tough resins	+ Can impart special properties such as lubricity, conductivity, flame retardance
Modified PPO		
Polystyrene	+ Can provide 1000 ohm-cm resistivity	
Polysulfones		+ Reduce and balance shrinkage
	+ Reduce shrinkage	
	− Reduce melt flow	− Reduce melt flow
	− Raise cost	+ Can lower cost
Crystalline:	+ Can more than triple tensile strength	− Lower tensile strength
Aceals		+ Can more than triple flexural modulus
Nylon 6, 6/6	+ Can raise flexural modulus sevenfold	
6/10, 6/12, 11, 12		+ Raise HDT slightly
Polypropylene	+ Can nearly triple HDT	− Embrittle resins
Polyphenylene sulfide	± Toughen brittle resins, embrittle tough resins	+ Can impart special properties such as lubricity, conductivity magnetic properties, flame retardance
Thermoplastic		
polyesters	+ Can provide 1 ohm-cm resistivity	
Polyethylene		+ Reduce shrinkage
	+ Reduce shrinkage	+ Reduce distortion
	− Cause distortion	
	− Reduce melt flow	− Reduce melt flow
	− Raise cost	+ Can lower cost

configuration. Its properties after quenching can be signficantly different than those it would have had if it had been cooled properly/slowly and allowed to recrystallize. (During processing it becomes amorphous.) The effects of time are similar to those of temperature, in the sense that any given plastic has a preferred, or an equilibrium, structure in which it is more likely to arrange itself. However, it may be prevented from doing so instantaneously or by a shortening of the available time. Afterwards, if given enough time, the molecules will rearrange themselves into the preferred pattern. Heating causes molecular rearrangements to occur more rapidly. During the heating action, severe shrinkages and property changes can occur in all directions of the processed plastic.

These morphological changes of TPs do not occur with TSs. As previously explained, when TSs are processed, their individual chain segments are strongly bonded together in an irreversible chemical reaction.

Heat Profile

To obtain the best processing melts for any plastics, one starts with the plastic manufacturer's recommended heat profile and/or one's own experience. These are starting points for various types of plastics, as shown in Fig. 1–12 and Table 1–5. The time and effort spent on startup make it possible to achieve maximum efficiency of performance vs. cost for the processed plastics. By the application of logic, the information gained can be stored and applied to future setups. One must recognize that in all probability similar machines (even from the same manufacturer) will not permit duplication of a process, but knowledge thus gained will guide the processor in future setups.

An amorphous material usually requires a fairly low initial heat in a screw plasticator; its purpose is to preheat material but not melt it in the feed section before it enters the compression zone of the screw (see, for example, Chapter 2). On the other hand, crystalline material requires a higher heat initially to ensure that it melts prior to reaching the compression zone; otherwise satisfactory melting will not occur. Careful implementation of these procedures results in the best melt, which in turn produces the best part. (Filled plastics, particularly those with thermally conductive fillers, usually require different heat profiles, i.e., a reverse profile where the feed throat area is better than the front zone.)

THERMAL PROPERTIES

Processes are influenced by the thermal characteristics of plastics, such as melt temperature (T_m), glass transition temperature (T_g), thermal conduc-

Example of a Thermoset
Processing Heat-Time Profile Cycle

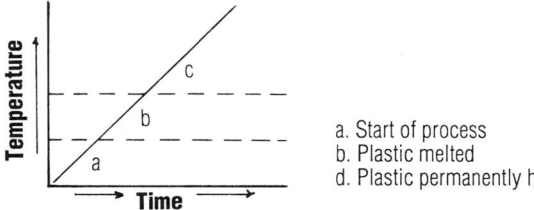

a. Start of process
b. Plastic melted
d. Plastic permanently hard

Example of a Thermoplastic
Processing Heat-Time Profile Cycle

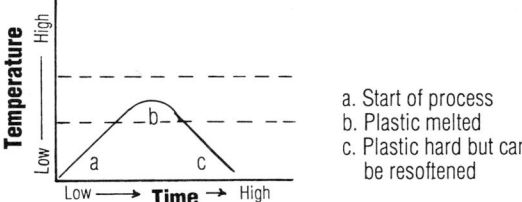

a. Start of process
b. Plastic melted
c. Plastic hard but can
   be resoftened

Fig. 1-12. Melting characteristics of TPs and TSs.

Table 1-5. Melt Processing Temperatures for TPs.

MATERIAL	PROCESSING TEMPEATURE RATE	
	°C	°F
ABS	180–240	356–464
Acetal	185–225	365–437
Acrylic	180–250	356–482
Nylon	260–290	500–554
Polycarbonate	280–310	536–590
LDPE	160–240	320–464
HDPE	200–280	392–536
Polypropylene	200–300	392–572
Polystyrene	180–260	356–500
PVC, rigid	160–180	320–365

Note: Values are typical for injection molding and most extrusion operations. Extrusion coating is done at higher temperatures (i.e., about 600°F for LDPE).
See the appendix for English to metric conversions.

tivity, thermal diffusivity, heat capacity, and coefficient of thermal expansion (Table 1-6). All these properties relate to the selection of the optimum processing conditions. There is a maximum processing temperature—or, to be more precise, a maximum time to temperature cycle—for all materials prior to their decomposition or destruction (Table 1-6). If certain additives or reinforcements are included, such as metal powders or glass fibers, thermal properties change (1-4, 8, 9, 11-26).

Melt Temperature

The melt temperature (T_m) occurs at a relatively sharp point for crystalline materials (Table 1-4b). The amorphous materials basically do not have a T_m; they start melting as soon as the heat cycle begins. In reality there is no single T_m point, but a range. It is often taken as the peak of a DSC curve (see Chapter 10).

The T_m is dependent on the processing pressure and time at heat, particularly during a slow temperature change for relatively thick melts. Also if T_m is too low, the melt's viscosity is high, and more power is required to process it. If the viscosity is too high, degradation will occur.

Glass Transition Temperature

The glass transition temperature (T_g) is the temperature below which the polymer behaves like glass, being strong and very rigid but brittle. Above this temperature it is not as strong or as rigid as glass, but it is not brittle. At T_g the plastic's volume or length increases (Fig. 1-13). The amorphous TPs have a more definite T_g.

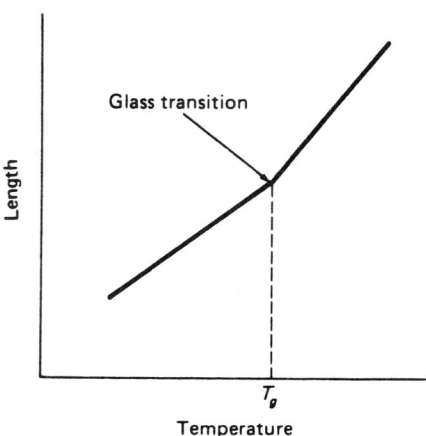

Fig. 1-13. Effect of T_g on volume or length of plastics.

Table 1-6. Examples of Thermal Properties of TPs. (Properties of some common materials included for comparison.)

PLASTICS (MORPHOLOGY*)	DENSITY, LB/FT³ (g/cm³)	MELT TEMPERATURE T_m, °F (°C)	GLASS TRANSITION TEMPERATURE T_g, °F (°C)	THERMAL CONDUCTIVITY, BTU/LB°F (10^{-4}CAL/S CM°C)	HEAT CAPACITY, BTU/LB°F (CAL/G°C)	THERMAL DIFFUSIVITY, 10^{-3} ft²/hr (10^{-4} CM²/S)	THERMAL EXPANSION 10^{-6} IN./IN. °F (10^{-6} CM/CM°C)
PP (C)	56 (0.9)	334 (168)	41 (5)	0.068 (2.8)	0.004 (0.9)	1.36 (3.5)	45 (81)
HDPE (C)	60 (0.96)	273 (134)	−166 (−110)	0.290 (12)	0.004 (0.9)	5.4 (13.9)	33 (59)
PTFE (C)	137 (2.2)	626 (330)	−175 (−115)	0.145 (6)	0.001 (0.3)	3.53 (9.1)	39 (70)
PA (C)	71 (1.13)	500 (260)	122 (50)	0.140 (5.8)	0.003 (0.75)	2.64 (6.8)	44 (80)
PET (C)	84 (1.35)	490 (250)	158 (70)	0.087 (3.6)	0.002 (0.45)	2.29 (5.9)	36 (65)
ABS (A)	66 (1.05)	221 (105)	215 (102)	0.073 (3)	0.002 (0.5)	1.47 (3.8)	33 (60)
PS (A)	66 (1.05)	212 (100)	194 (90)	0.073 (3)	0.002 (0.5)	2.2 (5.7)	28 (50)
PMMA (A)	75 (1.20)	203 (95)	212 (100)	0.145 (6)	0.002 (0.56)	3.45 (8.9)	28 (50)
PC (A)	75 (1.20)	510 (266)	300 (150)	0.114 (4.7)	0.002 (0.5)	3.0 (7.8)	38 (68)
PVC (A)	84 (1.35)	390 (199)	194 (90)	0.121 (5)	0.002 (0.6)	2.4 (6.2)	128 (50)
Aluminum	167 (2.68)	1,100		72.5 (3000)	0.23	1900 (4900)	10.6 (19)
Copper/bronze	549 (8.8)	1,800		109 (4500)	0.09	2200 (5700)	10 (18)
Steel	493 (7.9)	2,750		21.3 (880)	0.11	338 (1000)	6.1 (11)
Maple wood	28.1 (0.45)	400 (burns)		0.073 (3)	0.25	10.5 (27)	33 (60)
Zinc alloy	418 (6.7)	800		60.4 (2500)	0.10	1430 (3700)	15 (27)

*C = Crystalline resin. A = Amorphous resin.

The T_g influences the rate of change in temperature when it is increasing or decreasing, particularly with crystalline materials. For the crystalline type, the processor must make sure that a large amount of heat is applied in the T_g range; otherwise, it will not melt properly (or at maximum efficiency). If insufficient heat is used, an extended time period will be required, usually increasing production time and cost. With amorphous types, the application of heat is gradual and easier to control. Additional heat application just wastes energy.

The thermal properties of plastics, particularly T_g, influence processability in many different ways. Plastic selection should take these properties into account. A more expensive plastic could cost less to process because of a shorter processing time, requiring less energy per pound.

Thermal Conductivity

Thermal conductivity is the rate at which heat can be transferred through a material. It is an important parameter in evaluating the rate of heating and cooling in melts. Thermal conductivity is not strongly temperature-dependent. Values for amorphous types are in the range of 2×10^{-4} to 5×10^{-4} cal/s cm°C. Because of the high degree of molecular order for crystalline types, thermal conductivity values tend to be twice those of the amorphous types, except for PP (Type 1–6).

The heat capacity is an indicator of how much heat has to be added to a unit mass of plastic in order to raise its temperature 1°C, and it is readily measured by ASTM tests.

Thermal Diffusivity

Whereas heat capacity is a measure of the energy, thermal diffusivity is a measure of the rate at which energy is transmitted through the plastic; it relates to processability. In contrast, metals have values that are hundreds of times larger than those of plastics.

Heat Capacity

The heat capacity, or specific heat, is the amount of energy required to raise the temperature of a unit mass of a material one degree. In metric units it is expressed in cal/g°C. It can be measured at constant pressure or constant volume; at constant pressure, it is larger than at constant volume because additional energy is required to bring about the volume change against external pressure. The specific heat of amorphous polymers increases with temperature in an approximately linear fashion below and above T_g. A step-like change occurs near the T_g.

Coefficient of Thermal Expansion

This property measures expansion upon heating and contraction upon cooling. It is expressed as the relative change in length per degree of temperature change. The design of equipment (plasticator screws, molds, dies, etc.) must allow for these changes, or the equipment will operate inefficiently, and processed products will not meet at least dimensional requirements. Thermal expansion is reduced by orientation, cross-linking, rigid fillers, and so on, and is enhanced with plasticizers, lubricants, solvents, and so forth. Table 1-6 lists some of these values, but with certain additives/reinforcements the values could be zero.

ORIENTATION

Plastic molecular orientation can be accidental or deliberate. (Here, accidental refers to orientations that occur in processing plastics that may be acceptable. However, excessive frozen-in stresses can be extremely damaging if parts are subject to environmental stress cracking or crazing in the presence of chemicals, heat, etc.) As previously noted, inintially the molecules are relaxed; molecules in amorphous regions are in random coils, and those in crystalline regions relatively straight and folded. During processing, the molecules tend to be more oriented than relaxed, particularly when sheared, as during injection molding and extrusion. After temperature–time–pressure is applied, and the melt goes through restrictions (molds, dies, etc.), molecules tend to be stretched and aligned in a parallel form. The result is a change in directional properties and dimensions. The amount of change depends on the type of thermoplastic, the amount of restriction, and, most important, the rate of cooling. The faster the rate is, the more retention there is of the frozen orientation. After processing, parts could be subject to stress relaxation, with changes in performance and dimensions. With certain plastics and processes, there is an insignificant change. If changes are significant, one must take action to change the processing conditions, particularly increasing the cooling rate.

By deliberate stretching, the moelcular chains of a plastic are drawn in the direction of the stretching, and inherent strengths of the chains are more nearly realized than they are in their naturally relaxed configuration. Stretching can take place with heat during or after processing (blow molding, extruding film, thermoforming, etc.). Products can be drawn in one direction (uniaxial) or in two opposite directions (biaxial), in which case many properties significantly increase uniaxially or biaxially (see Table 1-7 and Fig. 1-14, and references 1-3, 8, 11-13, 22-24).

PROCESSING DIAGRAMS

The processor setting up a machine, regardless of the type of controls available, uses a systematic approach that should be outlined in the machine and/or control manuals. Once the machine is operating, the processor methodically makes one change at a time to determine the result. Two basic examples are presented in the following paragraphs to show a logical approach to evaluating changes made with any processing machine. As the injection molding machine is very complex with all the controls required to set it up, these examples refer to the injection molding process (1-4, 30, 31).

Figure 1-15 shows what happens when changing mold temperature and injection (ram) pressure. This molding area diagram (MAD) provides information on the best combination. Any setting outside the diagram results in defective molded parts that will not meet the required performances. The size of the diagram shows a molder the latitude available to produce good parts. However, to mold at the fastest cycle and/or lowest cost, the machine would be set at the lowest temperature and highest pressure, or the upper

Table 1-7. Effects of Orientation of Polypropylene Films.

	STRETCH (%)				
PROPERTIES	NONE	200	400	600	900
Tensile strength, psi	5,600	8,400	14,000	22,000	23,000
Elongation at break, %	500	250	115	40	40

PROPERTIES	AS CAST	UNIAXIAL ORIENTATION	BALANCED ORIENTATION
Tensile strength, psi			
MD*	5,700	8,000	26,000
TD*	3,200	40,000	22,000
Modulus of elasticity, psi			
MD	96,000	150,000	340,000
TD	98,000	400,000	330,000
Elongation at break, %			
MD	425	300	80
TD	300	40	65

*MD = Machine direction.
** TD = Transverse direction, and direction of uniaxial orientation.
See the appendix for English to metric conversions.

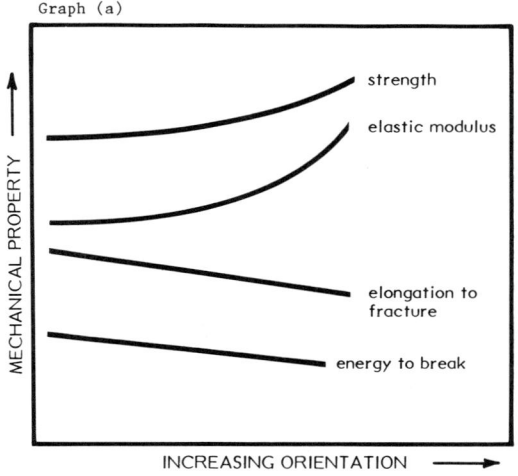

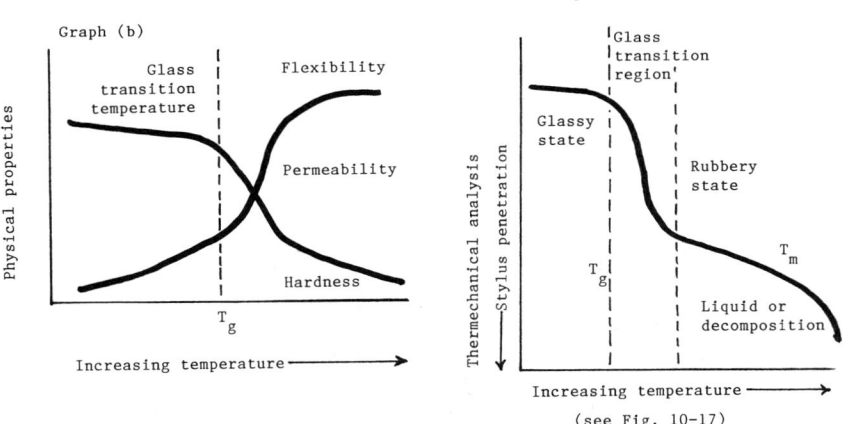

Fig. 1-14. Effect of orientation on properties of plastics.

left corner of the diagram. One must be aware that there could be variations in the material (thermoplastic), basic machine functions, and/or controls; so even with operation in that corner of the diagram, defective parts can occur. Unless the original cycles were run with all these variables at their worst, all parts should be okay. It is statistically probable that rejects will occur; so one must determine how far away from optimum and in what

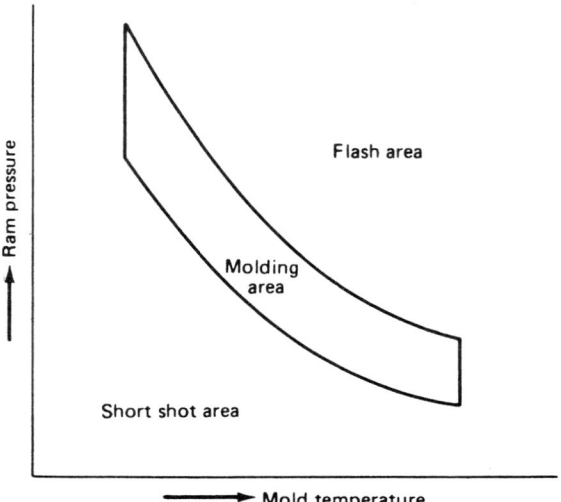

Fig. 1-15. Two-dimensional molding area diagram (MAD) that plots injection molding ram pressure vs. mold temperature.

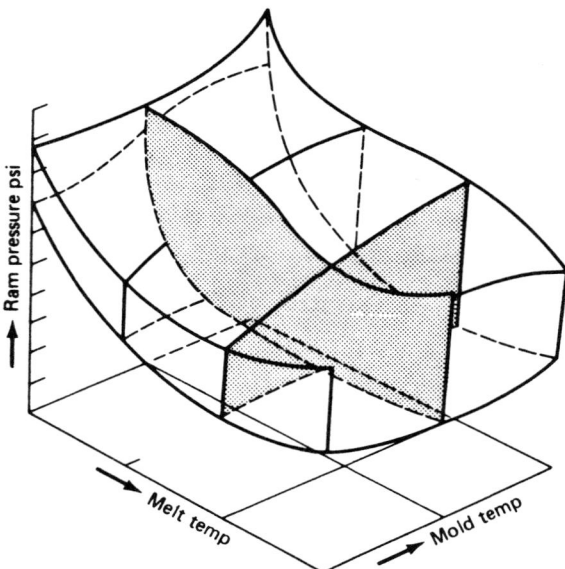

Fig. 1-16. Three-dimensional molding volume diagram (MVD) that plots melt temperature, mold temperature, and ram pressure.

directions the molding conditions can deviate and still provide good parts. If a thermoset was used and a diagram was developed, its best operation would be at high pressure and high temperature, or the upper right corner of the diagram. In spite of the different shapes developed for a TP and a TS, the approach is similar for both.

The second example is the use of a three-dimensional diagram (Fig. 1-16). This molding volume diagram (MVD) compares the behavior of a thermoplastic in the mold based on varying melt temperature, mold temperature, and ram pressure. Thus all parts molded within the volume produce good parts. To operate with maximum efficiency, one should plan to operate the machine at the highest pressure, lowest melt temperature, and lowest mold temperature possible. This approach can be continued with any controls available in the machine. True, it is easier said than done. Initially, a costly amount of time will be required; but eventually this becomes second nature so that many shortcuts can be taken. The fact remains that some logical approach has to be used to determine the best setting for the machine. Proper record keeping is vital. With the computer control systems that are available, a systematic approach can be readily conducted, process parameters can be stored, and an "expert system" can be developed.

Chapter 2

INJECTION MOLDING

INTRODUCTION

The injection molding (IM) process is greatly preferred by designers because the manufacture of parts of complex shape and three-dimensions can be more accurately controlled and predicted with IM than with other processes. As its method of operation is much more complex than others, IM requires a thorough understanding. Figures 2–1 and 2–2 show schematics of the load profile and the molding cycle that highlight: the way in which the melt is plasticized (softened) and forced into the mold; the clamping system for opening and closing the mold under pressure; the type of mold used; and the machine controls (1, 3).

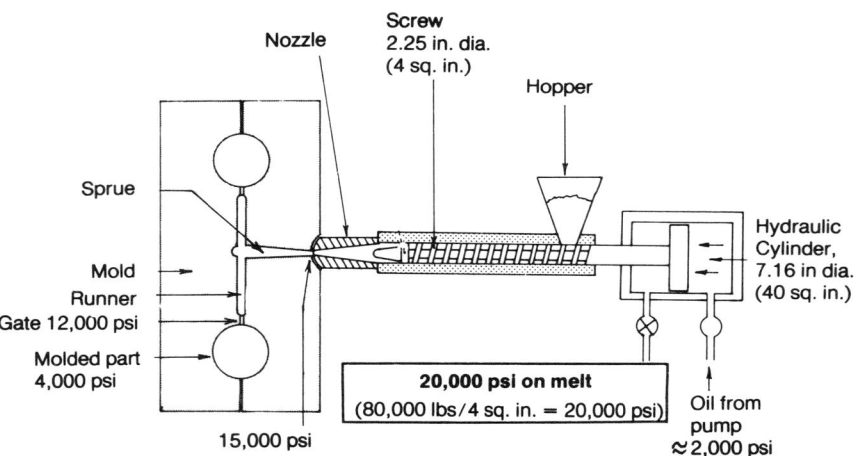

Fig. 2–1. Example of pressure loading on plastic melt during injection molding. (See the appendix for English to metric conversions.)

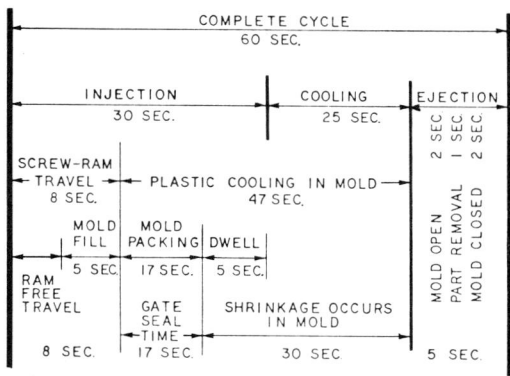

Fig. 2-2. Example of injection molding cycle processing thermoplastics (TPs).

Plastic moves from the hopper onto the feeding portion of the recipro-cating extruder screw (Fig. 2-3). The flights of the rotating screw cause the material to move through a heated extruder barrel where it softens (is made fluid) so that it can be fed into the shot chamber (front of screw). This motion generates pressure (usually 50–300 psi), which causes the screw to retract. When the preset limit switch is reached (or a position transducer on newer machines), the shot size is met and the screw stops rotating; and at a preset time the screw acts as a ram to push the melt into the mold. Injection takes place at high pressure (up to 30,000 psi melt pressure in the nozzle). Adequate clamping pressure must be used to eliminate mold open-ing (flushing). The melt pressure within the mold cavity ranges from 1–15 tons/sq. in., and is dependent on the plastic's rheology/flow behavior (see Chapter 1).

Time, temperature, and pressure controls indicate whether performance requirements of a molded part are met. Time factors include rate of injec-tion, duration of ram pressure, time of cooling, time of plastication, and screw RPM (Fig. 2-2). Pressure factors are injection high and low pressure, back pressure on the extruder screw, and pressure loss before the plastic enters the cavity, which can be caused by a variety of restrictions in the mold (Fig. 2-4). Temperature factors are mold (cavity and core), barrel, and nozzle temperatures, as well as the melt temperature due to back pres-sure, screw speed, frictional heat, and so on (see Fig. 2-5 and references 1-4 and 6-13).

The large number of variables summarized in Fig. 2-6 will cause part changes if not controlled properly. Basic settings for these variables are provided by the plastic producer: the injection barrel temperature (Table 2-1), the cavity melt pressure, and so on. However, the final settings are

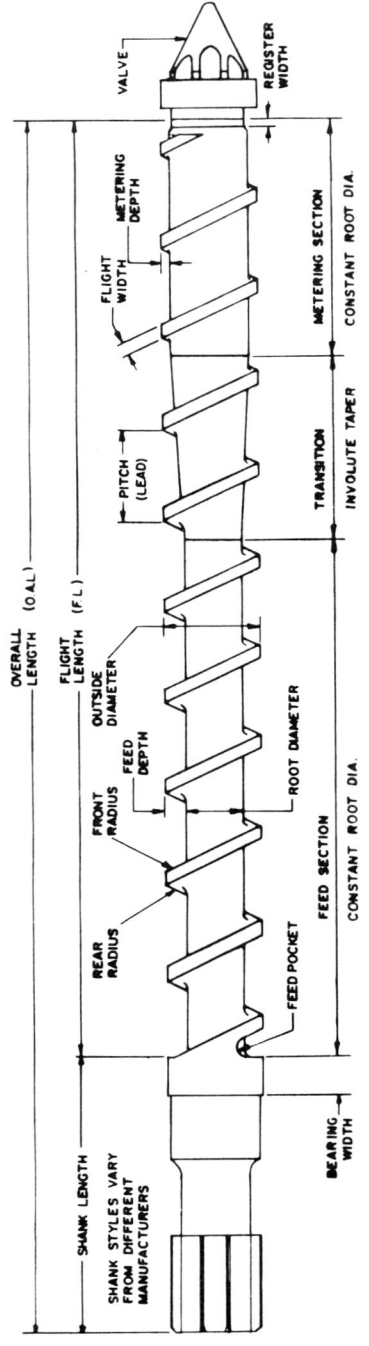

Fig. 2-3. Screw nomenclature.

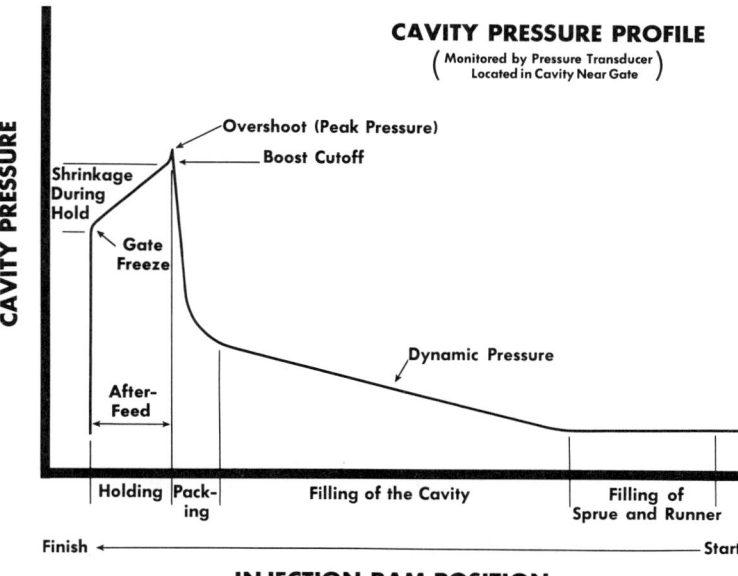

Fig. 2-4. Cavity pressure profile.

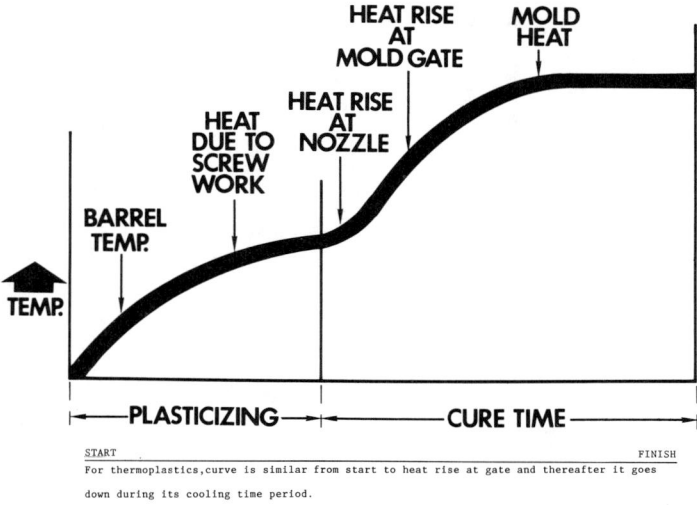

Fig. 2-5. Thermal load profile during injection molding for thermosets (TSs). (For TPs the curve is similar from the start to the heat rise at the gate; thereafter it descends during its cooling time period.)

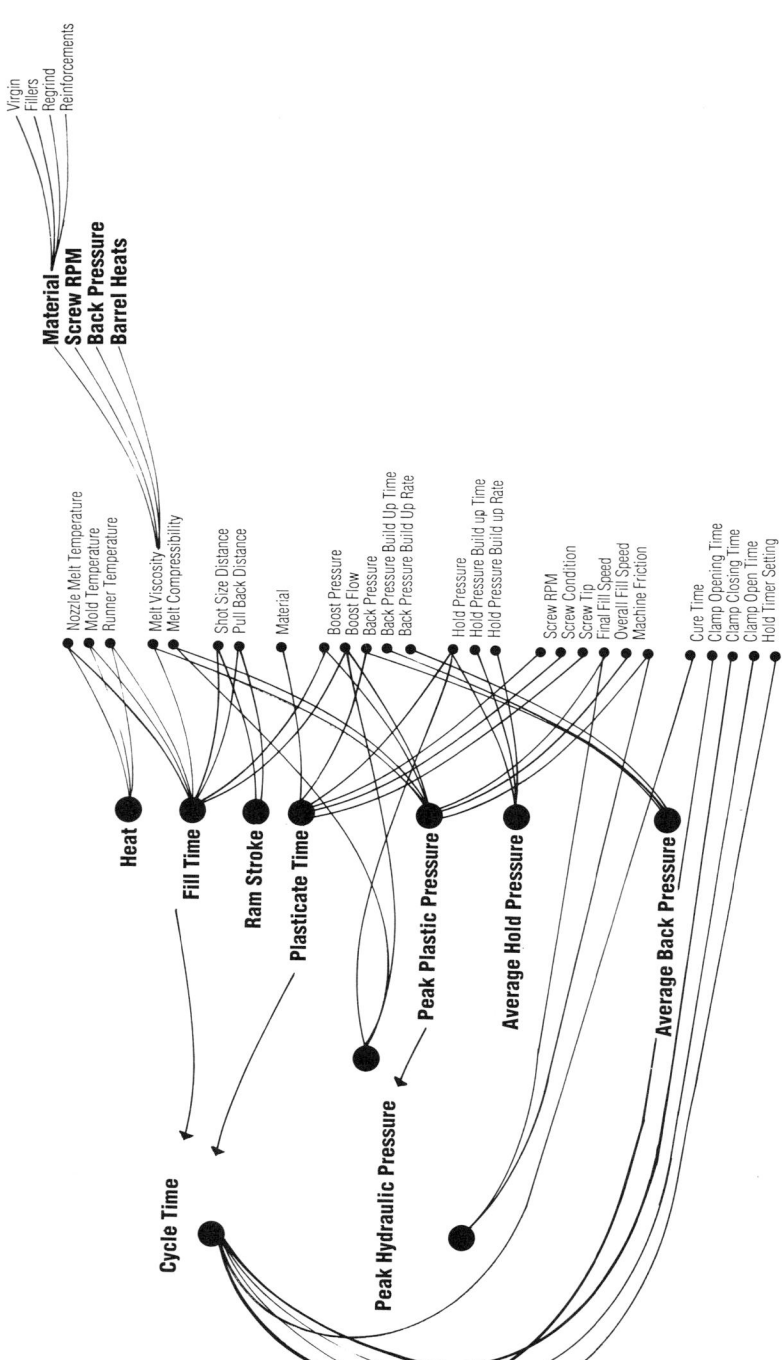

Fig. 2-6. Injection molding machine controls.

42

Table 2-1. Guide for Injection Molding and Extrusion Machine Settings. Specific Information on all Machine Settings and Plastic Properties Initially is Acquired by Using the Resin Supplier's Data Sheet on the Particular Compound/Resin to be Used.

RESIN DATA — These are strictly typical, average values for a resin class. Consult your resin supplier for values and more accurate information.	SPECIFIC GRAVITY G/CM³	DENSITY LB/FT³	SPECIFIC VOLUME IN³/LB	SPECIFIC VOLUME CM³/G	EXTRUSION TEMPERATURE °F	INJECTION TEMPERATURE °F	LINEAR MOLD SHRINKAGE IN./IN.	SPECIFIC HEAT BTU/LB/°F	WATER ABSORPTION % IN 24 HOURS	MAXIMUM WATER CONTENT ALLOWABLE FOR MOLDING
ABS—EXTRUSION	1.02	64.0	27.0	.980	435		.005	.34	.25	
ABS—INJECTION	1.05	65.0	26.0	.952		500	.005	.40	.40	.20
ACETAL—INJECTION	1.41	88.0	19.7	.709		390	.020	.35	.25	
ACRYLIC—EXTRUSION	1.19	74.3	23.3	.839	375		.004	.35	.30	
ACRYLIC—INJECTION	1.16	72.0	24.1	.868		450	.005	.35	.20	.08
CAB	1.20	74.6	23.1	.833	380	440	.004	.35	1.50	.15
CELLULOSE ACETATE—EXTRUSION	1.28	80.2	21.6	.781	380		.005	.40	2.50	
CELLULOSE ACETATE—INJECTION	1.26	79.0	21.9	.794		450	.005	.36	2.40	.20
CELLULOSE PROPRIONATE—EXTRUSION	1.22	76.1	22.7	.821	380		.004	.40	1.70	
CELLULOSE PROPRIONATE—INJECTION	1.22	75.5	22.9	.828		425	.004	.40	2.00	.25
CTFE	2.11	134.0	13.1	.473		550	.008	.22	.01	
FEP	2.11	134.0	12.9	.465	600	600	.010	.28	<.01	

(continued)

Table 2-1. Continued

RESIN DATA — These are strictly typical, average values for a resin class. Consult your resin supplier for values and more accurate information.	SPECIFIC GRAVITY G/CM³	DENSITY LB/FT³	SPECIFIC VOLUME IN.³/LB	SPECIFIC VOLUME CM³/G	EXTRUSION TEMPERATURE °F	INJECTION TEMPERATURE °F	LINEAR MOLD SHRINKAGE IN./IN.	SPECIFIC HEAT BTU/LB/°F	WATER ABSORPTION % IN 24 HOURS	MAXIMUM WATER CONTENT ALLOWABLE FOR MOLDING
IONOMER—EXTRUSION	.95	59.6	29.0	1.050	500		.007	.54	.07	
IONOMER—INJECTION	.95	59.1	29.2	1.060		420	.007	.54	.20	
NYLON 6	1.13	70.5	24.5	.886	520	550	.013	.40	1.60	.15
NYLON 6/6	1.14	71.2	24.3	.878	510	510	.015	.40	1.50	.15
NYLON 6/10	1.08	67.4	25.6	.927	475	450	.011	.40	.40	.15
NYLON 6/12	1.07	66.8	25.9	.935	460	500	.011	.40	.40	.20
NYLON 11	1.04	64.9	26.6	.962	450	450	.005	.40	.30	.10
NYLON 12	1.02	63.7	27.1	.980	450	445	.003	.47	.25	.10
PHENYLENE OXIDE BASED	1.08	67.5	25.6	.926	480	525	.006	.32	.07	
POLYALLOMER	.90	56.2	30.7	1.110	405	405	.015	.50	.01	
POLYARYLENE ETHER	1.06	66.2	30.7	.940	460	535	.006		.10	
POLYCARBONATE	1.20	74.9	23.1	.832	550	575	.006	.30	.20	.02
POLYESTER PBT	1.34	83.6	20.7	.746		460	.020		.08	.04
POLYESTER PET	1.31	81.8	21.1	.746	480	490	.002	.40	.10	.005
HD POLYETHYLENE—EXTRUSION	.96	59.9	28.8	1.040	410	480	.025	.55	<.01	
HD POLYETHYLENE—INJECTION	.95	59.3	29.1	1.050		480	.025	.55	<.01	

Material										
HD POLYETHYLENE—BLOW MOLDING	.95	56.9	28.8	1.040	410		.025	.55	<.01	
LD POLYETHYLENE—FILM	.92	57.44	30.1	1.090	350		.032	.55	<.01	
LD POLYETHYLENE—INJECTION	.92	57.4	30.1	1.090		400	.032	.55	<.01	
LD POLYETHYLENE—WIRE	.92	57.4	30.1	1.090	400		.025	.55	<.01	
LD POLYETHYLENE—EXT. COATING	.92	57.1	30.0	1.090	600		.025	.55	<.01	
LLD POLYETHYLENE—EXTRUSION	.92	57.4	30.1	1.087	500					
LLD POLYETHYLENE—INJECTION	.93	58.0	29.8	1.075		425				
POLYPROPYLENE—EXTRUSION	.91	56.8	30.4	1.100	450		.005	.50	.03	
POLYPROPYLENE—INJECTION	.90	56.2	30.7	1.110		490	.018	.50	<.01	
POLYSTYRENE—IMPACT SHEET	1.04	64.9	26.6	.963	450		.005	.34	.10	
POLYSTYRENE—G.P. CRYSTAL	1.05	65.5	26.2	.943	410	425	.004	.32	.03	
POLYSTYRENE—INJECTION IMPACT	1.04	64.9	26.6	.968		440	.006	.34	.10	.05
POLYSULFONE	1.25	77.4	22.3	.807	650	680	.007	.28	.30	
POLYURETHANE	1.20	74.9	23.1	.834	400	400	.020	.40	.10	.03
PVC—RIGID PROFILES	1.39	86.6	19.9	.720	365		.025	.25	.02	
PVC—PIPE	1.44	87.5	19.7	.714	380		.025	.25	.10	
PVC—RIGID INJECTION	1.29	83.6	21.0	.756		380	.025	.25	.10	.07
PVC—FLEXIBLE WIRE	1.37	85.5	20.2	.731	365		.025			
PVC—FLEXIBLE EXTRUDED SHAPES	1.23	76.8	22.5	.814	350		.025			
PVC—FLEXIBLE INJECTION	1.29	80.5	21.4	.776		300	.025			
PTFE	2.16	134.8	12.9	.464				.25	<.01	
SAN	1.08	67.4	25.6	.927	420	470	.005	.31	.03	.02
TFE	1.70	106.1	16.3	.589		610	.040	.46	.01	
URETHANE ELASTOMERS	.83	51.6	33.5	1.210	390	400	.001	.46	.07	.03

See the appendix for English to Metric conversions.

determined by the processor on a specific machine and mold. Applicable process control techniques will be reviewed in this text.

Even though most of the literature on processing (as in this book) specifically identifies or refers to thermoplastics (TPs), some thermosets (TSs) are used (TS polyesters, phenolics, epoxy, etc.). TPs reach maximum heat prior to entering "cool" mold cavities, whereas TSs reach maximum temperature in "hot" molds (see Fig. 2–5 and Chapter 1).

Productivity

Productivity is directly related to cycle time. There usually is considerable common knowledge about part geometry and process conditions that will provide a minimum cycle time (1–7, 32–52). Practices such as using thinner wall sections, cold vs. hot runners for TPs or hot vs. cold for TSs (Fig. 2–7), narrow sprues and runners, optimal size and location of coolant (or heat) channels, and lower melt/mold heat, when possible, will decrease the solidification time.

Numerous factors affect the elapsed time required to eject a part, as different plastics can have dramatically different melt behaviors. Many of these influences are poorly, if at all, understood. Some critical ones are the coefficient of expansion, melt rheology, thermal diffusivity, and the thermo-mechanical spectrum (see Chapter 1). Although the usual and important ways to optimize time are based on part design and process conditions, it can be shown that additional and significant decreases occur by using modified molding compounds via additives, alloying ratios, molecular weight distribution, and so on (see Chapters 1, 9, and 10 and references 33–96).

Mathematical models of part geometry and melt flow within the mold cavity, which are available for mold design, are useful tools for optimizing cycle times. They allow a wide variety of plastics materials and process parameters to be evaluated in a convenient and cost-effective manner (1–4, 9, 36, 85, 97–99). CAD (computer-aided design) and engineering algorithms offer a continually higher level of sophistication in determining the best heat distribution throughout the part. In actual practice, parts are not ejected according to a measurement of the internal or wall temperature. Ejection times are set through secondary characteristics of part heat, such as the ability of the part to withstand the forces of ejection, the occurrence of sink marks or other thermal warpage, and the overall gloss or appearance. The prediction of sink marks appears amenable to CAD/CAE (computer-aided engineering). Part appearance still plays a larger role in the art of IM.

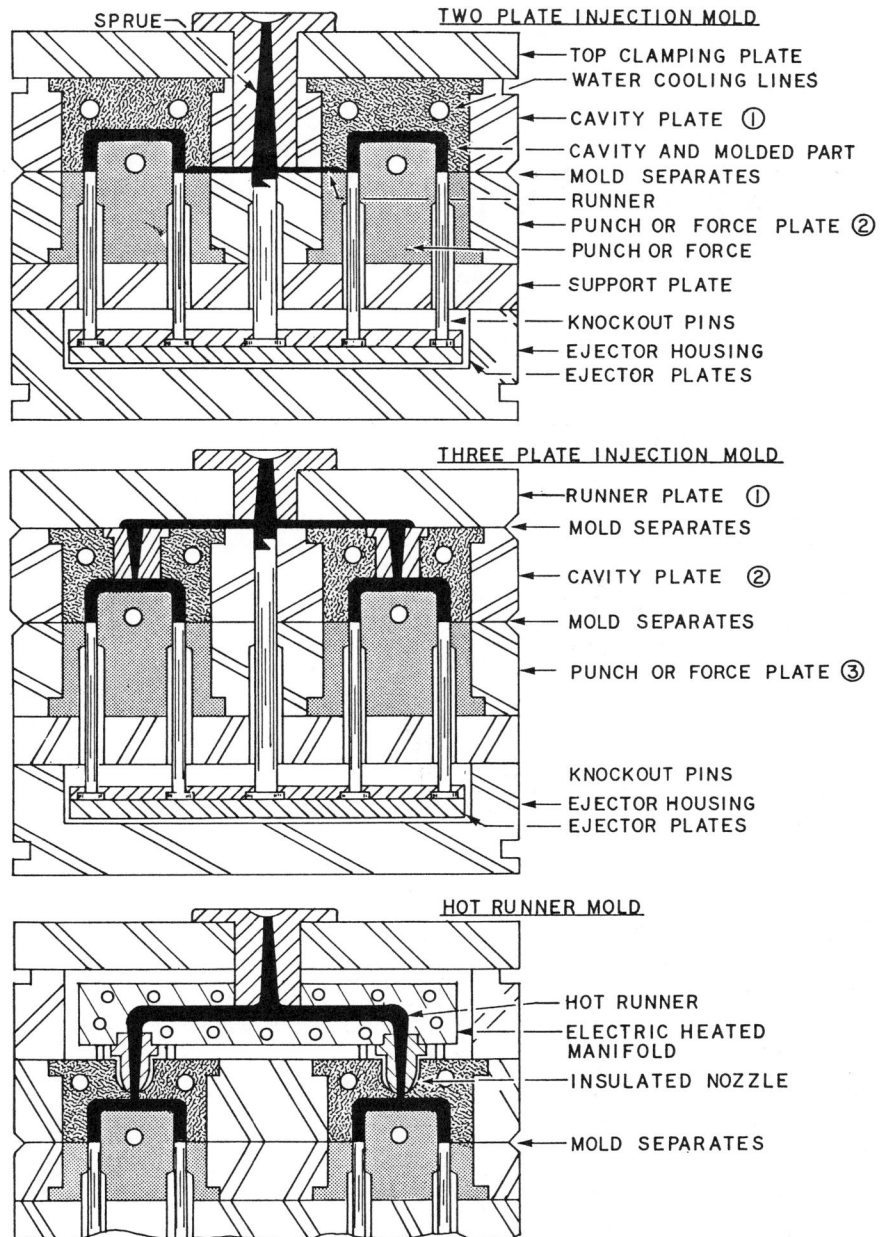

Fig. 2-7a. Types of injection molds illustrated here are cold runner two-plate, cold runner three-plate, and hot runner (1).

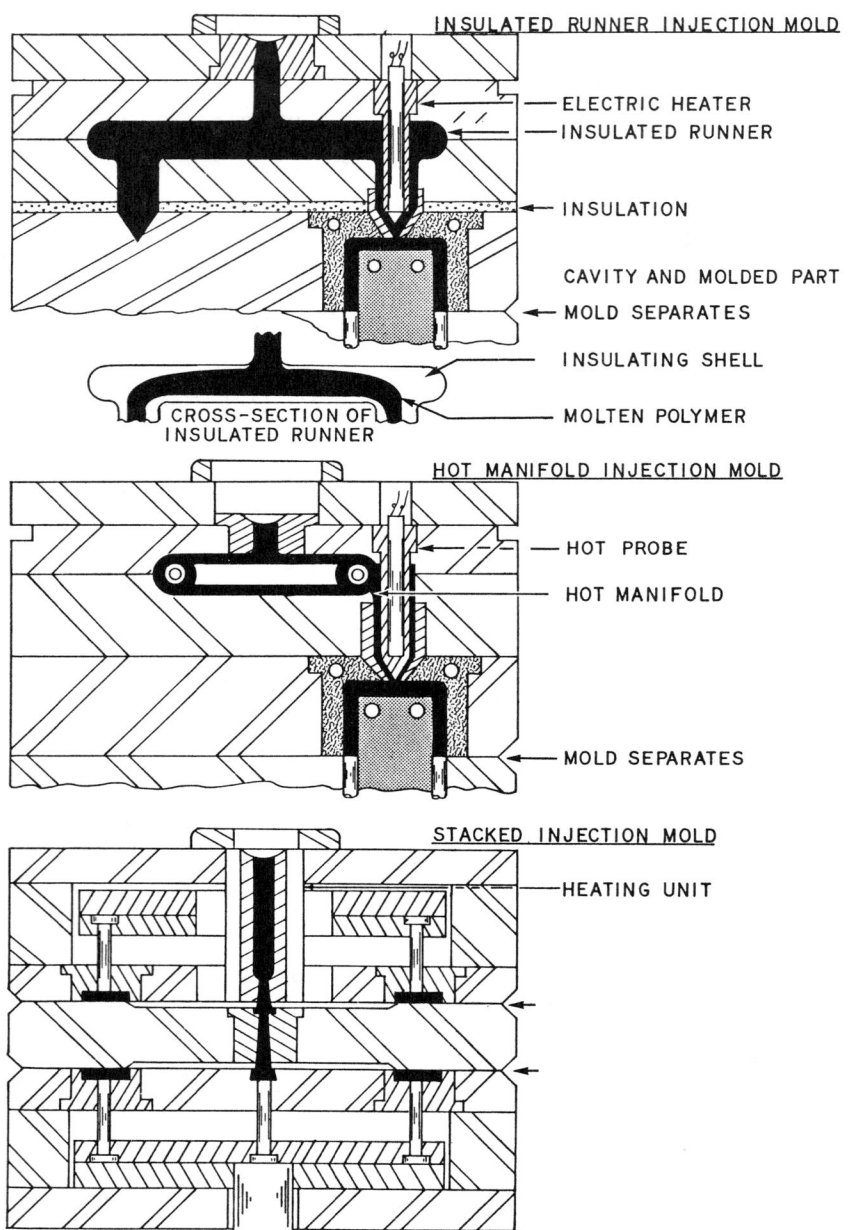

Fig. 2-7b. Types of injection molds illustrated here are insulated runner, hot manifold, and stacked (1).

SCREW DESIGN

With practically all machines, only the cylinder temperature is directly controlled (see Chapter 1). The actual heat of the melt, within the screw and as it is ejected from the nozzle, can vary considerably, depending on the efficiency of the screw design and the method of operation. Factors affecting melt heat include: the time plastic remains in the cylinder; the internal surface heating area of the cylinder and the screw, per volume of material being heated; the thermal conductivity of cylinder, screw, and plastic (Table 1-6); the heat differential between the cylinder and the melt; and the amount of melt turbulence in the cylinder. In designing the screw, a balance must be maintained between the need to provide adequate time for heat exposure and the need to maximize output most economically.

In general, heat transfer problems have led screw designers to concentrate on making screws more efficient heat transfer devices. As a result, the internal design and performance of screws vary considerably to accommodate the different plastics that are used (1–4, 85, 100–102). Most machines are single, constant-pitch, metering-type screws to handle the majority of plastics (Fig. 2–3).

Plastic in the screw channel is subject to changing experiences as the screw operation changes during the cycle. Each operation of the screw, whether it is moving forward, rotating, and retracting during shot preparation or static during an idle period, subjects the plastic to different thermal and shear situations. Consequently, the IM plasticating process becomes rather complex, but it is controllable and repeatable within the limits of equipment capability. At a fixed screw speed, the screw pitch, diameter, and channel depth determine output. A deep-channel screw is much more sensitive to pressure changes than a shallow channel screw. In the lower pressure range, a deep channel will provide more output; however, the reverse is true at high pressures. Shallower channels tend to give better mixing and flow patterns. Thus, although the screw is usually a simple-looking device, it accomplishes many different operations at the same time with its three sections: (1) solids conveying or feeding; (2) compressing, melting, and pressurizing the melt; and (3) mixing, melt refinement, and pressure–temperature stabilization.

Hypothetical data on screws are given in Table 2–2, which provides some examples of variations on the same length-to-diameter screw processing different plastics. In reality, the L/D ratios (flight length/outside diameter) vary according to the rheology of the resin. Advantages of a short L/D are: less residence time is necessary in the barrel, so heat-sensitive resins are kept at melt heat for a shorter time, lessening the chance of degradation; the design occupies less space; it requires less torque, making the screw strength and amount of horsepower less important; and it requires a smaller in-

Table 2-2. Hypothetical Screw Designs for General Types of Plastic.

DIMENSIONS, IN.	RIGID PVC	IMPACT POLYSTYRENE	LOW-DENSITY POLYETHYLENE	HIGH-DENSITY POLYETHYLENE	NYLON	CELLULOSE ACET/ BUTYRATE
diameter	4½	4½	4½	4½	4½	4½
total length	90	90	90	90	90	90
feed zone (F)	13½	27	22½	36	67½	0
compression zone	76½	18	45	18	4½	90
metering zone (M)	0	45	22½	36	18	0
depth (M)	.200	.140	.125	.155	.125	.125
depth (F)	.600	.600	.600	.650	.650	.600

vestment cost, initially and for replacement parts. Long *L/D* advantages are: the screw can be designed for a greater output or recovery rate, provided that sufficient torque is available, and it can be designed for more uniform output and greater mixing; also it will pump at higher pressures and give greater melting with less shear, as well as providing more conductive heat from the barrel.

The compression ratio (C/R) relates to compression that occurs on the resin in the transition section; it is the ratio of the volume at the start of the feed section divided by the volume in the metering section (determined by dividing the feed depth by the metering depth). The C/R should be high enough to compress the low bulk unmelted resin into a solid melt without air pockets. A low ratio will tend to entrap air bubbles. High percentages of regrind, powders, and other low bulk materials will be achieved by a high compression ratio. A high C/R can overpump the metering section. A common misconception is that engineering and heat-sensitive resins should use a low C/R. This is true only if it is decreased by deepening the metering section, and not by making the feed section shallower. The problem of overheating is more related to channel depths and shear rates than to C/R. As an example, a high C/R in polyolefins can cause melt blocks in the transition section, leading to rapid wear of the screw and/or barrel. In the processing of TS material, the C/R is usually zero so that accidental overheating does not occur and cause the melt to solidify in the barrel. With overheating, melt solidification occurs, and the zero ratio permits ease of removal—remove the nozzle, and it can be "unscrewed." Zero ratios are also used for TPs when the rheology so requires.

The output of a metering screw is fairly predictable, provided that the melt is under control. With a square pitch screw (conventional screw where distance from flight to flight is equal to the diameter), a simplified formula for output is:

$$R = 2.3 \, D^2 \, h \, g \, N$$

where R = rate or output in lb/h, D = screw diameter in inches, h = depth of the metering section in inches (for two-stage screw use the depth of the first metering section), g = gravity of the melt, and N = screw RPM.

This formula does not take into account back flow and leakage flow over the flights. These flows are not usually a significant factor unless the resin has a very low viscosity during processing, or the screw is worn out. The formula assumes pumping against low pressure, giving no consideration to melt quality and leakage flow of severely worn OD screws. With all these and other limitations, the formula can still provide a general guide to output. If the output is significantly greater, that is caused by a high C/R that overpumps the metering section, which sometimes is desirable but can lead to surging and rapid screw wear if it is excessive; and if the output is a lot less than estimated, that usually indicates a feed problem or a worn screw and/or barrel. A feeding problem sometimes can be corrected by changes in the barrel heat. More often, the probem is caused by such factors as screw design, shape and bulk density of the feedstock, surface conditions of the screw root, the screw heat, and so forth. The problem of screw/barrel wear can be assessed by measuring the screw/barrel.

An accurate method used to determine output loss due to screw wear is to compare the worn screw's current output with the initial production benchmark, originally determined by shooting into a "bucket" to check the shot weight. Another approach is to measure the worn screw's clearance to the barrel wall (W), which is used along with the original measured screw clearance (O) and the metering depth (M) from the screw root to the barrel wall. Here the approximate percentage output loss (OL), with RPM being constant, is calculated from the formula:

$$OL = (W - O)/M \times 100$$

Mixing Devices

Processors have developed an almost universally accepted model for melting in a single screw (Fig. 2–3). This model, which is the basis for most computer simulations, has been demonstrated to be correct by many "freeze tests" (1, 2, 100). The model shows how the plastic goes from the solid state to a melt as it moves through the action of the controlled screw and barrel. The results describe such relationships as high output via deep screws, low melt temperatures via deep screws, and melt quality via shallow screws. For these different situations, solutions have been developed to provide good mixing and product uniformity at high production rates without excessive melt heat.

A variety of different mixing and barrier screws, designed to improve melt processability with high output, can be used to meet the different flow requirements of the various plastics. The Dulmage Mixer (Dow Chemical), one of the first, usually is located at the end of the screw. It has a series of semicircular grooves cut on a long helix in the direction of the screw flights. These grooves interrupt laminar flow by dividing and recombining the melt many times, like a static mixer. Mixing pins placed radially in the screw root, using different patterns and shapes, improve performance. The Maddock Mixer (Union Carbide) consists of a series of opposed, semicircular grooves along the screw exit. Alternate grooves are open to the upstream entry, and the other grooves are open to the downstream discharge. Its ribs and flutes, which divide the alternating entry and discharge grooves, also alternate. This mixer does an effective job of mixing and screening unmelted plastics. Unmelted material gets trapped in grooves and mixed so that it is melted. In the Pulsar Mixer (Spirex Corp.) the metering section is divided into constantly changing sections, which are either deeper or shallower than the average metering depth. The resin alternates many times between the shallower depth with a somewhat higher shear and deeper channels with a lower shear. This tumbling and massaging action produces excellent mixing and melt uniformity without high shear (1, 2, 100) (see Chapter 3).

Barrier Screws

The original barrier screw was a very important design. The first patent, in 1959 (Maillefer and Geyer of Uniroyal), had broad claims; so all the barrier screw designs patented thereafter, pay a royalty under the original patent (1, 2, 100). Barrier screws have two channels in the barrier section, which are located mainly in the transition section (Fig. 2–8a). A secondary flight is started, usually at the beginning of the transition, to create two distinct channels, a solids channel and a melt channel. The barrier flight is undercut below the primary flight, allowing melt to pass over it.

There are many different barrier screw designs, each offering unique advantages. They include: the Uniroyal (the original barrier screw); the MC-3 of Hartig; the Efficient of New Castle Ind.; the Barr 2 of R. Barr Inc.; the VPB of Davis Standard; the Willert II of W. H. Willert Inc.; and the Double Wave of HPM Corp.

Custom-Designed Screws

The specially designed barrier screws, as noted, meet special requirements. Plastics tailored for specific applications usually have a high or very high molecular weight, and are heat- and shear-sensitive. Traditionally single-flighted screws provide melts via relatively high shear rates. As the solid plastic moves down the screw, there is a thin melt film between the barrel and the solid plastic bed. Due to screw rotation, the shear on the film is quite high, generating a great amount of shear-energy heat. This heat is transferred to the solid bed, promoting additional melting. The film becomes thicker as it moves to the rear of the channel, where it is scraped off the barrel by the screw flight. It then can only go down into the channel, and create the usual flow profile in the screw (Fig. 2–8b) (103).

The melting barrier type of screw is now a well known means of improving melt processability. However, in spite of its advantages, it has dif-

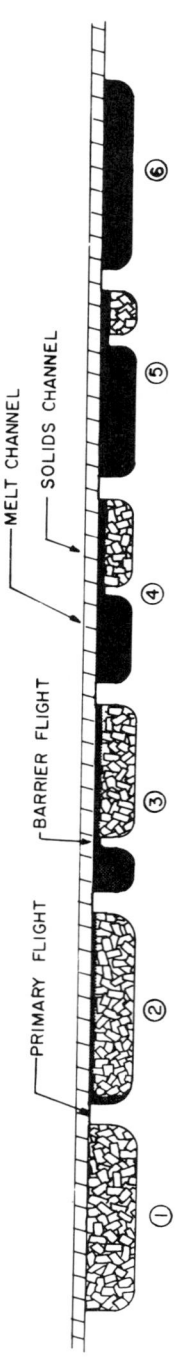

PRIMARY FLIGHT BARRIER FLIGHT MELT CHANNEL SOLIDS CHANNEL

① ② ③ ④ ⑤ ⑥

(1) The feed section establishes the solids conveying in the same way as a conventional screw.

(2) At the beginning of the transition (compression), a second flight is started. This flight is called the barrier or intermediate flight, and it is undercut below the primary flight OD. This barrier flight separates the solids channel from the melt channel.

(3) As melt progress down the transition, melting continues as the solids are pressed and sheared against the barrel, forming a melt film. The barrier flight moves under the melt film and the melt is collected in the melt channel. In this manner, the solid pellets and melted polymer are separated and different functions are performed on each.

(4) The melt channel is deep, giving low shear and reducing the possibility of overheating the already melted polymer. The solids channel becomes narrower and/or shallower forcing the unmelted pellets against the barrel for efficient frictional melting. Break up of the solids bed does not occur to stop this frictional melting.

(5) The solids bed continues to get smaller and finally disappears into the back side of the primary flight.

(5) The solids bed continues to get smaller and finally disappears into the back side of the primary flight.

(6) All of the polymer has melted and gone over the melt flight. Melt refinement can continue in the metering section. In some cases mixing ssections are also included downstream of the barrier section. In general, the melted plastic is already fairly uniform upon exit from the barrier section.

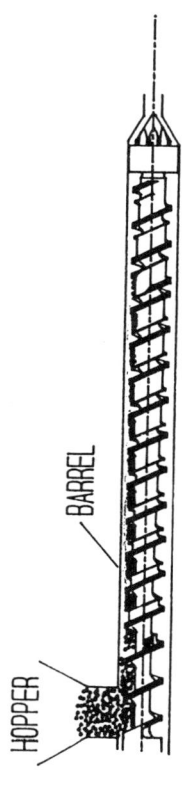

HOPPER BARREL

BARRIER TYPE SCREW (BARR-2 SHOWN)

Fig. 2-8a. Melt model of the barrier screw.

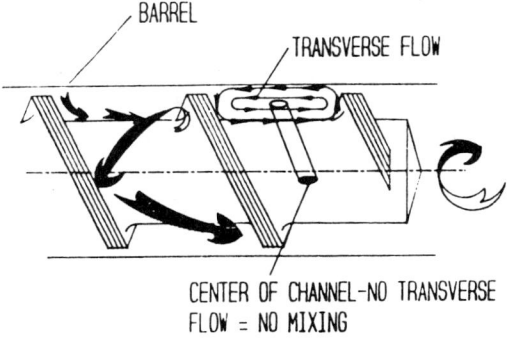

Fig. 2-8b. Typical flow profile in conventional screw.

ficulty running at a low-enough melt heat for high molecular weight plastics because it does rely on high shear melting. The Chung Melt Simulator was developed to directly apply melting to an improved screw design (104). An example of its use was in the Barr ET barrier screw for GE's Xenoy 1100 and 1500 (high performance plastics) (105). Simulator test results produced a screw that met IM requirements, as shown in Table 2-3. The ET screw initially follows the conventional solid bed mechanism, but the ET section is located at the end of the relatively short transition (compression) section. The result is melting rates that are up to 30 percent higher. The melting mechanism involved, which is similar to that seen in many twin screw extruders, is referred to by Nichols (106) as "dissipative mix melting." In addition to its higher melting capacity, the ET is transferring energy from the molten polymer to the unmelted bed, with consequent lower melt heat, a distinct advantage in most processes (IM, extrusion, blow molding, etc.).

Mechanical Requirements

(The following calculations were presented in reference 100.)

Screws always run inside a stronger more rigid barrel. For this reason,

Table 2-3. Custom-Designed Barrier Screw; Barr Et Screw.

RPM	HP	MELT °F	BACK PRESSURE PSI	RECOVERY OZ/S
50	43	426	30	1.63
70	65	435	75	2.16
120	115	448	120	3.69

Screw diameter = $4\frac{1}{2}$ in., L/D = 21, barrel temperature profile = 370, 420, and 420°F, and processing GE's Xenoy 1500 pellets.

they are not subjected to high bending forces. The critical strength requirement is resistance to torque. This is particularly true of smaller screws with
diameters of $2\frac{1}{2}$ in. and less. Unfortunately, the weakest area of a screw is
the portion subject to the highest torque. This is the feed section, which
has the smallest root diameter. A rule of thumb is that a screw's ability to
resist twisting failure is proportional to the cube of the root diameter in the
feed section. The maximum torque that can be applied to a screw can be
calculated by the following formula:

$$T = \frac{5{,}250 \ (HP)}{(RPM)}$$

where:

T = Torque in foot pounds
HP = Horsepower of the drive motor
RPM = Lowest screw speed in revolutions per minute at which the drive
can deliver full horsepower. (Be careful here because some
variable speed drives can deliver full horsepower at less than
full speed.)

Once we have the maximum torque, we can determine whether the screw
can take the torque, or can use the information to design the feed depth
section. Of course, other feeding characteristics are also important. Here
is an approximate formula used to determine the minimum root diameter
of the screw:

$$D_{\min.} = \sqrt{3 \ \frac{102 \ T}{S_t} + d^3}$$

where:

D = Root diameter of the screw in the deepest section (usually the feed).
d = Diameter of the cooling hole if there is one; a very minor factor in
extrusion screws. Use $d = 0$ when there is no hole.
T = Torque in foot pounds.
S = Tensile yield strength in psi. (Obtain from the screw material chart,
Table 2–4. The assumption has been made that the torsional strength
is 60% of the tensile yield strength.)

This formula is intended strictly as a guide, and is not exact for a number
of reasons. (1) We assume that torsional shear strength is 60% of the tensile
yield. Torsional shear strength is inexact at best, and there are very few
places where you can find these data. (2) We do not take into consideration

Table 2-4. Common Materials Used in Screw Construction (100).

	TENSILE YIELD STRENGTH PSI	HARDNESS AS MACHINED ROCKWELL	AVAILABILITY OF CASE HARDNESS ROCKWELL	O.D. WEAR RESIST.	ROOT WEAR RESISTANCE	CORROSION RESISTANCE	MATERIAL AVAILABILITY	EASE OF MACHINING	APPROX. COST PER POUND $
ALLOY STEELS									
AISI 4140	100,000	28–32Rc[1]	48–55Rc[1]	Fair[1]	Poor	Poor	Excellent	Fair	.90
AISI 4340	110,000	28–32Rc[1]	48–55Rc[1]	Fair[1]	Poor	Poor	Fair	Fair	1.30
Stressproof[2]	100,000	30Rc	48–55Rc[1]	Fair[1]	Poor	Poor	Fair	Fair	.90
Nitralloy[3] 135M	85,000	33Rc	60–70Rc[4]	Good[5]	Good	Poor	Poor	Fair	1.65
STAINLESS STEELS									
304	335,000	80Rb	90Rb	Poor[5]	Poor	Good	Good	Fair-Poor	3.40
316	35,000	95Rb	90Rb	Poor[5]	Poor	Good	Good	Fair-Poor	4.15
416	115,000	30Rc	30Rc	Poor[5]	Poor	Fair-Good	Poor	Fair-Poor	2.95
17-4PH[6]	175,000	38Rc	42Rc[5]	Poor[5]	Fair-Poor	Good	Fair	Fair-Poor	2.90
15-5PH[6]	175,000	38Rc	42Rc[5]	Poor[5]	Fair-Poor	Good	Poor	Fair-Poor	4.40

(continued)

Table 2-4. Continued

	TENSILE YIELD STRENGTH PSI	HARDNESS AS MACHINED ROCKWELL	AVAILABILITY OF CASE HARDNESS ROCKWELL	O.D. WEAR RESIST.	ROOT WEAR RESISTANCE	CORROSION RESISTANCE	MATERIAL AVAILABILITY	EASE OF MACHINING	APPROX. COST PER POUND $
TOOL STEELS									
CPM-10V[7]	275,000	98Rb	56–58Rc	Excellent	Excellent	Fair	Fair	Fair	15.00
D-2	>300,000	96Rb	58–60Rc	Good	Good	Poor	Good	Fair	4.50
H-13	>300,000	96Rb	50–60Rc	Good	Good	Poor	Good	Fair	3.95
SPECIALTY MATERIALS									
Duranickel[8] 301	125,000	30–38Rc	32Rc[5]	Poor[5]	Poor	Excellent	Very poor	Poor	25.00
Hastelloy[9] 276	80,000	86Rb	86Rb[5]	Poor[5]	Poor	Excellent	Very poor	Poor	25.00

(1) Flame or induction hardened. (2) Trademark of La Salle Co. (3) Trademark of Joseph T. Ryerson & Son Inc. (4) Nitrided. (5) Usually improved by hardsurfacing. (6) Trademark of Armco Steel Corp. (7) Trademark of Crucible Specialty Metals. (8) Trademark of Huntington Alloys Inc. (9) Trademark of Cabot Corp.

the added strength provided by the flight. We merely calculate the section as a cylinder. This gives a built-in safety factor. (3) There is no consideration given to rapid shock loads, which can be a significant factor. (4) Stress concentration factors can vary widely, depending on the radius at the bottom of the flight. (5) All strengths given in the tables and in most other places are at room temperature. They are bound to be less at operating temperatures.

If you do not wish to go through the step of calculating torque, the following formulas can be used to calculate: (1) minimum root diameter; (2) minimum screw RPM at full HP; (3) maximum HP that can be applied to a screw with known dimensions. These formulas have the same limitations as the above formula for torque (100).

$$D_{min.} = \sqrt[3]{\frac{535,000\ (HP)}{(RPM)S_t} + d^3} \qquad RPM_{min.} = \frac{535,000\ (HP)}{(D^3 - d^3)S_t}$$

$$HP_{max.} = \frac{(D^3 - d^3)(RPM)\ S_t}{535,000}$$

By interrelating mechanical factors and obtaining the best combination of screw geometry, power input, and speed of rotation, the processor directly affects an important IM characteristic: screw recovery rate and plasticizing capacity. Energy consumption vs. plasticizing capacity is shown in Fig. 2–9, using the SPI-SPE plasticizing standard.

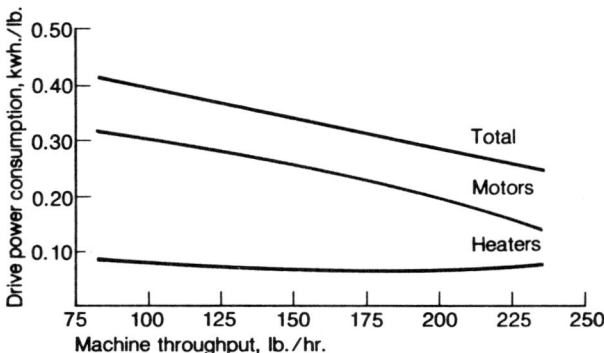

Fig. 2–9. Example of energy consumption vs. melt throughput using general purpose polystyrene with screw running continuously.

Valves/Screw Tips

When the melt is forced into the mold, the plunger action could cause the melt to flow back into the screw flights. Generally, with heat-sensitive resins such as PVC and thermosets, a plain or smearhead screw tip is used. For other resins this is not adequate, and a number of different check valves are used. These devices work in the same manner as a check valve in a hydraulic system, allowing fluid to pass only in one direction. These check valves, which are basically a sliding ring or ball check design (Fig. 2–10) are supplied by many manufacturers. Here are some comparisons (100):

	Advantages	Disadvantages
Sliding ring valve:	• Greater streamlining for less degrading of materials. Best for heat-sensitive materials. • Less barrel wear. • Less pressure drop across valve. • Best for vented operation. • Easier to clean. • Less expensive than side discharge ball check.	• Less positive shut off, especially in 4½″ dia. and larger sizes. Less shot control. • More expensive than front discharge ball check.
Ball check valve:	• More positive shutoff. Better shot control. • Front discharge ball check less expensive than sliding ring valve.	• Less streamlined. More degrading of heat-sensitive materials. • More barrel wear. • Side discharge type more expensive than sliding ring type. • Greater pressure drop, creating more heat. • Poor for vented operation. • Harder to clean.

Table 2–5 shows the influence of check valves on plastics performance.

VENTED BARRELS

Moisture retention in and on plastics has always been a problem for all processors. Surface moisture or moisture absorbed within the plastic can cause splay, an unsightly surface defect of the molded part, and reduce mechanical properties. The increased use of hygroscopic plastics (see Chapter 9) also requires care and the assurance of proper drying of material via the usual technique, using dryers and/or vented barrels (Fig. 2–11). There are advantages of using vented barrels as opposed to the more familiar dryers (100):

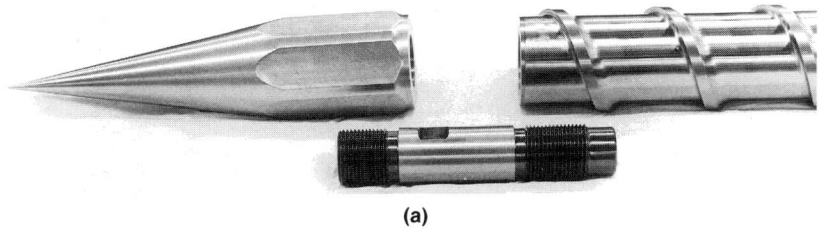

(a)

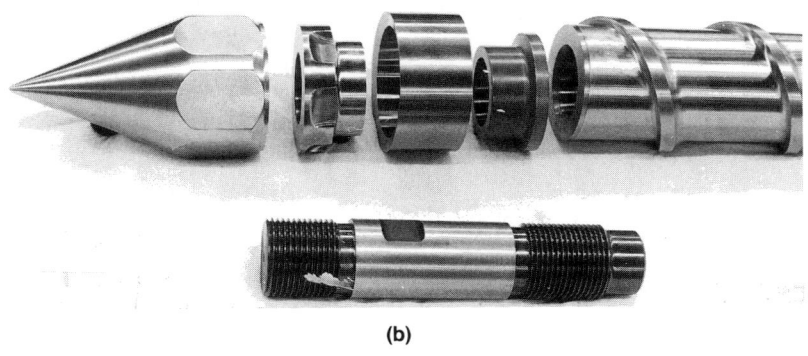

(b)

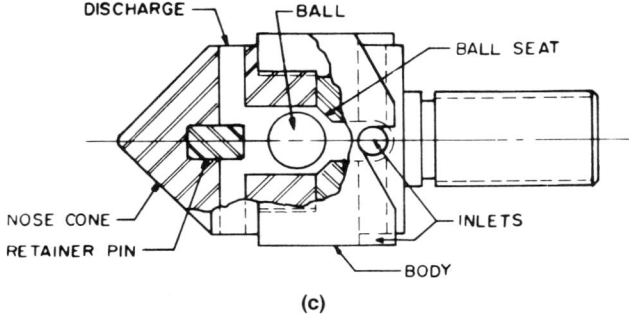

(c)

Fig. 2-10. Basic screw tip designs. (a) Plain or smearhead (dissaembled). (b) Sliding ring (disassembled). (c) Ball check (assembled).

Table 2-5. Effect of Varying Check Valves and Processing Conditions on Final Physical Properties of Dry Blended Polystyrene Copolymer.

	CONTROL	CHECK VALVE	GATE SIZE	BACK PRESSURE	SCREW SPEED	FILL TIME
Check valve	Ring	Ball	Ring	Ring	Ring	Ring
Gate size, in.	0.13×0.25	0.13×0.25	0.062×0.063	0.13×0.25	0.13×0.25	0.13×0.25
Back pressure, psi	0	0	0	125	0	0
Screw speed, RPM	73	73	73	73	52	73
Fill time, s	1	1	1	1	1	4
Notched Izod impact strength, ft/lb/in.	3.2	2.6	1.9	1.5	4.0	3.9
Flexural strength, 10^3 psi	17	17	17	17	19	18
Flexural modulus, 10^6 psi	0.98	0.98	0.98	0.98	1.00	0.98

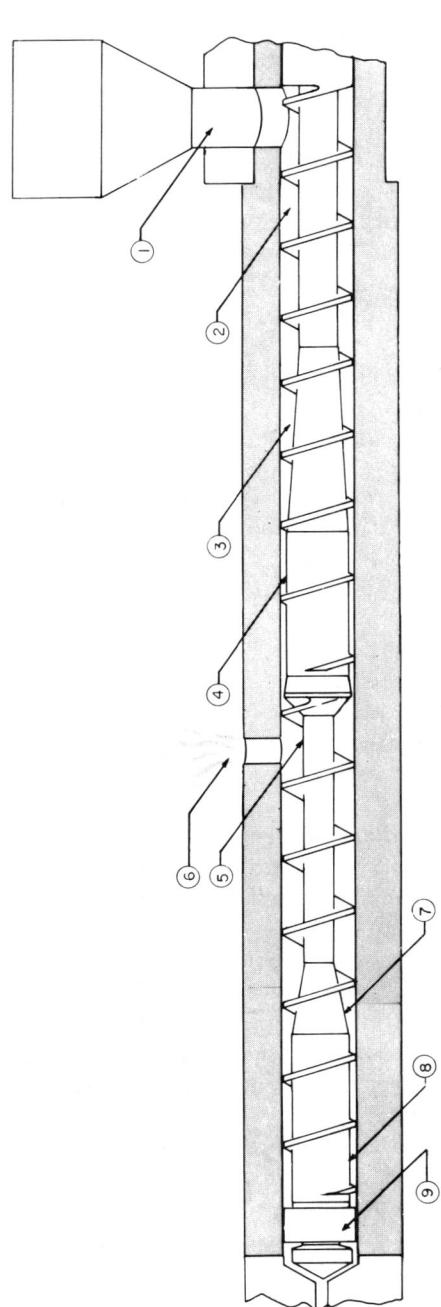

(1) Wet material enters from a conventional hopper. (2) The pellets are conveyed forward by the screw feed section, and are heated by the barrel and by some frictional heating. Some surface moisture is removed here. (3) The compression or transition section does most of the melting. (4) The 1st metering section accomplishes final melting and evens flow to the vent section. (5) Resin is pumped from the 1st metering section to a deep vent or devolitizing section. This vent section is capable fo moving quantities well in excess of the material delivered to it by the 1st metering section. For this reason, the flights in the vent section run partially filled and at zero pressure. It is here that volatile materials such as water vapor, and other nondesirable materials, escape from the melted plastic. The vapor pressure of water at 500°F is 666 psi. These steam pockets escape the melt, and travel spirally around the partially filled channel until they escape out the vent hole in the barrel. (6) Water vapor and othe volatiles escape from the vent. (7) The resin is again compressed and pressure is built in the 2nd transition section. (8) The 2nd metering section evens the flow and maintains pressure so that the screw will be retracted by the pressure in front of the non-return valve. (9) A low resistance, sliding ring, non-return valve works in the same manner as it does with a nonvented screw.

Fig. 2-11. Basic operation of a vented barrel.

63

1. *Elimination of predrying.* A vented injection unit removes moisture more completely without a dryer. Often a dryer cannot do the job completely in a reasonable time period.

2. *Rapid startup and color or material changes.* You do not have to wait for hours when starting up or when changing colors or materials. This increases machine and personnel utilization.

3. *Superior parts.* The improved melt, free of volatiles, renders higher-quality parts with excellent appearance and better physical properties. Splay marks are eliminated from appearance parts and parts to be plated.

4. *Energy efficiency.* The vented machine uses less energy. Btu's are not lost while material stands in large hoppers at elevated temperatures for long periods. Dryers are large users of energy.

5. *Removal of other volatiles.* Water vapor is not always the only volatile contaminant that should be removed. The vent removes other undesirable materials that come off at temperatures not possible in a dryer. Of course, escape of volatiles is easier from a melted and agitated plastic. This has been very effective in solving mold and ejector pin plate-out problems.

6. *Elimination of dryer maintenance.* Dryers are high maintenance items with clogged filters, heater element burnout, and contaminated desiccant beds. Even in shops with good routine maintenance programs, it is common to operate with ineffective dryers for long periods before they are noticed. When this happens, quality goes down, and scrap accumulates.

7. *Contamination and material handling cut.* There is no need to clean out larger complicated hopper dryer systems on every material or color change. The simple, lightweight, standard hopper is easier to clean.

8. *Less space required.* The hopper dryer requires a large volume in order to get up to 5 hours of drying time. This means a heavy, large, and high hopper that may not fit into the space available.

9. *Elimination of dryer variability.* The variation in part quality and appearance due to changes in dryer performance is eliminated. The vent operates the same all the time.

10. *Greater use of regrind.* The improved moisture removal ability of the vent allows the use of percentages of regrind. The vent also allows the storage of materials in open containers.

11. *Reduced mold venting.* The removal of volatiles from the vent reduces the mold venting problem. It can also eliminate the problem of clogged mold vents.

Care is required when a plug is placed in the vent opening so that a vented barrel can be used in the same way as a solid nonvented barrel. Internal barrel pressure can develop that exceeds the strength of the bolt retaining the plug, causing the plug to be released with an explosive action. Operators may be hit by the plug and/or the hot melt. One safety precaution is to rotate the barrel so that the plug is pointing downward or away from the operator. The most important safety measure is to use retaining bolts (or another method of plug attachment) that will provide more than enough strength to prevent plug release. The injection molding or extrusion barrel manufacturer should be able to supply the information needed to ensure safe operation. Another precaution is to install a pressure gauge at the head of the barrel and establish a maximum pressure at which the barrel can be safely operated. To eliminate any run-away situation, shear bolts or a rupture disc can be installed near the plug.

During start-up and operation of a vented barrel, make sure that the barrel in front of the vent is at a temperature above the "freezing" point of the plastic. To help retain heat in the metal barrel, thermal insulation should be used to cover exposed metal surfaces (this action should be taken even if the barrel is not vented).

Air Entrapment

Air can be entrapped in the melt during processing. This can happen when plastic (pellets, flakes, etc.) is melted in a normal air environment (as in a plasticating extrusion process or in an injection barrel, compression mold, casting form, spray system, etc.), and the air cannot escape. Generally the melt is subject to a compression load, or even a vacuum, which causes release of air; but in some cases the air is trapped. If air entrapment is acceptable, no further action is required. However, it is usually unacceptable, for reasons of both performance and aesthetics.

Changing the initial melt temperature in either direction may solve the problem. With a barrel and screw, it is important to study the effects of temperature changes. Another approach is to increase the pressure in processes that use process controls. Particle size, melt shape, and the melt delivery system may have to be changed or better controlled. A vacuum hopper feed system may be useful. With screw plasticators, changes in screw design may be helpful. Usually, a vented barrel will solve the problem.

The presence of bubbles could be due to: air alone or moisture, plastic surface agents or volatiles, degradation, or the use of contaminated regrind. With molds such as those used for injection, compression, casting, or reaction injection, air or moisture in the mold cavity will be the culprit. So the first step to resolving a bubble or air problem is to be sure what problem exists. A logical troubleshooting approach can be used.

SHOT-TO-SHOT VARIATION

During IM, shot-to-shot variations can occur. Major causes of inconsistency are worn nonreturn valves, bad seating of a nonreturn valve, a broken valve ring, a worn barrel in the valve area, or a poor heat profile. To identify the cause, one follows a logical procedure (1-3, 107). Any problem caused by the valve will cause the screw to rotate in the reverse direction during injection. To locate the trouble, one must pull and inspect the valve, and check the OD of the ring for wear. The inspector looks for a broken valve stud (caused by cold startup when the screw is full of plastic), bad seating of the ring or ball (angles of the ring ID and the seat must be different, in order to ensure proper shutoff action at the ID of the ring), and a broken ring. One checks the dimensions of the valve and compares them with those determined before using the machine.

A poor heat profile for crystalline resins can cause unmelted material to be caught between the ring and the seat, holding the valve open and allowing leakage. A change in the heat profile (see Chapter 1) or the machine's plasticizing capacity is not sufficient to correct the problem. For any resin, if the problem does not occur with every shot, the cause may be improper adjustment or damaged barrel heat controls.

Nonuniform melt density could be caused by nonuniform feeding to the screw and/or the regrind blend, which could have a different bulk density. Increasing the back pressure may help (107). This throughput condition, the residence time of the plastic in the barrel, and the barrel heat profile are all important in obtaining the best melt quality. The heat profile is the most important parameter and varies from resin to resin, as well as with different cycle times and shot sizes. As the following example shows, a screw operating under two different conditions will produce different results.

Consider a screw that is a 2-in. diameter, $20/1$ L/D with 20 oz melt screw capacity. With a 15 s cycle and a shot size of 2 oz, it operates as follows:

20 oz (screw capacity) $\div$ 2 oz = 10 cycles.
15 s cycle = 4 cycles/min.
10 cycles $\div$ 4 = 2.5 min. of residence time, from the time plastic starts through the screw until it enters the mold.

Another set of requirements uses a 6 oz shot size with the same 15-second cycle:

20 oz ÷ 6 = 3.33 cycles.

3.33 cycles ÷ 4 = 0.83 min. of residence time.

In the second case, a higher rate of melting will be required, with the probability that the screw will be inadequate for the melt, and problems will develop.

The inventory in a screw will run between 1½ and 2 times the maximum shot size rating in polystyrene. With other resins, calculate the differences in density to arrive at the maximum shot size and the expected inventory (108).

PURGING

Purging has always been a necessary evil, consuming substantial amounts of materials, labor, and machine time, all nonproductive. In IM (extrusion, blow molding, etc.), sometimes it is necessary to run hundreds of pounds of resin to clean out the last traces of a dark color before changing to a lighter one. Sometimes there is no choice but to pull the screw for a thorough cleaning. Although there are few generally accepted rules on how to purge, the following tips should be considered: (1) try to follow less viscous with more viscous resins; (2) try to follow a lighter color with a darker color resin; (3) maintain the equipment; (4) keep the materials handling equipment clean; and (5) use an intermediate resin to bridge the temperature gap (such as that encountered in going from acetal to nylon), and use a PS as a purge.

Ground/cracked cast acrylic and PE-based materials are the main purging agents, but others also are commercially available for certain machines and materials. Cast acrylic, which does not melt completely, is suitable for virtually any resin. About one pound for each ounce of injection capacity will be needed (5–10 lb/in. of screw diameter in an extruder). With extruders, special conditions and preparations are required, which suppliers of the purging compound can recommend (remove dies, screen packs, etc.).

PE-based compounds usually contain abrasive and release agents. They are used to purge the "softer" TPs (polyolefins, styrenes, some PVCs, etc.). With extruders, many of the requirements/restrictions do not apply.

These purging agents function by mechanically pushing and scouring residue out of the machines. Others also apply chemical means. Tables 2-6 and 2-7 provide information on purging.

CLAMPING FORCE

The clamping force required to keep the mold closed during injection must exceed the force given by the product of the live cavity pressure and the

Table 2-6. Guidelines for Purging Agents.

MATERIAL TO BE PURGED	RECOMMENDED PURGING AGENT
Polyolefins	HDPF
Polystyrene	Cast acrylic
PVC	Polystyrene, general-purpose, ABS, cast acrylic
ABS	Cast acrylic, polystyrene
Nylon	Polystyrene, low-melt-index HDPE, cast acrylic
PBT polyester	Next material to be run
PET polyester	Polystyrene, low-melt-index HDPE, cast acrylic
Polycarbonate	Cast acrylic or polycarbonate regrind; follow with polycarbonate regrind; do not purge with ABS or nylon
Acetal	Polystyrene; avoid any contact with PVC
Engineering resins	Polystyrene, low-melt-index, HDPE, cast acrylic
Fluoropolymers	Cast acrylic, followed by polyethylene
Polyphenylene sulfide	Cast acrylic, followed by polyethylene
Polysulfone	Reground polycarbonate, extrusion-grade PP
Polysulfone/ABS	Reground polycarbonate, extrusion-grade PP
PPO	General-purpose polystyrene, cast acrylic
Thermoset polyester	Material of similar composition without catalyst
Filled and reinforced materials	Cast acrylic
Flame-retardant compounds	Immediate purging with natural, non-flame-retardant resin, mixed with 1% sodium stearate

Table 2-7. Guidelines for Resin Changes.

MATERIAL IN MACHINE	MATERIAL CHANGING TO	MIX WITH RAPID-PURGE AND SOAK	TEMPERATURE BRIDGING MATERIAL	FOLLOW WITH
ABS	PP	ABS	—	PP
ABS	SAN	SAN	—	SAN
ABS	Polysulfone	ABS	PE	Polysulfone
ABS	PC	ABS	PE	PC
ABS	PBT	ABS	PE	PBT
Acetal	PC	Acetal	PE	PC
Acetal	Any material	PE	—	New material
Acrylic	PP	Acrylic	—	PP

Table 2-7. Continued

MATERIAL IN MACHINE	MATERIAL CHANGING TO	MIX WITH RAPID-PURGE AND SOAK	TEMPERATURE BRIDGING MATERIAL	FOLLOW WITH
Acrylic	Nylon	Acrylic	—	Nylon
TPE	Any material	PE	—	New material
Nylon	PC	PC	—	PC
Nylon	PVC	Nylon	PE	PVC
PBT	ABS	PBT	PE	ABS
PC	Acrylic	PC	—	Acrylic
PC	ABS	PC	PE	ABS
PC	PVC	PC	PE	PVC
PE	Ryton	PE	PE	Ryton
PE	PP	PP	—	PP
PE	PE	PE	—	PE
PE	PS	PS	—	PS
PETG	Polysulfone	PETG	—	Polysulfone
Polysulfone	ABS	Polysulfone	PE	ABS
Polysulfone	ABS	Cracked acrylic	—	ABS
PP	ABS	ABS	—	ABS
PP	Acrylic	Acrylic	—	Acrylic
PP	PE	PE	—	PE
PP	PP	PP	—	PP
PS	PP	PP	—	PP
PVC	Any material	LLDPE or HDPE	—	New material
PVC	PVC	LLDPE or HDPE	—	PVC
PPS	PE	PPS	PE	PE
SAN	Acrylic	Acrylic	—	Acrylic
SAN	PP	SAN	—	SAN

total projected area of all impressions and runners. The projected area can be defined as the area of the shadow cast by the molded shot when it is held under a light source, with the shadow falling on a plane surface parallel to the parting line.

With cold-runner systems for TPs (or hot-runner systems for TSs), the projected areas include runners and sprues. For hot-runner TPs (cold-runner TSs), runners and sprue are not included. As an example, if the projected area is 132 sq. in., and a pressure of 4,000 psi is required, the clamp force is:

132 sq. in. × 4,000 psi = 528,000 lb
528,000 lb ÷ 2,000 lb = 264 tons

One should consider including a safety factor of 10 to 20 percent to ensure sufficient clamping force, particularly when not familiar with the operation and/or material. Then the clamping force would be 290 to 317 tons in the cavity. However, because of partial hardening of plastic as it flows through the relatively restrictive sprue and runners in a cold-runner TP system (or a hot-runner TS system), the actual pressure in the cavity is less than the applied plunger pressure (Fig. 2–1).

The actual pressure developed within a mold cavity varies directly with the thickness of the molded section and inversely with the melt viscosity. Thick sections require greater clamping force than thin sections because the melt in the thick sections stays semifluid longer. Similarly, a higher stock heat, hotter mold, larger gates, or a faster rate of injection will require a higher clamping force (1, 2).

Whereas the projected area determines the clamping force, the weight or volume of a shot determines the capacity of the IM machine required. For the hot runners of TPs, the shot size includes the gate and runners. Capacities of machines are generally rated in ounces of general purpose PS; with other resins, convert to the correct capacity by relating the resin densities to that of PE. If the shot size is based on volume, densities are not involved.

Using too much clamping force has drawbacks: (1) a slower cycle time; (2) possible damage to the mold; (3) reduced venting; (4) possible damage to platens if a small mold is used in a large-platen machine (the machine builder should provide information); and (5) extra energy consumption in hydraulics.

To determine the proper clamping force, one must start at low force and begin molding at a reduced injection pressure, gradually building up melt pressure until the mold fills without flashing. If flash occurs prior to mold fill, it is necessary to increase the clamp tonnage gradually until no flash exists. Then the mold lockup is set correctly. If the melt temperature is lowered, the injection pressure and time will have to be changed, with a possible increase of clamp force.

If the mold still flashes, usually in the center, the mold clamping area may be too large. To determine if this is the problem, one rubs machinist bluing on the face of one mold half, and then clamps and opens the mold to see if bluing was transferred. With incomplete transfer, the processor puts the mold in a machine with higher clamp tonnage, and operating costs go up. Running the mold in the original machine would require removal of probably 0.010 in. of metal in the cavity areas that were not blued (1–3, 6, 11, 33–36, 100, 108).

Tie-Bar Growth

One problem that most controls do not consider involves the effect of heat on tie-bars, which can directly influence mold performance, particularly at start-up. If the heat differs from top and bottom bars, it is necessary to insulate the mold from the platens. The insulator pad used also confines heat more to the mold, producing savings in heat and/or better heat control. The following paragraphs describe calculation of the elongation of these usually evenly heated bars, and of thermal mold growth (1).

Tie-Bar Elongation. The change in tie-bar length, *e,* can be calculated as follows:

$$e = \frac{F \times L}{E \times A}$$

where F = force per tie bar, L = bar length, E = modulus of elasticity, and A = cross-sectional area of bar.

At maximum die height (178 in.) on a 500-ton injection-molding machine with a tie-bar diameter of 6 in. (or a cross-sectional area of 28.27 sq. in.), tie-bar elongation equals:

$$e_{max} = \frac{250,000 \text{ lb} \times 178 \text{ in.}}{30 \times 10^6 \text{ psi} \times 28.27 \text{ in.}^2} = 0.0524 \text{ in.}$$

At minimum die height (146 in.), the elongation is:

$$e_{min} = \frac{250,000 \text{ lb} \times 146 \text{ in.}}{30 \times 10^6 \text{ psi} \times 28.27 \text{ in.}^2} = 0.0430 \text{ in.}$$

To calculate the effect of a small change in elongation on the force on a tie-bar, one solves for *F:*

$$F = \frac{eEA}{L}$$

At maximum die height, the change in force per 0.001 in. elongation equals:

$$F_{max} = \frac{0.001 \text{ in.} \times 30 \times 10^6 \text{ psi} \times 28.27 \text{ in.}}{178 \text{ in.}} = 4,764 \text{ lb}$$

At minimum die height, the change in force for the same elongation is:

$$F_{max} = \frac{0.001 \text{ in.} \times 30 \times 10^6 \text{ psi} \times 28.27 \text{ in.}}{146 \text{ in.}} = 5,808 \text{ lb}$$

Thermal Mold Growth. Uneven mold growth can occur with a temperature differential across the mold. Mold growth, *G,* is calculated by the following formula:

$$G = \text{Mold length} \times \text{Coefficient of linear expansion} \\ \times \text{Mold temperature}$$

In a 20-in.-long mold, where the temperatures are 100 and 120°F, mold growth equals:

$$G_{100} = 20 \text{ in.} \times 6 \times 10^{-6} \text{ in. per in. per degree} \times 100°F = 0.0120 \text{ in.}$$
$$G_{120} = 20 \text{ in.} \times 6 \times 10^{-6} \text{ in. per in. per degree} \times 120°F = 0.0144 \text{ in.}$$

The difference in growth on different sides of the mold, then, is:

$$G_{120} - G_{100} = 0.0144 - 0.0120 = 0.0024 \text{ in.}$$

MOLDS

A mold must be considered as one of the most important pieces of production equipment in the plant. It is a controllable complex device (Fig. 2–12) that must be an efficient heat exchanger. If not properly handled and maintained, it will not be an efficient operating device. Hot melt, under pressure, moves rapidly through the mold. Water or some other medium circulates in the mold to remove heat (for TPs) or add heat (for TSs). Air is released from cavities to eliminate melt burning and/or voids in the part. All kinds of action operate, including sliders and unscrewing devices (1). Parts are ejected (knockout pins, air, etc.) at the proper time. These basic operations in turn require all kinds of interaction, including such param-

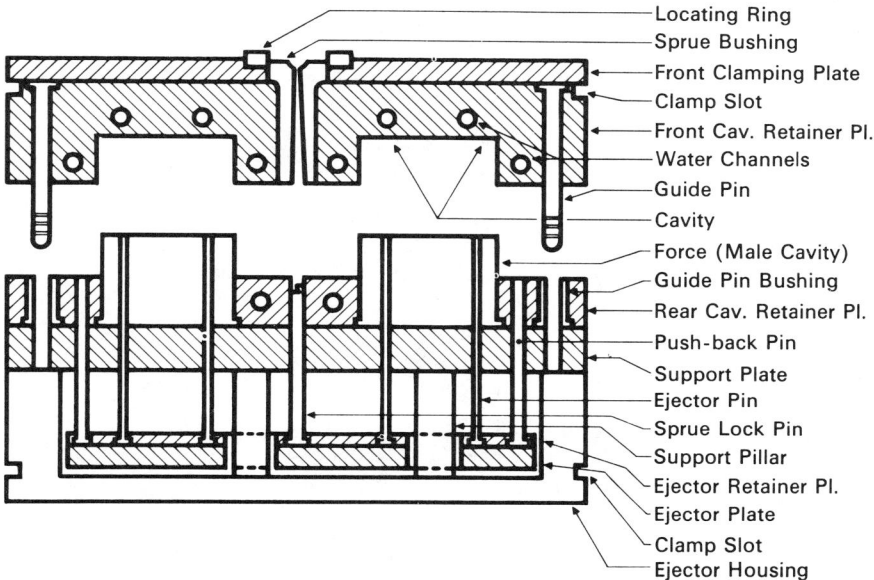

Locating Ring
Sprue Bushing
Front Clamping Plate
Clamp Slot
Front Cav. Retainer Pl.
Water Channels
Guide Pin
Cavity
Force (Male Cavity)
Guide Pin Bushing
Rear Cav. Retainer Pl.
Push-back Pin
Support Plate
Ejector Pin
Sprue Lock Pin
Support Pillar
Ejector Retainer Pl.
Ejector Plate
Clamp Slot
Ejector Housing

Fig. 2–12. Illustration of a two-part standard mold.

eters as fill-time, hold pressure, and other variables, as shown in Figs. 2–6 and 2–13. Typical mold steels are shown in Table 2–8.

Each of the plastics used has special distinctive properties. Some are abrasive or corrosive; others require very tight heat control and pressure (Fig. 2–6). Settings that work for one resin probably will not work for another; or machine change to a "duplicate" probably will require different settings. Shrinkage requires special attention. Crystalline resins shrink more than amorphous resins (see Chapter 1). Differential shrinkage can cause warpage. With tight part tolerances, it is necessary to leave more, rather than less, steel in the mold so that corrections requiring metal "cutting" can correlate processing with tolerances.

CAD and CAE programs are available that can aid in mold design and in setting up the complete process (1, 2, 4, 9, 29, 99). These programs are concerned with melt flow to part solidification, and the meeting of performance requirements. Many different factors are incorporated, including heat transfer, thermal conductivity, thermal expansion, coefficients of friction, machine and mold operating setup, and so on.

Some plastics, particularly filled ones, can be very abrasive. This necessitates use of abrasion-resistant metals in the mold or the application of special coatings (Table 2–9a). Some resins degrade during processing and

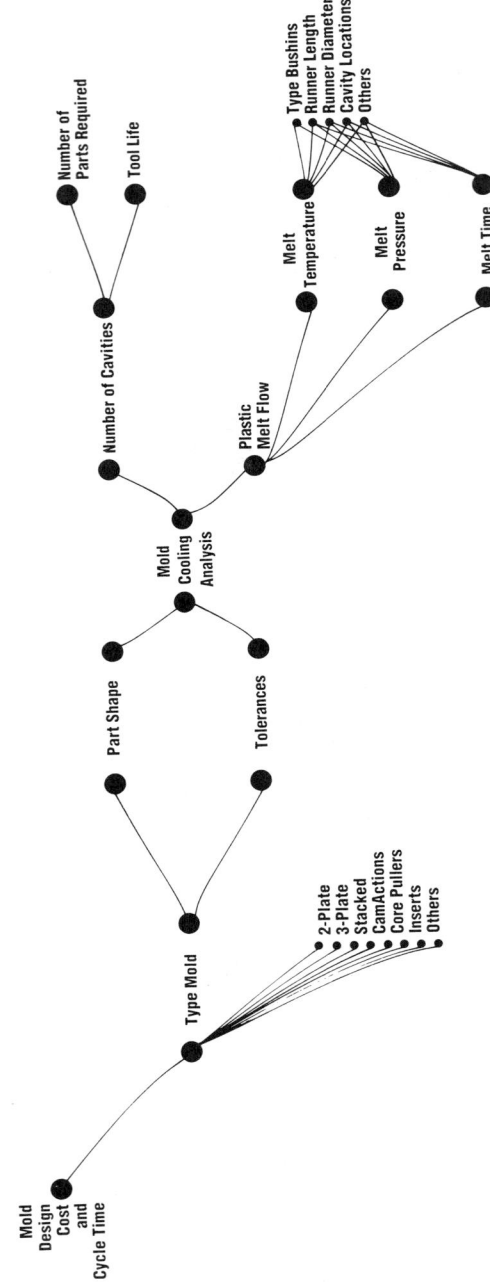

Fig. 2-13. Control factors important to mold operation.

Table 2–8. Examples of Steels Used in Different Parts of a Mold.

TYPE OF STEEL	TYPICAL USES IN INJECTION MOLDS
4130/4140	General mold base plates
P-20	High-grade mold base plates, hot-runner manifolds, large cavities and cores, gibs, slides, interlocks
4414 SS, 420 SS (pre-hardened) P5, P6	Best grade mold base plates (no plating required), large cores, cavities and inserts. Hobbed cavities
01	Gibs, slides, wear plates
06	Gibs, slides, wear plates, stripper rings
H-13	Cavities, cores, inserts, ejector pins and sleeves (nitrided)
S7	Cavities, cores, inserts, stripper rings
A2	Small inserts in high-wear areas
A6	Cavities, cores, inserts for high-wear areas
A10	Excellent for high-wear areas, gibs, interlocks, wedges
D2	Cavities, cores, runner and gate inserts for abrasive plastics
420 SS	Best all-around cavity, core, and insert steel; best polishability
440C SS	Small to medium-size cavities, cores, inserts, stripper rings
250, 350	Highest touchness for cavities, cores, small unsupported inserts
455M SS	High toughness for cavities, cores, inserts
M2	Small core pins, ejector pins, ejector blades (up to $\frac{5}{8}$ in. diam.)
ASP 30	Best high-strength steel for tall, unsupported cores and core pins

are corrosive; so proper materials are required for construction and coatings. To keep molds operating properly, mold cleaning is necessary. Certain plastics and mechanical mold operations may require daily cleaning (as recommended by resin suppliers) (Table 2–9b).

Runner System

Even though a molder has received a mold intended to meet part performance requirements at the lowest cost, it may not be cost-efficient because of a poor runner design. One can simulate the gate and runner design on commercial CAD programs such as Moldflow® before the runner system is built. Once it is built, very limited rework can be done.

Traditionally, there have been a number of misconceptions about proper runner design—for example, the larger the runner, the faster melt enters the cavity. Here larger actually means a longer chill-cycle time; a larger shot size and larger machine capacity are needed, and more scrap and a higher reprocessing cost with potential increase of contamination; there is a greater projected area, requiring a higher clamping force, and so on. As impossible as it appears, with high production it may pay to have extensive rework on the mold, runner, and cavity done to obtain higher profits. It is easy to

Table 2–9a. Protective Coatings for Molds/Dies.

MATERIAL	METHOD OF APPLICATION
Chromium	Plating
Nickel	Plating
Electroless nickel	Solution treatment
Nedox electroless nickel	Solution treatment followed by TFE impregnation; used on copper and ferrous alloys
Tufram TFE aluminum	Deep anodizing process followed by TFE impregnation; used on aluminum alloys
TFE ceramic	Spray and bake application; used for all die materials that can withstand 250°C bake
Tungsten silicide	Solution treatment; used on steel and ferrous alloys
Tungsten carbide	Explosion impact or flame spray with plasma arc; used for all high-melting metals to improve abrasion resistance
Aluminum oxide	Plasma flame spray; used for extreme abrasion resistance; used on steel dies but usually limited to small dies because of expansion problems; works best on 18-8 stainless
PTFE	Spray and bake application; used for low-friction and low-adhesion application; poor abrasion resistance
Polyimide, aramid	Straight organic coatings with high softening points (450–550°C), which are applied by spray and baked; low friction characteristics against some resins (for example, PVC); moderate abrasion resistance
Filled polyimide, aramid	Aramid and polyimide systems containing TFE and other fluorocarbon resins to improve the friction properties

have the mold properly designed initially with all the tools available (1, 34–36, 109–114).

PROCESS CONTROL

Process controls can range from unsophisticated to very sophisticated devices. Their cost includes the equipment and using them correctly (they take time, patience, and a willingness to learn new ways of molding). Figure 2–14 provides a simplified approach to understanding controls, and Fig. 2–15 reviews some variables that influence part performance. Figures 2–6 and 2–13 show the many parameters that are interfaced to develop the most efficient machine operation. Tradeoffs are inevitable in a complex operation such as IM. Many of these variables influence end results, and some of the variables interact.

The development of powerful computer controls and programs has greatly accelerated integrating process variables with a goal of zero defects at the

Table 2-9b. Guide to Mold/Die Cleaning Methods.

METHOD	MOLD WEAR	EXPENSE	CLEANING SPEED	CLEANING DEGREE	HAZARD WASTE	DISPOSAL PROBLEM	OPERATE HAZARD	DAMAGE CAUSED
Chemical tank	2	4	5	1	1	Yes	Yes*	a
Ultrasonic chemical	2	10	3	5	1	Yes	Yes*	a
Wet-blast glass bead	5	7	8	1	10	No	No	b, c
Dry-blast glass bead	8	1	10	10	10	No	No	b, c, d
Plastic dry-blast	1	1	10	10	10	No	No	None
Hand cleaning	2	5	5	10	5	No	No	e
Dry-blast with sand or other abrasive	10	1	10	10	10	No	No	b, c, d

No. 1 low/slow to No. 10 fast/best.
*Use proper precautions.
a = surface etching; b = surface pitted; c = round edge; d = removes chrome; e = filing/grinding premature surface/edge wear.

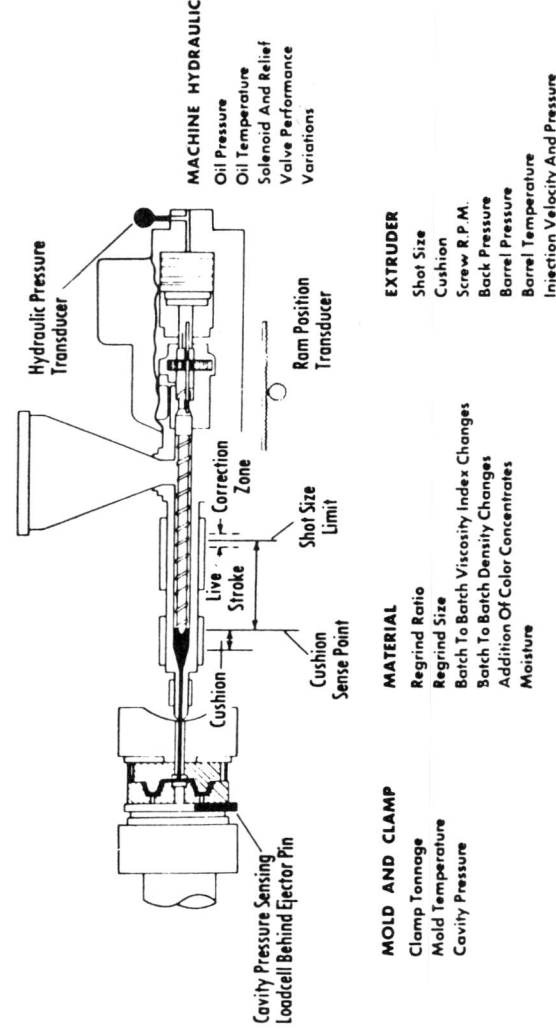

MACHINE HYDRAULICS

Oil Pressure
Oil Temperature
Solenoid And Relief
Valve Performance
Variations

EXTRUDER

Shot Size
Cushion
Screw R.P.M.
Back Pressure
Barrel Pressure
Barrel Temperature
Injection Velocity And Pressure

MATERIAL

Regrind Ratio
Regrind Size
Batch To Batch Viscosity Index Changes
Batch To Batch Density Changes
Addition Of Color Concentrates
Moisture

MOLD AND CLAMP

Clamp Tonnage
Mold Temperature
Cavity Pressure

Hydraulic Pressure Transducer

Ram Position Transducer

Correction Zone

Shot Size Limit

Live Stroke

Cushion Sense Point

Cushion

Cavity Pressure Sensing
Loadcell Behind Ejector Pin

Fig. 2-14. Examples of need for process controls.

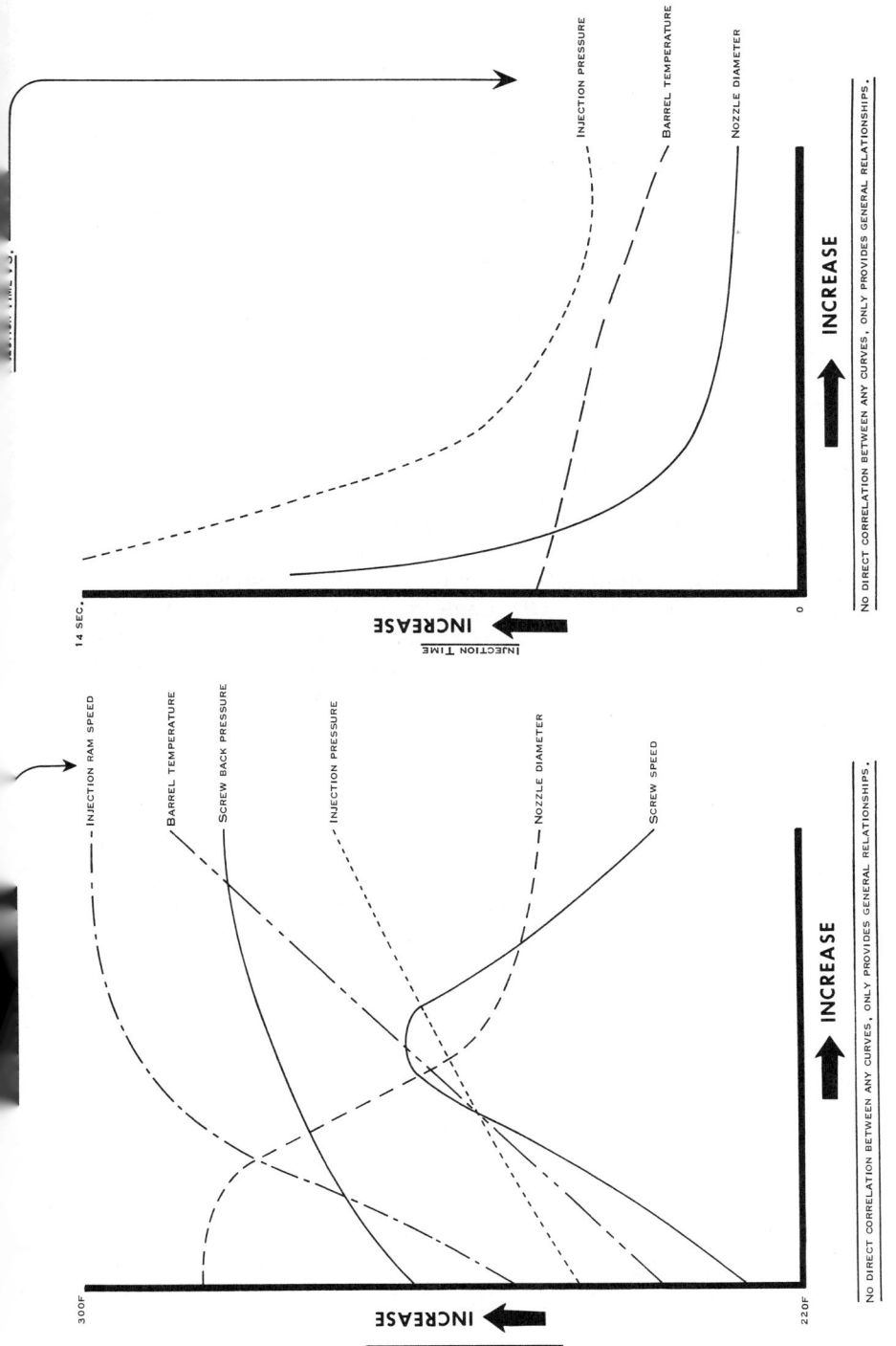

Fig. 2-15. (a) Injection stock temperature *vs.* injection ram speed, barrel temperature, screw back pressure, injection pressure, nozzle diameter, and screw speed. (b) Injection time *vs.* injection pressure, barrel temperature, and nozzle diameter.

79

lowest cost. The computer simplifies the fine-tuning of machine settings with molding variables. Examples include establishing melt conditions for mold filling and packing, which involves the simultaneous measurement and control of two or more critical variable parameters (Fig. 2–13). It is during this phase of operation that most variations make themselves evident and can be easily detected. A change in melt viscosity is reflected as a change in ram speed and can be detected by measuring the ram position with respect to time.

A change in resin viscosity is reflected as a change in melt pressure and can be detected by measuring mold or cavity pressure with respect to time. Other variations that similarly display themselves and can be detected include melt temperature, hydraulic pressure, oil temperature, and so on.

COMPUTER INTEGRATED INJECTION MOLDING

The ultimate result of Computer Integrated Injection Molding (CIIM) in software packages is to translate the results of computer simulation of the molding of a specific part into machine settings for specific microprocessor-controlled machines (Fig. 2–16). CIIM automates the entry of a large number of set points (Figs. 2–2 and 2–13) in microprocessor-controlled machines and maximizes their efficiency, on the basis of extensive development by Ernest C. Bernhardt (Plastics & Computer Inc., Montclair, NJ 07042, USA), a world leader on this subject (110, 111).

Microprocessor Control Systems

Microprocessor control systems (MCS) make it possible to completely automate an IM plant. They control machines, automatically, enabling them to achieve high quality and zero defects. These systems readily adapt to enhancing the ability of processing machines. There are many moldings that would be difficult, if not impossible, to produce at the desired quality level without this feature.

Once processing variables are optimized through computer simulation, these values are entered in computer programs in the form of a large number of machine settings. Establishing the initial settings during start-up is inherently complex and time-consuming. The many benefits of these systems are well recognized and accepted, but it is evident that self-regulation of IM can be effective only when the design of the part and the mold is optimized, and when the correct processing conditions for the operation have been predetermined. Otherwise, a self-regulating machine is confused and can provide conflicting instructions. The results could be disastrous, damaging the machine or the mold. Therefore, the efficient utilization of

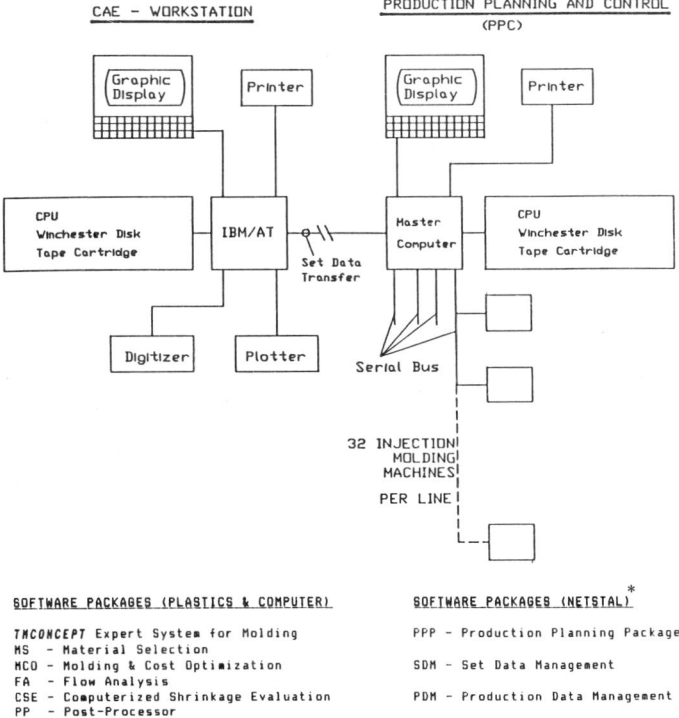

CAE – WORKSTATION

PRODUCTION PLANNING AND CONTROL
(PPC)

32 INJECTION
MOLDING
MACHINES
PER LINE

SOFTWARE PACKAGES (PLASTICS & COMPUTER)

TMCONCEPT Expert System for Molding
MS – Material Selection
MCO – Molding & Cost Optimization
FA – Flow Analysis
CSE – Computerized Shrinkage Evaluation
PP – Post-Processor

SOFTWARE PACKAGES (NETSTAL)*

PPP – Production Planning Package
SDM – Set Data Management
PDM – Production Data Management

Fig. 2–16. Overview of Computer Integrated Injection Molding (CIM).
*Swiss machine manufacturer.

microprocessor control systems depends on the success of utilizing correct and optimum programs (111).

Computerized Process Simulation of IM

The simulation approach replaces the traditional trial-and-error method. Programs, such as the Bernhardt TMConcept, are packaged for the complete molding process, including materials selection, molding and cost optimization, flow analysis, computerized shrinkage evaluation, and mold thermal analysis. The programs include mold filling, packing, and so forth, which accurately model the performance of microprocessor programmed injection. Major 3D CAD systems for part and mold design, as well as for structural and flow analysis, are integrated with these systems.

SHRINKAGE/PART TOLERANCE

Certain IM parts can be molded to extremely close tolerances of less than a thousandth of an inch, or down to 0.0 percent, particularly when TPs are filled with additives or TS compounds are used. To practically eliminate shrink and provide a very smooth surface, one should use a small amount of chemical blowing agent (< 0.5 percent, by weight) and a regular packing procedure. For conventional molding, tolerances can be met of ±5 percent for a part 0.020 in. thick, ±1 percent for 0.050 in., ±0.5 percent for 1.000 in., ±0.25 percent for 5.000 in., and so on. Thermosets generally are more suitable than TPs for meeting the tightest tolerances.

Economical production requires that tolerances not be specified tighter than necessary. However, after a production target is met, one should mold "tighter" if possible for greater profit. Table 2-10 reviews factors affecting tolerances. Many plastics change dimensions after molding, principally because molecular orientations/molecules are not relaxed (see Chapter 1). To ease or eliminate the problem, one can change the processing cycle so that the plastic is "stress-relieved," even though that may extend the cycle time, and/or heat-treat per the resin supplier's suggestions.

Table 2-10. Parameters that Influence Part Tolerance.

PART DESIGN:	Part configuration (size/shape). Relate shape to flow of melt in mold to meet performance requirements that should at least include tolerances.
MATERIAL:	Chemical structure, molecular weight, amount and type of fillers/additives, heat history, storage, handling.
MOLD DESIGN:	Number of cavities, layout and size of cavities/runners/ gates/cooling lines/side actions/knockout pins/etc. Relate layout to maximize proper performance of melt and cooling flow patterns to meet part performance requirements; preengineer design to minimize wear and deformation of mold (use proper steels); lay out cooling lines to meet temperature to time cooling rate of plastics (particularly crystalline types).
MACHINE CAPABILITY:	Accuracy and repeatability of temperature/time/velocity/ pressure controls of injection unit, accuracy and repeatability of clamping force, flatness and parallelism of platens, even distribution of clamping on all tie rods, repeatability of controlling pressure and temperature of oil, oil temperature variation minimized, no oil contamination (by the time you see oil contamination damage to the hydraulic system could have already occurred), machine properly leveled.
MOLDING CYCLE:	Set up the complete molding cycle to repeatedly meet performance requirements at the lowest cost by interrelating material/machine/mold controls.

Mold Shrink Allowances

An easy method for estimating shrink allowance is as follows:

$$M = (1 + S) L$$

where M = mold dimension, S = plastic shrinkage (in./in. or mm/mm), and L = part dimension.

If parts are small and have thin walls, this estimate is the best guide. If parts are larger (> 10 in.) and/or use rather high-shrink plastics, consider using:

$$LM = L/(1 - L)$$

where LM = largest mold dimension.

Computerized Shrinkage Analysis

As reviewed in the section on CIIM and included as CSE in Fig. 2–16, there is a computerized shrinkage analysis program (112). This program takes into account processing variables that affect shrinkage (Table 2–10). To do the job, it is designed as part of the total molding concept (TMConcept).

STATISTICAL PROCESS CONTROL

Statistical process control (SPC) seeks to closely control the manufacturing process and permits the manufacture of tighter-tolerance parts by indicating when a process is starting to drift away from the ideal set point. There are basically two possible approaches for real-time SPC. The first, done on-line, involves the rapid dimensional measurement of a part or a nondimensional "bulk" parameter such as weight, and is the more practical method. In contrast to weight, other dimensional measurements of the precision needed for SPC are generally done off-line, the second approach, and result in a response that is too slow. Also obtaining the final dimensional stability needed to measure a molded part may take time; amorphorus resins require at least 30 to 60 min. to cool and stabilize (113–114).

There are four common features of SPC:

1. *Raw material characterization.* A simple, rapid single-point melt index (MI) test is used (see Chapter 10). Although the MI does not completely characterize a resin, for purposes of SPC it does not need to. Also used is a simple time-dependent sampling method. The resin supplier should sequentially number each box/container in the preliminary runs. A sample from each box is measured, and results are plotted using a standard control chart format.

2. Internal material handling (drying, blending, etc.). The importance of drying resins varies with the type used, but improper control of drying will make SPC impossible (see Chapter 9). Minor blend ratio shifts (virgin/additive or concentrate/regrind) are easily discernible on weight response control charts. Regrind requires rigorously measured (controlled) conditions, as its weight is significantly different from that of virgin material.

3. Injection molding: use of part or shot weight. The person who physically operates the machine must be an integral participant in the SPC procedure, so that real-time requirements are met in setting up the machine process controls. The primary operating problem that usually emerges is operator overcontrol; the usual range of changes that an operator makes to "improve" a process will push it out of control. This is not the operator's fault, but is a result of his or her inability to quantitatively determine the results of control actions in a timely manner. When the operator is put in the "process loop" by weighing the parts, he or she receives instant feedback as to the effect, direction, and magnitude of control changes.

4. Implementation. The complete support and commitment of management are required, as SPC initially requires significant funds. It cannot be done quickly; data gathering requires time. SPC may not be practical for every product because of the high cost in time and personnel. New products being prepared for production are often excellent candidates.

With the use of weight for SPC, the data developed can go directly into an operators statistical display or even be used in simple manual plotting. However, weight is a variable property, and confusion over the weight response can lead to the rejection its use as a reliable indicator of quality. There are two common reasons why the use of weight is rejected. First, it is not likely that an absolute weight correlation will always occur. However, even if it is not perfect, it can still be a useful characteristic unless "absolute perfection" is needed. Second, the weighing method may not be reliable. Modern scales, with powerful internal error reduction algorithms, must be used. The most common error is simply improper selection of the scale's resolution: for a shot weight of 100 g, weigh to 0.001 g; for a shot weight of 100 to 999 g, weigh to 0.01 g.

TROUBLESHOOTING

Processing of all types (IM, extrusion, etc.) has become more sophisticated, particularly with regard to process and power controls; so troubleshooting requires a thorough, logical understanding of the complete process (Fig. 1-1) and continues to be a very important function. Problems are presented throughout this book, with suggested approaches to solutions. One must assemble information of this type as the basis for a troubleshooting guide.

Each problem will have its own solution or solutions (Fig. 2–17). Simplified guides to troubleshooting granulators, conveying equipment, metering/proportioning equipment, chillers, and dehumidifiers are available (1–3, 6–8, 33–35).

Recognize that no two similar machines (from one or more suppliers) will operate in exactly the same manner, and that plastics do not melt or soften as perfect blends, but they do all operate within certain limits.

A simplified approach to troubleshooting is to develop a checklist that incorporates the basic rules of problem solving: (1) have a plan and keep updating it based on experience gained; (2) watch the processing conditions; (3) change one condition/control at a time; (4) allow sufficient time for each change, keeping an accurate log of each; (5) check housekeeping, storage areas, granulators, etc.; and (6) narrow the range of areas in which the problem belongs—that is, machine, mold/dies, operating controls, material, part design, and management. To accomplish item (6), several steps may be taken:

(a) Change the resin. If the problem remains the same, it probably is not the resin.
(b) Change the type of resin used, as that may pinpoint the problem.
(c) If the trouble occurs at random, it is probably a function of the machine or the heat control system. Change the mold/die to another machine to determine if it is the machine. Also consider changing the operator.
(d) If the problem appears, disappears, or changes with the operator, observe the differences in the operators' actions.
(e) If the problem always appears in about the same position of a single-

Fig. 2–17. Call your supervisor if anything even looks wrong!

cavity mold, it is probably a function of the flow pattern due to unsatisfactory cooling, and requires readjustments (1, 2).

(f) If the problem appears in the same cavity or cavities of a multicavity mold, it is in the cavity or gate and runner system.

(g) If a machine operation malfunctions, check the hydraulic or electric circuits. As an example, a pump makes oil flow, but there must be resistance to flow to generate pressure. Determine where the fluid is going. If actuators fail to move or move slowly, the fluid must be bypassing them or going somewhere else. Trace it by disconnecting lines if necessary. No flow, or less than normal flow in the system, will indicate that a pump or pump drive is at fault. Details on correcting malfunctions are in the machine instruction manual.

(h) Check for hydraulic contamination. Too little attention is paid to the cleanliness required of the oil used. Dirt is responsible for the majority of malfunctions, unsatisfactory component performance, and machine degradation—particularly with the increased use of electrohydraulic servosystems. Injection pressure, holding pressure, plasticating pressure, boost pressure, and boost cutoff are adversely affected by increased contamination levels in the fluid. Sources of contamination include: that introduced with new oil, that built into the hydraulic system (QC by builder poor), that introduced with air from the environment, that generated by hydraulic-component wear, and that introduced through leaking or faulty seals and introduced by shop maintenance activity. Contamination control is accomplished with the proper filters (such as 10 microns) (see suppliers), and with preventive maintenance procedures that are both correct and properly used.

(i) Set up a procedure to "break in" the new mold/die.

The procedure for setting up a mold/die is as follows: (1) Obtain samples and molding cycle information if the mold was used by others. (2) Clean a used mold. (3) Visually inspect the mold and make corrections if required. (4) Check out, on a bench, the actions of the mold/die-cams, slides, unscrewing devices, and so on. (5) Install safety devices. (6) Operate the mold/die in the machine, and move it very slowly under low pressure. (7) Open the mold/die and inspect it. (8) Dry-cycle the mold without injecting melt to check knockout stroke, speeds, cushions, and low pressure closing. (9) After the mold is at operating heat, dry-cycle it again; expansion or contraction of the mold parts may affect the fits. (10) Take a shot, using maximum mold lubrication and under conditions least likely to cause mold damage, usually low melt feed and pressure. (11) Build up slowly to operating conditions, and run the process until stabilized (usually 1 to 2 h). (12) Record operating information. (13) Take the part to quality control for approval. (14) Make required changes. (15) Repeat the process until it is approved by the customer.

Faulty or unacceptable parts usually result from problems in one or more of three areas of operation: (1) premolding—material handling and storage (see Chapter 9); (2) molding—conditions in the processing cycle; and (3) postmolding—parts handling and finishing operations (see Chapter 9).

Problems caused in premolding and postmolding may include those involving contamination, color, the static dust collector, and so on. In molding (item 2, above) the molder is required to produce a good-quality melt based on visual observation as it flows freely from the nozzle. Each mold and each material are unique; therefore, one cannot generalize about what typically makes a good melt. The experience of the molder and knowledge of process needs are the final determining factors.

There are several ways to determine the efficiency of the melt. One method is to observe the screw drive pressure; it should be about 75 percent of maximum. If it is less than that, lower the rear zone heat until the drive pressure starts to rise. With melt quality changing, raise the center zone to restore quality to what is required. Heat changes should be accomplished in 10 to 15 degree increments, with 10 to 15 min. of stabilization time allowed prior to the next change.

Once the rear zone is set, one should lower the front zones to whatever level will still give good molding conditions. With crystalline types such as nylon, PP, PE, and so on, the operator must watch the screw return. If the screw is moving backward in a jerky manner, there is insufficient heat in the rear zone; the unmelted resin is jamming or plugging the screw compression zone. The heat energy required to melt crystalline plastics is different from that needed for amorphous plastics (see Chapter 1).

Wear

All screws, barrels, molds or dies, and any device that handles melt will wear—but hopefully an insignificant amount that does not influence processability. The wear of screws (particularly on the flight OD) and barrels is a function of: (1) the screw–barrel–drive alignment; (2) the straightness of the screw and barrel; (3) the screw design; (4) the uniformity of barrel heating; (5) the material being processed; (6) abrasive fillers/reinforcing agents/pigments, and so on; (7) the screw surface material; (8) the barrel liner material; (9) a combination of the screw surface and the barrel liner; (10) improper support of the barrel; (11) excessive loads on the barrel discharge end/heavy molds or dies; (12) corrosion caused by additives such as flame retardants; (13) corrosion caused by certain polymer degradation; and (14) excessive back pressure on the injection recovery (2, 100).

Screws are usually aligned properly by the supplier prior to shipment, but can become misaligned during shipment, during installation, and by accidental impacts and other aspects of their use. An angular misalignment will generally cause wear uniformly around the screw in a fairly localized area. In that vicinity the barrel will be worn around the entire ID. If the barrel is bent, the screw will be worn all around near the center and near the discharge, whereas the barrel is usually worn on one side near the center. Wear on screws and barrels generally falls into three categories:

1. Abrasive wear, caused by abrasive fillers such as calcium carbonate, talc, glass fibers, barium sulfate (used in magnetic tapes, etc.), and even the titanium dioxide pigments used in all white and pastel shades. Glass fibers tend to abrade the root of the screw at the leading edge, and in severe cases can undermine the screw flight completely, usually leaving no flight in the compression/transition zone. This action occurs extensively when partially melted or unmelted plastic pushes the glass against the screw or barrel.
2. Adhesive wear, or galling, caused by metal-to-metal contact. Certain sensitive metals can momentarily weld to each other because of very high localized heating. As the screw rotates, the weld separates, and metal is pulled from the screw to the barrel or vice versa. Proper clearance usually eliminates this problem with proper alignment and hardness. With an improperly designed screw for a plastic operating at high output rates, an unmelted blockage will result, forcing the screw against the barrel and causing rapid adhesion wear.
3. Corrosion wear, caused by chemical attack in the melting of certain plastics such as PVC, ABS, PC, PUR, and others, as well as flame-retardant compounds, fiber sizing agents, and so on. Material suppliers can identify the offending agents. The wear usually shows a pitted appearance and usually is downstream, where it has a chance to overheat and degrade. This type of wear can be controlled by using proper operating procedures; do not let the machine stay at the operating heat for any length of time. Proper selection of the screw design and corrosion-resistant screw/barrel materials can help. Nonreturn valves and screw tips are also subject to wear; so it is important to use the best available material (Table 2–4).

Different coatings such as chrome and nickel plating are used to protect the screw surface (Table 2–8). Depending on the specific plastic being processed, a particular coating will be available. The wear surfaces, primarily of flight lands, usually are protected by welding special wear-resistant alloys over these surfaces. The most popular and familiar alloys are Stellite (trademark of Cabot Corp.) and Colmonoy (trademark of Wall Colomonoy Corp.); others are also used and are available from different suppliers. Similarly different heat treatments are used on the steels to increase wear resistance.

Inspection

Screws do not have the same outside continuous diameter. It is recommended that, upon receiving a machine or just a screw, one examine it in regard to its specified dimensions (diameters vs. locations, channel depths,

Table 2-11. Manufacturing Tolerances on Screws (100).

DIAMETERS			CHANNEL DEPTHS	
Outside diameter:	±	.001"	Depth	Tolerance
Shank diameter:	±	.005"	.000" .150"	± .002"
Injection registers:	±	.0005"	.151" .350"	± .003"
Clearance diameters:	+	.015"	.351"-.750"	+ .005"

LENGTHS				
Overall length (O.A.L.)	± $\frac{1}{64}$"		Hollow bore length:	± $\frac{1}{4}$
Transition zones:	± $\frac{1}{10}$ dia.		Flight Widths:	
Vent sections:	± $\frac{1}{10}$ dia.		0-.500":	± .010"
Shank lengths:	± $\frac{1}{32}$"		.501"-1.000":	± .0515"
Ring valve location:	+ $\frac{1}{32}$"		1.001" and over:	+ .020"

CONCENTRICITY			
T.I.R. of O.D.: to 100 length: .002"		Hollow bore to shank:	.015"
Over 100": .004"		Injection registers:	.001"

HARDNESS			
Base material 4140	28-32 Rc	Colmonoy[3] no. 5:	36-40 Rc
Flame-hardened flights:	48Rc min.	Colmonoy[3] no. 56:	46-50 Rc
Nitralloy[1] 135M		Colmonoy[3] no. 6:	50-55 Rc
(or equivalent):	60-70 Rc	Colmonoy 84:	36-42 Rc
Stellite[2] no. 6:	38-42 Rc	N-45[4]	40-44 Rc
Stellite[2] no. 12:	42-48 Rc	N-50[4]	44-48 Rc
		N-55[4]	46-50 Rc

FINISH
Unplated screws: 16 RMS max.
Plated screws: Root—8 RMS max.; Flight sides, O.D., and shank—16 RMS max.

1. Trademark of Joseph T. Ryerson & Son, Inc. 2. Trademark of Cabot Corp.
3. Trademark of Wall Colmonoy Corp. 4. Trademark of Metallurgical Industries Inc.

concentricity and straightness, hardness, spline/attachment dimensions, etc.) and make a "proper" visual inspection. This information should be recorded so that upon a later inspection comparisons can be made. The initial check also "guarantees" proper delivery. Some special equipment should be used for inspection other than the usual methods (micrometer, etc.) to ensure that the inspection is reproduced accurately. Such equipment is available from suppliers (100) and actually simplifies testing and takes less time, particularly for roller and hardness testing. It is important that screws be manufactured to "controlled" tolerances such as those given in Table 2–11.

Chapter 3

EXTRUSION

INTRODUCTION

The extruder, which offers the advantages of a completely versatile processing technique, is unsurpassed in economic importance by any other process. This continuously operating process, with its relatively low cost of operation, is predominant in the manufacture of shapes such as films, sheets, tapes, filaments, pipes, rods, and others. The basic processing concept is similar to that of injection molding (IM), in that material passes from a hopper into a cylinder in which it is melted and dragged forward by the movement of a screw. The screw compresses, melts, and homogenizes. When the melt reaches the end of the cylinder, it usually is forced through a screen pack prior to entering a die that gives the desired shape with no break in continuity (8–9, 85, 90–96, 100–106, 115–143) (Fig. 3–1).

A major difference between extrusion and IM is that the extruder processes plastics at a lower pressure and operates continuously. Its pressure usually ranges from 200 to 1,500 psi (1.4 to 10.4 MPa) and could go to 5,000 or possibly 10,000 psi (34.5 or 69 MPa) (Table 3–1). In IM, pressures

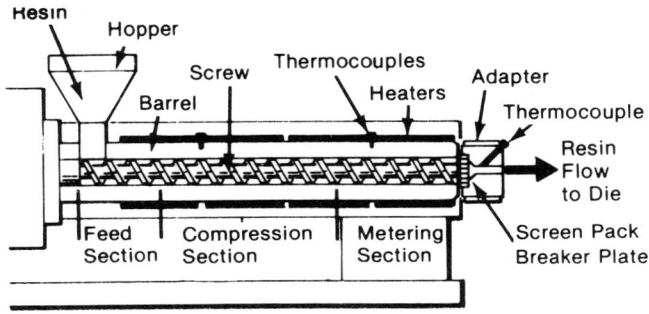

Fig. 3–1. Cross section of a single extruder.

Table 3–1. Typical Extruder Die Head Pressures.

	MELT PRESSURE AT THE DIE	
EXTRUDED SHAPE	PSI	MPA
Film—blown	1,000–5,000	6.9–34.5
Film—cast	200–1,500	1.4–10.4
Sheet	200–1,500	1.5–10.4
Pipe	400–1,500	2.8–10.4
Wire coating	1,000–5,000	6.9–34.5
Filament	1,000–3,000	6.9–20.7

go from 2,000 to 30,000 psi (14 to 210 MPa). However, the most important difference is that the IM melt is not continuous; it experiences repeatable abrupt changes when the melt is forced into a mold cavity (see Chapter 2). With these significant differences, it is actually easier to theorize about extrusion and to process plastics through extruders, as many more controls are required in IM.

Good-quality plastic extrusions require homogeneity in terms of the melt heat profile and mix, accurate and sustained flow rates, a good die design, and accurately controlled downstream equipment for cooling and handling the product. Four principal factors determine a good die design: internal flow length, streamlining, the materials of construction, and uniformity of heat control. Heat profiles, such as those in Fig. 3–2, are preset via tight controls (see Chapters 1 and 2). To accomplish this control, cooling systems are incorporated in addition to heater bands. Barrels use forced air and/or water jackets. In some machines water bubbler channels are located within the screws.

On leaving the extruder, the product is drawn by a pulling device, and in this stage it is subject to cooling, usually by water or blown air. This is an important aspect of control if tight dimensional requirements exist and/ or conservation of plastics is desired. Usually lines do not have adequate control of the pulling device. The processor's target is to determine the tolerance required for the pull rate and to see that the device meets the requirements. One should check with a supplier on the speed tolerances available. Even if tight dimensional requirements do not exist, the probability is that better control of the pull speed will permit tighter tolerances and reduce the material output. One should check the cost of replacing the puller.

As the molecules of the melt flow are aligned in the direction of the output from the die, the strength of the plastic is characteristically greater in that direction than at right angles. Depending on the product use, this may

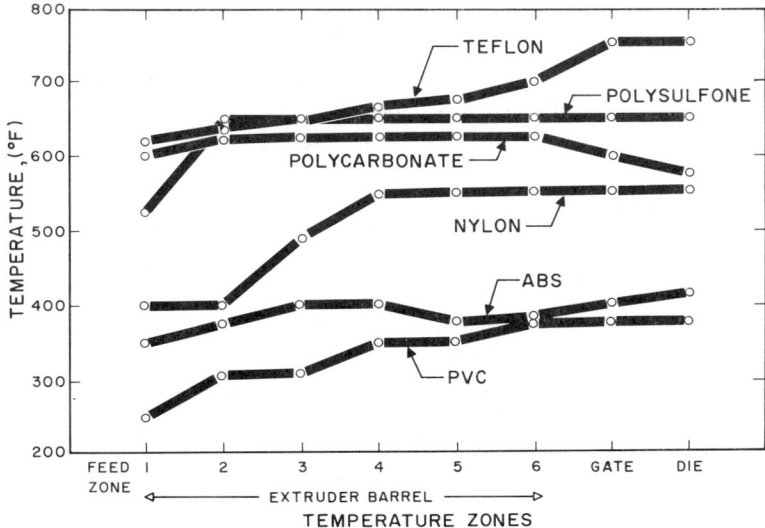

Fig. 3-2. Temperature profiles of different plastics going through an extruder.

or may not be favorable. The degree of orientation can be controlled (to be reviewed later; also see Chapter 1).

The success of any continuous extrusion process depends not only upon uniform quality and conditioning of the raw materials (see Chapters 1, 2, and 9) but also upon the speed and continuity of the feed of additives or regrind along with the virgin resin. Actually only thermoplastics go through extruders; markets have not developed to date for extruded thermosets. As reviewed in previous chapters, variations in the bulk density of materials exist in the hopper, requiring controllers such as weight feeders and perhaps requiring some type of packing feed, such as rams, screw packers, and so on (see an auxiliary equipment supplier).

Extrusion Instabilities or Variabilities

In extrusion, as in all other processes, an extensive theoretical analysis has been applied to facilitate understanding and maximize the manufacturing operation. However, the "real world" must be understood and appreciated as well. The operator has to work within the many limitations of the materials and equipment (the basic extruder and all auxiliary upstream and downstream equipment). The interplay and interchange of process controls can help to eliminate problems and/or aid operation with the variables that exist. The greatest degree of instability is due to improper screw design (or using the wrong screw). Proper instrumentation, particularly barrel heat, is important to diagnosis of the problem(s) (see Chapter 1 on heat control). For uniform/stable extrusion, it is important to periodically check the drive system, the take-up device, and other equipment, and compare it to its original performance. If variations are excessive, all kinds of problems will develop. An elaborate process control system can help, but it is best to improve stability in all facets of the extrusion line. Examples of instabilities and problem areas include: (1) nonuniform plastics flow in the hopper; (2) troublesome bridging, with excessive barrel heat that melts the solidified plastic in the hopper and feed section and stops plastic flow; (3) variations in (a) barrel heat, (b) screw heat, (c) screw speed, (d) screw power drive, (e) die heat, (f) die head pressure, and the (g) take-up device; (4) insufficient melting and/or mixing capacity; (5) insufficient pressure generating capacity; (6) wear and/or damage of the screw and/or the barrel; (7) melt fracture/sharkskin (see Chapter 1); and so on.

Finally one must check the proper alignment of the extruder and the downstream equipment. Proper alignment and isolation of the vibrators is a must for high-quality and high-speed output.

Plastics Handling

Care should be taken to prevent conditions that promote surface condensation of moisture on the plastic and moisture absorption by any existing pigments in color concentrates. As explained throughout this book, surface condensation can be avoided by proper storage of the plastic and keeping it in an area at least as warm as the operating temperature for at least 24 h prior to its use. If moisture absorption by a color concentrate is suspected, heating for 8 to 24 h in a 250 to 300°F (120 to 150°C) oven should permit sufficient drying. With hygroscopic resins, special precautions and drying are required (see Chapter 9). A hopper dryer's heat can be used to improve melt performance and extruder output capacity (Fig. 3–3). When the dryer preheat is insufficient, heat can be applied in the screw's solids conveying zone and/or the barrel feed throat (assuming the capability exists).

A typical heat profile that is adequate at low screw speed may be inadequate when speeds are increased, because of the greater rate of material flow through the extruder. (In fact, the heat profile may even be inadequate at low speed.) The output may not increase at a linear rate with increasing speed unless the heat profile increases (Fig. 3–4). With a proper heat profile, a linear relationship between speed and output can be maintained through a wider operating range. However, too high a heat at low screw

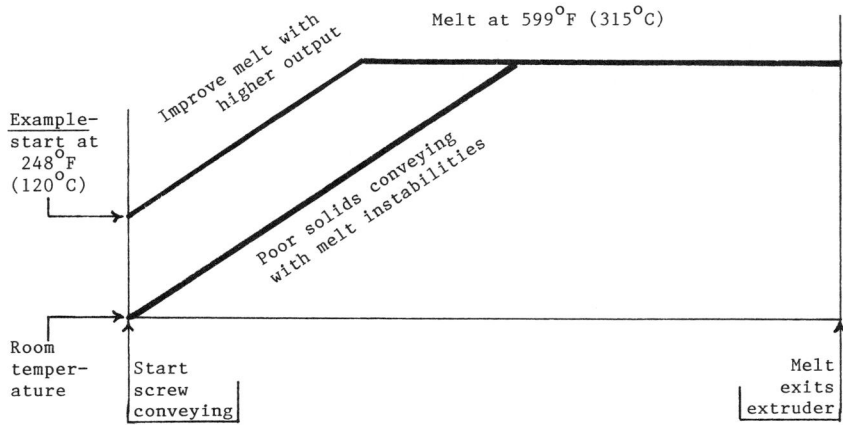

Fig. 3–3. Technique for improving material and/or machine performance by preheating "solids" entering the extruder's feed throat.

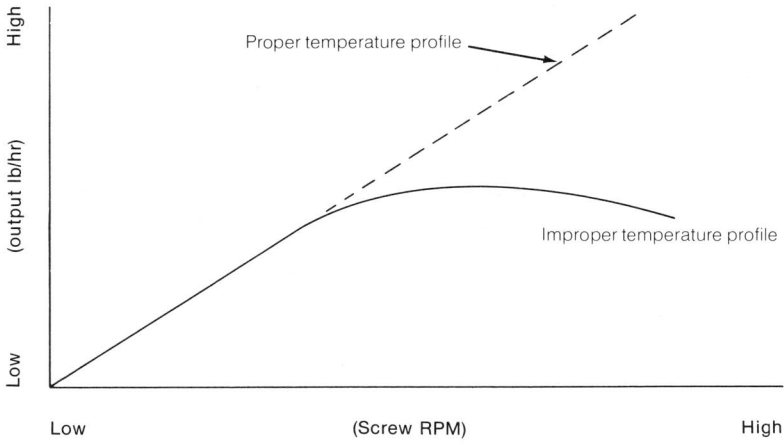

Fig. 3–4. Illustration showing that a higher temperature input for plastics entering the barrel throat will increase the extruder output capacity with an increase in RPM.

speed is likely to cause melting all the way back to the barrel throat, causing resin bridging and degradation.

Different techniques are used to improve feeding from the hopper—a ram stuffer, a tapered (larger) screw in the throat section, a grooved feed section, starved feeding, and so on. All have advantages and disadvantages, based on the capabilities of the machines and of the resin being processed. What may help with one machine in the plant may be useless in another machine using the same resin. Use of a barrel feed with an intensively water-cooled grooved feed section basically avoids early melting. The grooves impart better solids conveying for resins with a low coefficient of friction such as UHMWPE and other resins where additives make them extremely slippery. The grooves usually have right-angle profile and axial arrangements, although helical grooves are used for certain materials. Optimum sizing and dimensioning of the feed section and the correct control of cooling are necessary, as well as proper design of the feed screw, to achieve a high-output revolution with a low-friction load.

If a resin exhibits a feeding inefficiency (as the screw speed is increased, the output of the extruder does not measurably increase), grooved feed sections should be considered. Most materials feed well, as seen through moderate or high pressure levels early in the barrel, and the state of screw design has led to efficient "melting rate limited" extrusion. A grooved feed that is already limited by the melting rate will overdrive the screw and cause melt quality effects that must be corrected through the use of shallower screws. Should a material not be "melting rate limited" but feed reasonable well, a grooved feed will increase the output. Probably the next larger screw will accomplish the same result at a lower cost. One must consider economics, including the added maintenance and operational costs of grooves.

Starve or metered feeding devices do not allow equipment to run with feed flights completely filled; so this approach is best avoided, although some vented screws do use these devices. Machines will run most efficiently and with the best cleaning when feed flights are full. Starve feeding leads to flights that are partially empty for some distance, an undesirable situation (wear problems, etc.). Some two-stage screws that handle a variety of resins may not be versatile enough, and then a feeder gives added control. When machines are having a feed problem, and no other approach is possible, starving can help.

As the feeder is adjusted, the screw's performance will at some point deteriorate so that the output consistency or the melt efficiency is threatened. Adjusting the screw design to preclude the need to starve is a better alternative, which usually is economically feasible.

Twin screw extruders producing profiles will often use starve feeders to reduce torque requirements. However, a better approach, when feasible, is to accomplish this via a screw design with adequate extruder torque power. Slight starving of twins may not seriously deteriorate the melt, but moderate to high levels of starving will reduce stability.

As explained in previous chapters and in Chapter 9, regrind can be a major problem. Many extruder lines, such as film trim and scrap, are reprocessed; the material is granulated and/or plasticized by a small extruder and fed back with virgin resin. The extrudability of the combination may be quite different from that of virgin resin. Problems also can develop in the blending of two or more different resins, such as LLDPE and LDPE, to improve specific performances. The LLDPE starts melting/breaking down at a lower heat than does LDPE, so their blend is more difficult to process than virgin/regrind blends. New screw designs overcome such problems (100, 102–106, 119–124).

MULTIPLE SCREW EXTRUDERS

Regardless of their particular designs, all extruders have the function of conveying plastic and converting it into a melt. For this purpose, both single and multiple screw extruders are suitable; but they all have individual characteristic features. Practical and theoretical data show that each type has its place (118, 124–126, 130–151). The single screw machine dominates and will be the focus of discussion in this chapter. However, other types are available, such as the twin screw extruders shown in Fig. 3–5, and they are often used to achieve improved dispersing and mixing, as in the compounding of additives.

Other claims for multiscrews, most often twin screw designs, include a high conveying capacity at low screw speed, a positive and controlled pumping rate over a wide range of temperature and coefficients of friction, low frictional heat generation which permits low-heat operation, low contact time in the extruder, relatively low motor power requirements, and the ability to feed normally difficult feeding materials such as powders. Twin screw types, because of their low heat extrusion characteristics, have found increasingly wide usage in heat-sensitive PVC processing. The most popular and functional multiscrews are the twin screw designs.

Twin screw extruders with nonintermeshing counter-rotating screws are mostly used for compounding by resin manufacturers, including situations where volatiles must be removed during extrusion. Meshing twin screws have found a substantial market in difficult compounding and devolatilization processes. To provide specialized compounding and mixing, particularly in the laboratory, different techniques are required, such as using interchangeable screw sections on a splined shaft (Fig. 3–6).

Most of the commercial machines on the market and in use today are

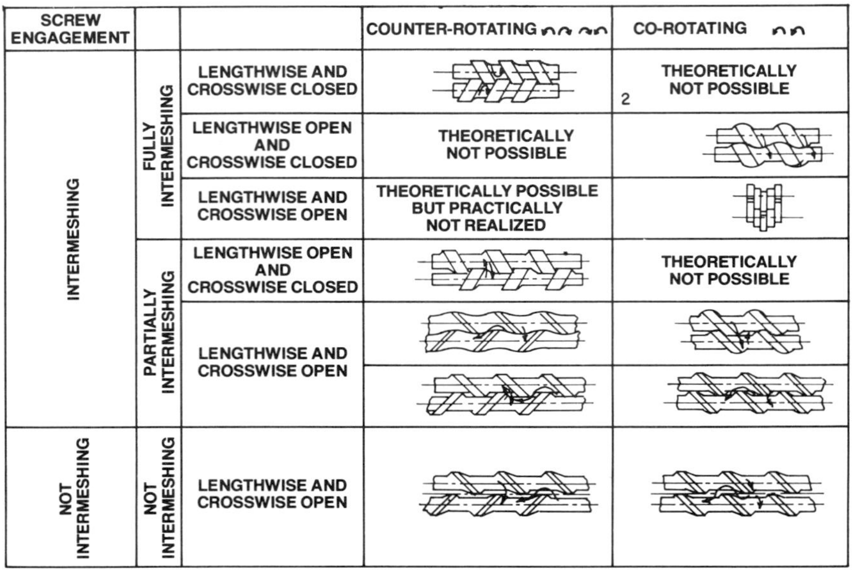

SCREW ENGAGEMENT			COUNTER-ROTATING	CO-ROTATING
INTERMESHING	FULLY INTERMESHING	LENGTHWISE AND CROSSWISE CLOSED	[diagram]	THEORETICALLY NOT POSSIBLE
		LENGTHWISE OPEN AND CROSSWISE CLOSED	THEORETICALLY NOT POSSIBLE	[diagram]
		LENGTHWISE AND CROSSWISE OPEN	THEORETICALLY POSSIBLE BUT PRACTICALLY NOT REALIZED	[diagram]
	PARTIALLY INTERMESHING	LENGTHWISE OPEN AND CROSSWISE CLOSED	[diagram]	THEORETICALLY NOT POSSIBLE
		LENGTHWISE AND CROSSWISE OPEN	[diagram]	[diagram]
			[diagram]	[diagram]
NOT INTERMESHING	NOT INTERMESHING	LENGTHWISE AND CROSSWISE OPEN	[diagram]	[diagram]

Fig. 3–5. Different types of commercial twin screw extruder mechanisms are used; they differ widely in operating principles and functions.

intermeshing. One interesting feature of nonintermeshing twins is the possibility of running the two screws at different speeds, thus creating frictional relationships between them, which in some instances can be exploited for the rapid melting of powders. In some twins, one screw is significantly shorter than the other. This design is used for resins that may be adequately conveyed by a single screw once in the form of a melt, but which, because of low bulk density or very low coefficients of friction of the solid against the surrounding walls, are difficult to feed into screw flights. Thus, after melting by the twins, the resin moves through the single screw.

Constant screw diameters can be used, as in single designs, or the screws can be conical. A conical twin with a large-diameter rear feed zone has a volume capacity greater than the compression and metering zones capacity. This design allows a greater amount of powder to be moved to the feed zone per RPM while using the least possible clearance between flights, to provide a higher output rate with minimum shear and frictional heat accumulation. Different variables are involved in multiscrew machines compared to singles, so multiscrews generally are sized on the basis of lb/h rather than L/D and/or diameters.

With intermeshing screws, the relative motion of the flight of one screw inside the channel of the other acts as a wedge that pushes material from

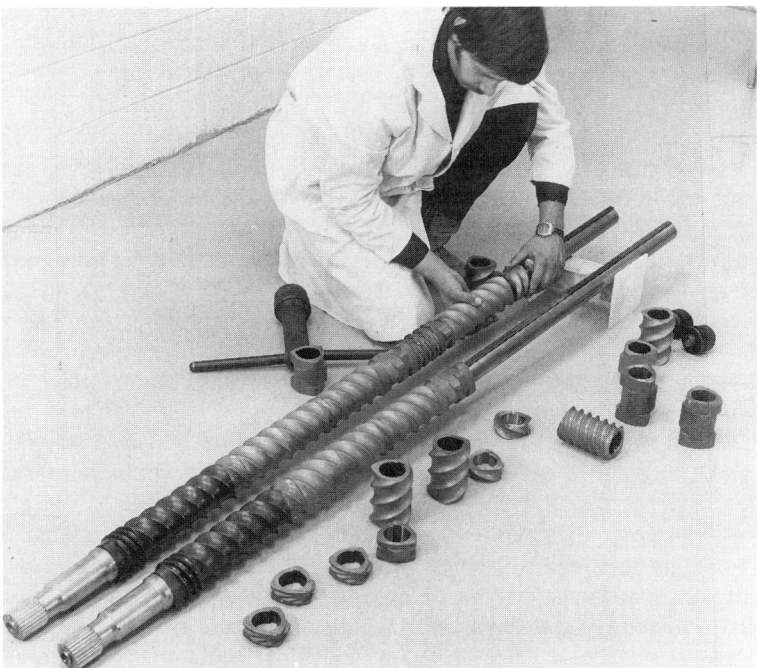

Fig. 3-6. Assembly and interchangeability of Werner and Pfleiderer screw assembly to provide specialized processing conditions.

the back to the front of the channel and from screw to screw. This pattern keeps the screw moving forward, almost as if the machine were a positive displacement gear pump, which conveys material at rather low RPM with low compression and very little friction. The friction in a single screw that causes material to move forward also generates heat. Twin screws do not have the problem of frictional heat buildup because heat is not influenced by friction. Heat is controlled from an outside source (barrels)—an action that becomes very critical in the processing of heat-sensitive plastics, such as PVC. The multiple screws provide the advantages of higher output rates and very tight heat control, as required, for example, to produce large PVC pipes. However, single screws are also used to produce PVC pipe.

Although multiple screws are more expensive than single screws, there are some advantages to using multiscrews. Twins can be used effectively in handling PVC dry blends that are compounded in a plant, potentially offering a significant costs savings as compared to buying compounded PVC. With a multiple and its lower operating heat profile, lower levels of heat stabilizers can be used, with potential savings in material costs.

Other uses for twin screws include the processing of extruded expandable PS (EPS) sheet and high molecular weight polymers. For EPS, there might be cost advantages in using one machine rather than the usual tandem single screw setup. One machine mixes while the other extrudes, with the second machine providing the required cooling time period. Products such as high molecular weight polyolefins or some of the TFE-fluoroplastics can be gently melted in twin screws, and these high-viscosity melts can be conveyed through a die without pulsations, a major advantage in processability.

In counter-rotating systems, the basic advantage is that the material that does pass through the nip of the two screws is subject to an extremely high degree of shear, just as if it were passing through the nip between two rolls of a two-roll mill. By varying the clearance (the free space that is left after the two screws have intermeshed), it is possible to vary the position of material that is carried through and just moves axially. The narrower the clearance between screws, the greater the shear force that will be exerted, and the larger the portion of material that remains in the "bank." With this action, it is easy to adjust the amount of shear to be applied.

In co-rotating twin screw extrusion, one screw transports the material around and up to the point where the screws intermesh. At that point two opposing and equal-velocity gradients exist, so nearly everything that one screw has carried is taken over by the other screw. In this system, plastics will follow an 8-shaped path along the entire barrel length. Advantages of co-rotating are: (1) chances are better statistically that all particles will be subjected to the same shear; (2) with the relatively long 8-shaped path, the chances of influencing the plastic in terms of the melt heat are good; (3) at the deflection point, the shear energy introduced can be regulated within very wide limits by adjusting the depths of the screw flights; and (4) the system allows for a much greater degree of self-cleaning or self-wiping than in other designs, as one screw completely wipes the other screw. This latter performance is important, not so much in terms of keeping screws clean, but in permitting greater control over the residence time distribution, which is very important with heat-sensitive plastics.

To summarize the comparison of screw types, in a single extruder the screw rotating inside the barrel is not able by itself to push the material forward. If for some reason the material filling the channels adheres to the screw, it becomes a rotating cylinder and provides no forward motion. Material pushed forward should not rotate, or at least should rotate at a slower rate than the screw. The only force that can keep the material from turning with the screw and make it advance along the barrel is the friction between the material and the inside surface of the barrel. The more friction, the less rotation of material with the screw; the less rotation, the more forward motion. To yield sufficient production with a low friction factor, the screw must have a large diameter and turn at high speed. However, a large-diameter screw rotating at high speed develops very high shear. The heat produced may increase the melt temperature over its limit, so cooling is necessary during the process (water jacket and/or blown air over the barrel). A nonintermeshing twin screw functions like a single screw extruder, with friction as the prime mover. Twins with intermeshing screws operate on a completely different principle. The direction of rotation, whether co-rotating or counter-rotating, has a great influence on the way they operate.

Counter-rotating twins tend to accumulate and compress the material where the screws meet and remove material where they part, thereby creating zones of high and low pressure around the screws. Material that is forced by this pressure through small passages at high speeds is subjected to high localized shear. In co-rotating twins, material is transferred from one screw to the other without such high shear action.

In co-rotating twins, the screws act like a positive displacement gear pump and do not depend on friction against the barrel to move material forward. As there is no relationship between the screw speed and the output rate, they can work with starve feeding. The channel depth is three to four times greater than in a single screw, and, more important, the screw speed is much lower (use 15–20 RPM); so the total shear rate is much less (can be 80 percent less in twin vs. single). The low screw speed accounts for the low shear rate, and this situation keeps material more viscous and the stock heat down. Consequently, any additional heat required has to be applied through the barrel heaters.

Unfortunately, there are losses as well as gains. The higher-cost multiple screw machines, with their more expensive and complicated drive systems, require constant preventive maintenance; otherwise, rather extensive down time (and repair costs) would occur. Regardless of the disadvantages, multiscrews have an important role in processing plastics. Their disadvantages should not influence their use if there are cost advantages.

SINGLE SCREW EXTRUDERS

The standard metering extrusion screw with its three zones (conveying, compression, and metering) basically operates like a conventional injection molding (IM) screw, as reviewed in Chapter 2. The nomenclature is the same for each (Fig. 2–3), except that no valve is used at the end of the extrusion screw (Fig. 3–7). As reviewed, extrusion screws operate at lower pressures and in a continuous mode (IM is repeatable with abrupt, completely on–off pressure changes and very fast cycles). Even though many

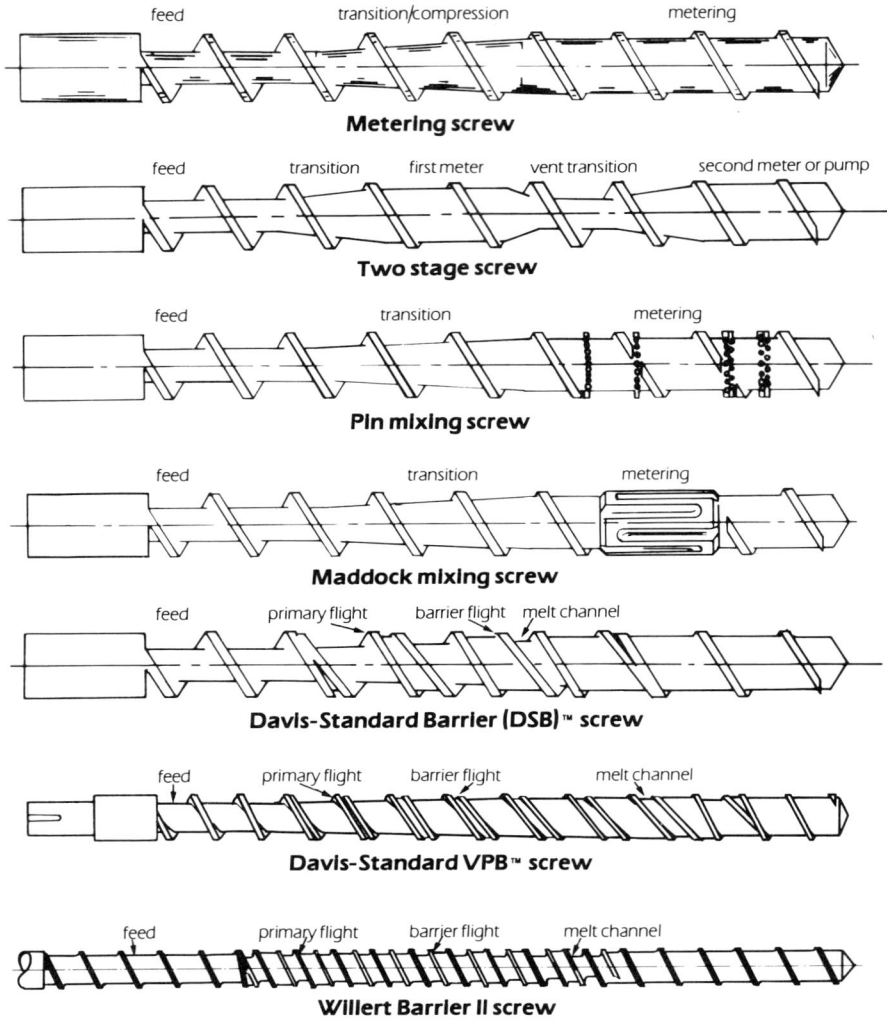

Fig. 3-7. Screw designs with different mixing sections.

variables must be considered, extrusion requires fewer controls and presents fewer problems than IM. Typical output rates for extruders are given in Table 3-2 (2-4, 8-9, 25-29, 57-58, 85-151).

Single screw extruders have changed greatly over the years. Today's functional modular concept developed mainly for reasons of effectiveness and favorable cost comparisons. Their output rates have significantly surpassed

Table 3-2. Guide for Extruder Plastics, Output Rates in Pounds Per Hour.*

MATERIAL	SCREW DIAMETER, INCHES					
	1½	2½	3½	4½	6	8
ABS	280	400	825	1,350	2,270	4,100
Acrylic	320	470	900	1,500	2,700	4,750
PC	210	320	680	1,025	1,850	3,200
PP	280	400	825	1,350	2,270	4,100
HIPS	340	560	1,100	1,800	3,250	5,750
PVC-flexible	300	450	900	1,500	2,700	4,750
PVC-rigid	180	250	500	800	1,450	2,300
LDPE	310	525	1,050	1,750	3,000	5,500
LLDPE	200	300	600	1,000	1,800	3,260
HDPE	215	325	725	1,175	2,150	3,750

*a. Output deviation from average is plus or minus 10 to 15 percent; output rates are based on different processing machine settings and the general composition of the plastics (as reviewed in the chapter).

b. To obtain the actual output rate, weigh the actual output based on machine settings and the specific plastic processed.

c. A "rough" estimate for output rate (OR) in lb/h can be calculated by using the barrel's inside dimension (ID) in inches in the following equation: $OR = 16\ ID^{2.2}$.

d. Pounds × 0.4536 = kilograms; see the appendix for English to metric conversions.

e. Standard barrel inside diameters, in inches (mm in parentheses), generally are: 1½ (38), 2 (50), 2½ (64), 3¼ (83), 3½ (89), 4½ (115), 6 (153), and 8 (204).

those of older designs. The output rates in Table 3–2 can be used as a guide to predict the output rate of a process. The performance of all machines and production lines (film, profile, etc.) will depend on the many factors that have to be controlled and syncronized going from upstream through the extruder and the downstream equipment. The type of screw used has always been a major influence in the complete line.

The blown film extruder has typified the new generation of extruders. The most effective screw design, in most cases, has been an L/D of 25. Longer machines with a 30 to 33 L/D are chosen for venting or special requirements. High outputs are obtained with LDPE blown film or PP cast film extrusion. The 20 L/D machines now are almost always used only for heat-sensitive plastics. The 25 L/D version offers exactly the right compromise for obtaining a high output and preventing overheating and damage of thermally sensitive plastics.

Even in today's high technology world, the art of screw design is still dominated by trial-and-error approaches. However, computer models (based on proper input and experience) play an important role. When new materials are developed or improvements in old materials are required, one must go to the laboratory to obtain rheological and thermal properties before computer modeling can be performed effectively (see Chapters 1 and 10). New screws improve one or more of the basic screw functions of: melt quality, mixing efficiency, melting performance along the screw, melt heat level, output rate, output stability, and power usage or energy efficiency.

As previously reviewed, heating can be controlled by using different machine settings, which involve various tradeoffs. For example, in choosing the optimum rotation speed, a slow speed places the melt in contact with the barrel and screw for a longer time via heat conduction, and the slower speed produces less shear, so that dissipative heating is reduced, and properties of the plastic (particularly of a film) are enhanced. Sometimes an internal heat control is used with a screw (Fig. 3–8). This type of screw is characterized by deeper channels, steeper helical angles, and an internal heating element. Its internal heating lowers the amount of viscous heating needed to process the material. As a result, the melt heat can be reduced by 50°F.

Mixing and Melting

In typical extrusion operations, mixing devices are used in the screws. Many dynamic mixers, such as those included in Fig. 3–7 and in Chapter 2, are used to improve screw performance. Static mixers are sometimes inserted at the screw end (Fig. 3–9) or at the end of the barrel. There are also mixing

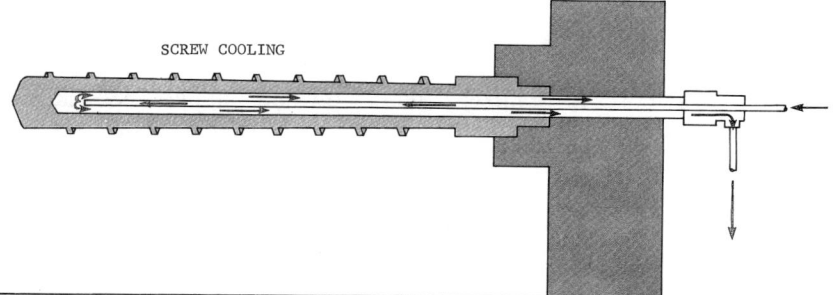

Fig. 3-8. Concept of screw with internal heat or cooling system.

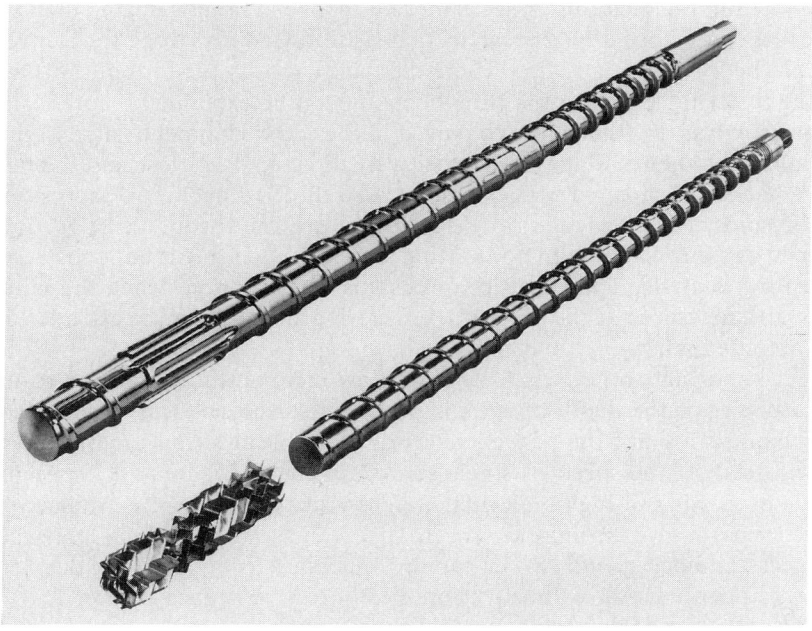

Fig. 3-9. Example of a static mixer located at the end of a screw.

devices that remain independent of the screw (2, 8–9, 85, 100, 121–124, 142).

Proof of the success and reliability of dynamic on-line mixers is shown by their extensive use. Each mixer offers its own advantages and disadvantages (see Chapter 2) with different machines and materials. Unfortunately there is no one system that solves all melting problems. The data available from the different equipment suppliers can be used in comparative studies.

Static mixers are successfully installed in the adapters between the barrel and the die for further thermal homogenization of the melt after it leaves the screw. This causes transitory fluctuations and temperature variations in the melt to decrease considerably. As a result an increased pressure drop occurs and the heat level increases a few degrees, while temperature peaks disappear. Thus in selecting a static mixer, one should be sure that its resistance to flow is as low as possible.

The dynamic mixing elements should be fitted as near the end of the metering zone as possible. For maximum performance, a standard screw can have the following sectional dimensions in respect to its diameter (D): feed zone = $5D$, compression zone = $3D$, metering zone = 12 to $13D$, and mixing zone 2 to $3D$. In principle, any mixing zone, wherever it is located within the metering zone, is more effective in homogenizing than the screw threads normally present in that position. Where practical, mixers should be located in a region where the melt viscosity is not too low. Only in the metering zone can the flow phenomenon be explained theoretically. In this analysis the melt flow conveyed in the screw channel to discharge is visualized as the resultant of the positively directed drag flow and the negative backward-directed pressure flow. Also directed backward, across the flight lands, is the leakage flow due to the pressure drop. Leakage flow, despite its significance with nonwetting materials, as a rule is not considered in a flow analysis. The extrusion process is best balanced when the initial pressures measured at the start of the metering zone and in the extruder die are about equal.

It is important to note that the back flow increases in proportion to the third power of the depth of the thread. For this reason, screws with deep cut channels are not the best choice for thermoplastics (but are okay with thermosets). If the speed of single screw extruders is increased, especially in the processing of high molecular weight viscous melts, the extrudate obtained may be rough, unattractive in appearance, and unsalable. Such results also can occur with slower-running machines using relatively deep cut screws in conjunction with extrusion dies of low resistance to flow.

This situation is due to an unstable combination of screw and die, and basically can be related to unsuitable pressure melt variations. Approaches to improving the situation include using longitudinally adjustable (regular mechanical) screws with tapered clearances as throttling sections, independent throttling valves in different positions of the die, and so on. The goal is to eliminate dead spots in which stagnating melt could be thermally degraded; so streamlining is in order, such as that provided by conical screw tips (Figs. 3-7 and 3-8).

In-line dynamic mixers, independently driven at optimum speeds, perform distributive mixing with moderate pressure losses, low power requirements, and small heat increases. The use of dynamic mixers mounted on the end of a screw may not be optimum because extruders may have to be driven at slower speeds to avoid problems such as surging; but independently driven mixers can be sized and run at optimum speeds to provide the best mixing. Other benefits of independently driven mixers involve feeding. For example, metering pumps can inject liquid additives directly into the mixer in the exact quantitites needed to modify a resin. In fact, an auxiliary extruder can add a secondary melt stream to the mixer, thereby allowing for such techniques as resin alloying while simultaneously coloring or stabilizing the resin.

A major and important use of mixers is to ensure minimum exposure of the melt to shear and high heat, so that, with heat-sensitive melts, less stabilizer is needed. With these devices, a moderate pressure drop and melt uniformity are achieved in a short process time. The mixers can be placed adjacent to a die, to provide the least flush time between material changes.

All the mixing devices discussed in this section make it possible to achieve better melts by literally breaking up the solid beds. Like an ice cube that more easily melts in water when it is cracked, the plastic "solids" break up, resulting in a better melt.

Venting

During extrusion, as in IM (see Chapter 2), melts must be freed of gaseous components (monomer, moisture, plasticizers, additives, etc.); so a vented screw is used (Figs. 2-11 and 3-10). It is especially difficult to completely remove air from some powdered materials unless the melt is exposed to vacuum venting (a vacuum is connected to the vent's exhaust). The standard machines operate on the principle of melt degassing. The degassing is assisted by a rise in the vapor pressure of volatile constituents, which results from the high melt heat. Only the free surface layer is degassed; the remainder of the plastic can release its volatile content only through diffusion (1-4, 8-9, 96, 100, 143). Diffusion in the nonvented screw is always time-dependent, and long residence times are not possible for melt moving through an extruder. Thus a vented extruder is used.

Most single screw vented extruders have two stages (Fig. 2-11); a few

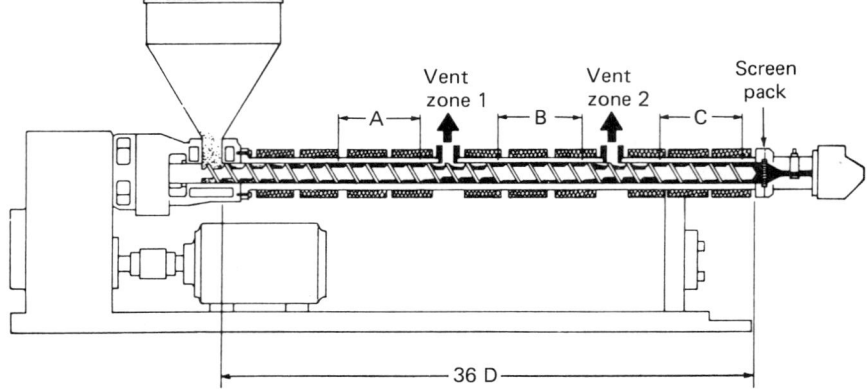

Fig. 3-10. A double-vented extruder.

have two vents and three stages (Fig. 3-10). The first stages of the transition and metering sections are often shorter than the sections of a single stage conventional screw. The melt discharges at zero back pressure into the second stage, under vacuum instead of pressure. The first stage extrudate must not be hot enough to become overheated in the second stage. Also, the first stage must not deliver more output per screw revolution at discharge pressure than the second stage can pump through the die under the maximum normal operating pressure, such as might occur just prior to a screen pack change. Usually this requirement means that the second stage metering section must be at least 50 percent deeper than the first stage.

In practice, the best metering section depth ratio (pump ratio) is about 1.8:1. The best ratio depends on factors such as screw design, downstream equipment, feedstock performance, and operating conditions. With a high compression ratio or metering depth ratio that is a little too low, melt flow through the vent is likely. If this ratio is moderately high, gradual degradation of the output occurs. If the screw channel in the vent area is not filled properly, the self-cleaning action is diminished, and the risk of plateout increases. In any case, sticking or smearing of the melt must be avoided, or degradation will accelerate.

Screen Packs

Melt from the screw usually is forced through a breaker plate with a screen pack. Extra heat develops when melt goes through the screens, so some heat-sensitive materials cannot use a screen pack. The function of a screen pack initially is to reduce rotary motion of the melt, remove large unmelted particles, and remove other contaminants. This situation can be related to improper screw design, contaminated feedstock, poor control of regrind, and so on. Sometimes screen packs are used to control the operating pressure of extruders. However, there are advantages in processing with matched and controlled back pressure, operating within the required melt pressure, as this can facilitate mixing, effectively balancing out melt heat. (Table 3–3a provides a troubleshooting guide to the use of screens.)

In operation, the screen pack is backed up by a breaker plate that has a number of passages, usually many round holes ranging from $\frac{1}{8}$ to $\frac{3}{16}$ in. in diameter. One side of the plate is recessed to accommodate round discs of wire screen cloth, which make up the screen pack (Table 3–3b). Pressure controls should be used on both sides of the breaker plate to ensure that the pressure on the melt stays within the required limits. Based on the processing requirements, manual to highly sophisticated screen changers are used. With limited runs or infrequent changes, manual systems are used. The packs are usually mounted outside the extruder between the head clamp and the die and can be changed via mechanical or hydraulic devices. Continuous screen changes also are used. The more sophisticated the system is, the higher its costs. One should consult suppliers about screen capabilities, disadvantages, and so forth.

Barrel and Screw Materials

The majority of barrels and screws are made from special steels (Tables 2-4 and 3-4), which are nitrided to a minimum depth by special techniques. Low alloy steels are sometimes used with wear-resistant liners. Usually the wear on these bimetallic cylinders is almost three times that of the others. In the processing of abrasive materials, feed sections are sometimes finished in hard metal or other special materials, and matched with the screws. If there is wear in the extruder, then the greatest damage is always on the screw. Often only a new screw is used as a replacement, as it is assumed that the barrel is not damaged. However, usually this assumption is a fallacy. If the screw is worn out, the barrel has been affected to some extent. It may well need complete replacement.

The rate of wear is increased considerably when the feed contains fillers such as titanium dioxide, glass fibers, and so on. As reviewed in Chapter 2, there are many variables that cause damage to the barrel and the screw. If a problem is likely to occur frequently, protect the screw and consider using barrels with replaceable inner liners.

Table 3-3a. Troubleshooting Guide for Screen Changers.

PROBLEM: Seal Leakage

POSSIBLE CAUSES:	REMEDY:
■ Head pressure above maximum set pressure	▲ Reduce head pressure
■ Excessive overhung load	▲ Support downstream die adaptor, etc.
■ Seals improperly installed	▲ Remove, clean, and re-install
■ Excessively scored seals or slide plate	▲ Correct cause of damage (see below) or replace
■ Disc spring relaxed	▲ Replace
■ Initial pressure setting of upstream body too low	▲ See procedure for setup in manual

PROBLEM: Excessively Scored Seals

POSSIBLE CAUSES:	REMEDY:
■ Maximum set pressure rating too high	▲ See procedure for setting U/S body
■ Wire strands protruding above face of slide plate	▲ Ensure correct screen size and installation
■ Screens tearing out or folding on shift	▲ Utilize positive screen retention
■ Tramp metal caught by screens	▲ Utilize hopper magnet or other metal-catching device

PROBLEM: No Slide Plate Movement

POSSIBLE CAUSES:	REMEDY:
■ Guards not closed	▲ Close guards
■ Pressure ready light not on (w/opt. solenoid valve)	▲ Check hydraulic power unit
■ Limit switches or interlock wiring faulty	▲ Repair or replace
■ Direction control valve not operating	▲ Repair or replace

PROBLEM: Erratic or Slow Slide Plate Movement

POSSIBLE CAUSES:	REMEDY:
■ Air in system	▲ Purge air
■ Low precharge in accumulator	▲ Recharge accumulator
■ Oil viscosity too high	▲ Replace with correct viscosity oil
■ Pressure drop across breaker plate too high	▲ Reduce pressure drop
■ Polymer viscosity too high	▲ Decrease viscosity
■ Initial pressure setting of upstream body too high	▲ Procedure for setting U/S body
■ Leakage past the directin control valve or changer cylinder	▲ Repair or replace

Table 3–3a. Continued.

PROBLEM: No Pressure Hydraulic Power Unit

POSSIBLE CAUSES:	REMEDY:
■ Motor not running	▲ Check disconnects, motor overloads, motor
■ Unload solenoid valve not closed	▲ Replace
■ Pump/shaft damaged	▲ Replace pump

PROBLEM: Low Pressure, Motor Stops

POSSIBLE CAUSES:	REMEDY:
■ Faulty pressure switch	▲ Replace switch

PROBLEM: Low Pressure, Motor Continues to Run

POSSIBLE CAUSES:	REMEDY:
■ Low fluid level	▲ Fill reservoir
■ Suction strainer blocked	▲ Clean or replace
■ Oil overheated	▲ See below
■ Air entrained in fluid	▲ Allow bubbles to subside, use anti-foam
■ Leakage past direction control valve	▲ Repair or replace
■ Leakage past screenchanger cylinder	▲ Repair or replace
■ Pump damaged/worn	▲ Replace pump

PROBLEM: Normal Pressure, Does Not Stop

POSSIBLE CAUSES:	REMEDY:
■ Faulty pressure switch	▲ Replace switch

PROBLEM: Rapid Pump/Motor Cycling

POSSIBLE CAUSES:	REMEDY:
■ Faulty pressure switch	▲ Replace switch
■ Leakage past check valve	▲ Replace pump
■ Pump damaged or worn	▲ Replace pump
■ Leaking past direction control valve or screen-changer cylinder	▲ Repair or replace

PROBLEM: Oil Overheating

POSSIBLE CAUSES:	REMEDY:
■ Rapid screenchanger cycling	▲ Shift less frequently
■ Restricted air movement or high temperature around reservoir	▲ Assure free movement of air around hydraulic unit
■ Fluid bypass through system causing excessive pump up time or rapid pump cycling	▲ See above

Table 3–3b. Screens Used Before the Breaker Plate to Filter Out Contaminants in the Melt.[a]

| | | WIRE MESH | | SINTERED |
CONTAMINANT	METAL FIBERS	SQUARE WEAVE	DUTCH TWILL[b]	POWDER
Gel captured	5	1	2	3
Contaminant capacity	6	2	3	3
Permeability	4	4	1	2

a. Range is from poorest (1) to best (6). Multiple screens usually are used; example screen pack has 20-mesh against breaker plate, followed by 40-, 60-, and 100-mesh (coarsest mesh has lowest mesh number)
b. Woven in parallel diagonal lines.

Table 3–4. Materials of Screw Construction.

| | | CORROSION | |
MATERIAL	STRENGTH	RESISTANCE	COST
4140 Tool steel	10	1	3
17-4PH Stainless steel	8	4	3
Hastalloy C-276	1	10	10
4140 Chrome-plated	10	5	3
4140 Electroless nickle-plated	7	8	3

4140 with hard facing
  Stellite #6
  Colmonoy #56
	10	5	6

Ratings: 1 = poorest to 10 = best.

ENERGY CONSUMPTION

Like the output capacity, the energy efficiency of an extruder is dependent on the torque available on the screw, screw RPM, heat control, and material being processed. Unfortunately, costly energy losses can occur, ranging from 3 to 20 percent and due to various factors, with the major loss occurring in the drive mechanism. The power for screw rotation is supplied by a variable speed motor drive system, and is transmitted through a gear reduction unit, a coupling, and a thrust bearing. Gear reducers impart the final speed and torque to the screw. Most gear reducers use double-reaction helical or herringbone gears for their ruggedness and to hold noise levels within acceptable limits. Worm and pinion gear combinations have been used on smaller extruders. The efficiency of the power transmission gear with the worm is a maximum of 85 percent; that of the helical gear reaches 95 percent, and the herringbone 97 percent.

Thrust bearings absorb the thrust force exerted by the screw as it turns against the material in the barrel. The size or rating of the thrust bearing provides an anticipated number of operating hours (if kept clean). The operating pressure, size of extruder, and operating speed are important factors in selection of a thrust bearing. The motor size must be sufficient to allow for the energy required to melt the plastic and to pump the melt at the desired output condition and rate. For example, higher outputs at higher screw speeds require more power than low outputs. The specific heat of the resin (Table 1-6) will also be a determining factor. As a guide, simple extrusions usually require 1 HP for every 10 to 15 lb/h output. For high energy or where high heat at low speeds is required, 1 HP may be required for every 3 to 5 lb/h. Plastics require different power or running torque on screws, based on the type of plastic and the screw size. Extrusion of rigid PVC requires about twice the power needed for LDPE. Increasing screw diameters could at least double torque requirements for every one inch of diameter increase. As previously observed, in comparing single extruders with intermeshing co-rotating twin screw extruders, the twins provide significant energy conservation and efficiency.

Energy consumption is a major factor in production costs, as well as all equipment efficiency. Many extruders, as well as other equipment, are usually overpowered. This situation may be better than using underpowered equipment; but processes should not waste energy. In an extruder, as in any other machine, the energy output is always equal to the energy input, regardless of the forms into which it may be converted. The energy may be furnished by mechanical or thermal sources. Mechanical energy is supplied by the drive motor; thermal energy is supplied "positively" by electrical heater bands or "negatively" by cooling devices. It is important to evaluate and compare equipment performance in order to minimize energy waste. Some machines (of the same type) are more wasteful, whereas others are more efficient. One should check the power consumption on incoming electrical lines going into equipment that operates from minimum to maximum load conditions. Perhaps all that is needed is an ammeter or watt meter physically placed on the line. The results will be obvious: use more efficient units, charge for operating costs based on efficiencies, and/or replace certain units (1-5, 8-9, 22).

GEAR PUMPS

Gear pumps, also called melt or metering pumps, have been standard equipment for decades in textile fiber production and in postreactor polymer finishing. In the 1980s they established themselves in all kinds of extrusion lines. They consist of a pump, a drive for the pump, and pump controls, located between the screen pack (or screw) and the die. Two counter-ro-

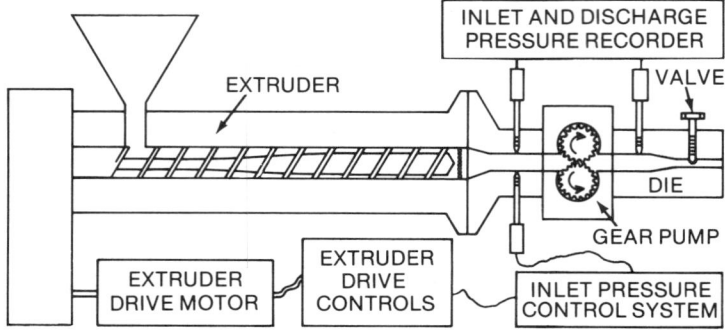

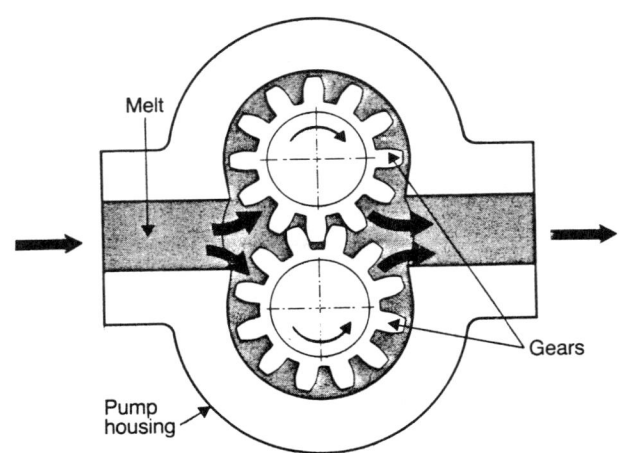

Fig. 3-11. Schematics of a gear pump used in a typical extrusion system.

tating gears will transport a melt from the pump inlet (extruder output) to the pump discharge outlet (die) (Fig. 3-11). Gear rotation creates a suction that draws the melt into a gap between one tooth and the next. This continuous action, from tooth to tooth, develops surface drag that resists flow; so some inlet pressure is required to fill the cavity (1-2, 124, 144-154).

The inlet pressure requirements vary with material viscosity, pump speed, and mixing requirements. These pressures are usually less than 1,000 psi but cannot go below certain specified pressures such as 300 psi. An extruder specifically designed for use with a pump only has to "mix," with no need to operate at high pressures to move the melt. It only has to generate the low pump inlet pressure; thus it can deliver melt at a lower than usual heat, requiring less energy and often yielding a higher output rate. *The positive displacement gear device pumps the melt at a constant rate.* It delivers the melt to the die with a very high metering accuracy and efficiency. Pressure differentials as high as 4,000 psi between the pump inlet and discharge are common.

The pump's volumetric efficiency is 85 to 98 percent. Some melt is deliberately routed across the pump to provide lubrication, and some slips past the gears. An incomplete fill on the inlet side will show up as a fast change in output and pressure at the exit. The extended loss of inlet pressure can damage the pump by allowing it to run dry. Overpressurization at the inlet, caused by the extruder's sudden surge, will at least change the melt conditions and in extreme cases can be dangerous to both equipment and operator. However, closed loop pressure controls are available for the inlet and exit, which eliminate the problem. To prevent overfeeding and overpressure, the screw metering section should have a larger than normal barrel clearance.

Melt pumps are most appropriate when the screw and die characteristics combine to give a relatively poor pumping performance by the total system. This can happen when die pressures are low but more often occurs when they are extremely high (5,000–8,000 psi), or when the melt viscosity is extremely low. When pumps are used to increase the production rate by reducing the extruder head pressure without a corresponding increase in the screw speed, the extrudate solids content often is increased. The result is an inferior product. This problem often necessitates additional filtration, which serves only to increase pressure and may counteract many of the benfits expected from the pump, as well as increasing the financial investment even further.

Depending on the screw design, the extruder often creates pulses, causing the production rate to fluctuate. Some products usually cannot tolerate even minor fluctuations, and a pump often can assist in removing these minor product nonuniformities. In general, a pump can provide output uniformity of $\pm$ 0.5 percent or better. Products include films (down to 0.75 mil thickness), precision medical tubing, HIPS with 3,500 lb/h output, fiber-optic sheathing, fibers, PET magnetic tape, PE cable jacketing (weight/ft variation reduced from 14 to 2.7 percent), and so on.

Pumps are very helpful to sheet extruders who also do in-house thermoforming, as they often run up to 50 percent regrind mixes. This normally variable-particle-size mix promotes surging and up to 2 percent gauge variation. Pumps practically eliminate the problem and make cross-web gauge adjustments much easier. Pumps are recommended in: (1) most two-stage vented barrels where output has been a problem, such as ABS sheet; (2) extremely critical-tolerance extrusions, such as CATV cable, where slight cyclic variations can cause severe electrical problems; (3) coextrusion, where precise metering of layers is necessary and low pressure differentials in the pump provide fairly linear outputs; and (4) twin screw extruders, where pumps permit long wear life of bearings and other components, thus helping to reduce their high operating costs.

Besides improving gauge uniformity, a pump can contribute to product quality by reducing the resin's heat history. This heat reduction can help blown film extruders, particularly those running high viscosity melts such as LLDPE and heat-sensitive melts such as PVC. Heat drops of at least 20 to 30°F will occur. In PS foam sheet extrusion, a cooling of 10 to 15°F occurs in the second extruder as well as a 60 percent reduction in gauge variation by relief of back pressure. One must be aware that all melts require a minimum heat and back pressure for effective processing.

Pumps cannot develop pressure without imparting some energy or heat. The melt heat increase depends on melt viscosity and the pressure differential between the inlet and the outlet (or ΔP). The rise can be 5°F at low viscosity and low ΔP, and up to 30°F when both these factors are higher. By lowering the melt heat in the extruder, there is practically no heat increase in the pump when ΔP is low. The result is a more stable process and a higher output rate. This approach can produce precision profiles with a 50 percent closer tolerance and boost output rates 40 percent. Better control of PVC melt heat could increase the output up to 100 percent. In one case, the output of totally unstabilized, clear PVC meat wrap blown film went from 600 to over 1,000 lb/h with the use of the gear pump.

With pump use, potential energy savings amount to 10 to 20 percent. Pumps are 50 to 75 percent energy-efficient, whereas single screw extruders are about 5 to 20 percent efficient.

Although they can eliminate or significantly improve many processing problems, gear pumps cannot be considered a panacea. However, they are worth examining and could boost productivity and profits very significantly. Their major gains tend to be principally in (1) melt stability, (2) temperature reduction in the melt, and (3) increased throughput with tighter tolerances for dimensions and weights.

DIES

The function of a die is to accept the "available" melt from an extruder and deliver it to takeoff equipment as a shaped profile (film, sheet, pipe, filament, etc.) with minimum deviation in cross-sectional dimensions and a uniform output by weight, at the fastest possible rate. A well-designed die should permit color and compatible resin changes quickly with little off-grade material. It will distribute the melt in the die flow channels so that it exits with a uniform density and velocity (Fig. 3–12).

The flow rate is influenced by all the variables that can exist in preparing the melt during extrusion—namely, die heat and pressure with time in the die. Unfortunately, in spite of all the sophisticated polymer flow analysis and the rather mechanical computer-aided design (CAD) capabilities, it is very difficult to design a die. An empirical approach must be used, as it is quite difficult to determine the optimum flow channel geometry from engineering calculations. It is important to employ rheological flow properties and other melt behavior (Chapter 1 and references) via the "applicable" CAD programs for the type of die required. The most important ingredient is experience, which, for the novice, is properly recorded in a computer program. Nevertheless, die design has remained more of an art than any other aspect of process design. Design experience can work only if the operator of the processing line has developed the important ability to debug it (2–5, 8–11, 32–37, 42–44, 49–58, 85, 93–98, 102, 116–117, 124, 127, 152–158).

The example presented below, using a simplified equation (G. P. Lahti) obtained through a high-speed computer study during the early 1960s, continues to be extremely useful in CAD programs. It provides an excellent empirical approach that pertains to extrusion channels and dies of several shapes. [Flow equations for dies of simple shapes, such as circular or rectangular channels, were known in the last century, and were first developed by M. J. Boussinesq in 1868 (*Journal de Mathematiques Pures et Appliquees*, 2, 13, 377–424, July 27, 1868). At that time formulas for pressure drop through more complex channels were not developed because of the extremely complicated mathematics.] As shown in Fig. 3–12d, the following equations can be used:

$$Q = \frac{1}{\mu} \frac{\Delta P}{L} \frac{BH^3}{12} F$$

or

$$\Delta P = \frac{12\,\mu\,QL}{BH^3} \frac{1}{F}$$

where dimensions are:

H = Minimum dimension of cross section, in. (cm)
B = Maximum dimension of cross section, $B \geq H$, in. (cm)

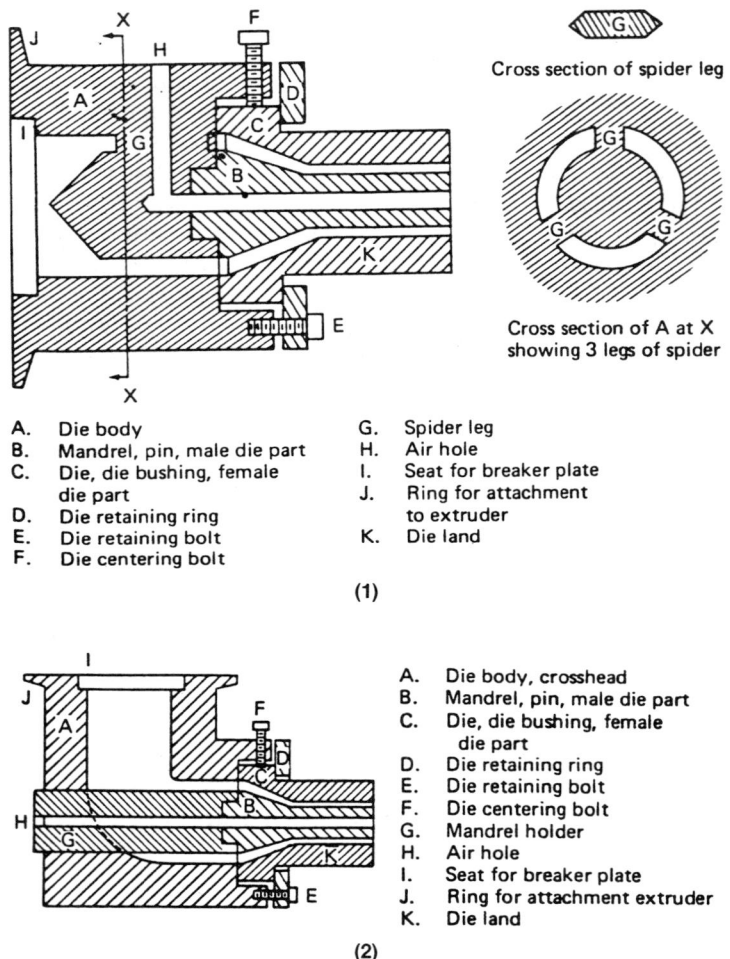

Cross section of spider leg

Cross section of A at X
showing 3 legs of spider

A. Die body	G. Spider leg
B. Mandrel, pin, male die part	H. Air hole
C. Die, die bushing, female die part	I. Seat for breaker plate
D. Die retaining ring	J. Ring for attachment to extruder
E. Die retaining bolt	K. Die land
F. Die centering bolt	

(1)

A.	Die body, crosshead
B.	Mandrel, pin, male die part
C.	Die, die bushing, female die part
D.	Die retaining ring
E.	Die retaining bolt
F.	Die centering bolt
G.	Mandrel holder
H.	Air hole
I.	Seat for breaker plate
J.	Ring for attachment extruder
K.	Die land

(2)

Fig. 3–12a. Examples of dies with nomenclature. (1) Pipe or tubing die for in-line extrusion. (2) Pipe or tubing die for crosshead extrusion. (3) Cast film die. (4) Sheet extrusion die.

H/B = Aspect ratio, dimensionless
μ = Viscosity, lb·s/in² (Pa·s)
Q = Volumetric flow rate, in³/s (cm³/s)
L = Length of channel, in. (cm)
ΔP = Pressure drop across L, lb/in² (Pa)
F = Flow coefficient (a function of shape and aspect ratio) from Fig. 3–12d, dimensionless

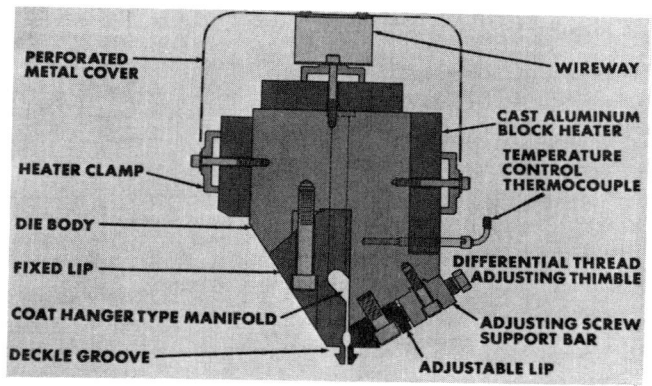

(3)

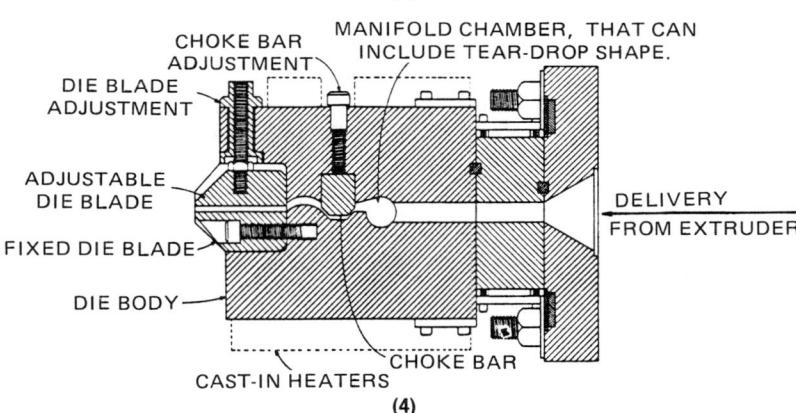

(4)

Fig. 3–12a. Continued

Regarding the length of the channel (L): to account for the entrance effect when a melt is forced from a large reservoir, a corrected channel length must be used, or the apparent viscosity obtained from shear rate–shear stress curves for the same L/H value as that of the existing channel must be used. The entrance effect becomes negligible for $L/H > 16$. This single equation can be used for a variety of flow channels, as shown in Fig. 3–12d (93).

A well-built die with adjustments—temperature changes (Fig. 3–12c), restricter/choker bars, valves, and/or other devices—may be used with a particular group of materials. Usually a die is designed for a specific resin. For example, conventional LDPE blown film dies with 0.030 in. (0.8 mm) die gaps will not process LLDPE satisfactorily at high output rates. The higher-viscosity LLDPE increases back pressure significantly, thereby decreasing

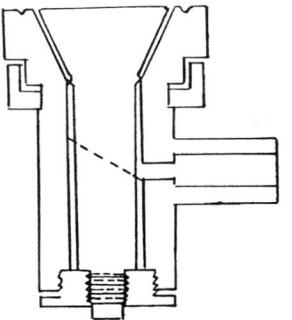

(1) *SIDE FEED DIE*
Advantages:
1. Low initial cost
2. Adjustable die opening
3. Will handle low flow
   materials
Disadvantages:
1. Mandrel deflects with
   extrusion rate, necessitating
   die adjustment
2. Die opening changes with
   pressure
3. Non-uniform melt flow
4. Cannot be rotated
5. One weld line in film

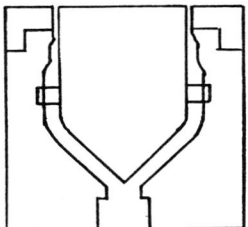

(2) *BOTTOM FEED SPIRAL DIE*
Advantages:
1. Positive die opening
2. Can be rotated
3. Will handle low flow resins
Disadvantages:
1. High initial cost
2. Very hard to clean
3. Two or more weld lines in
   film

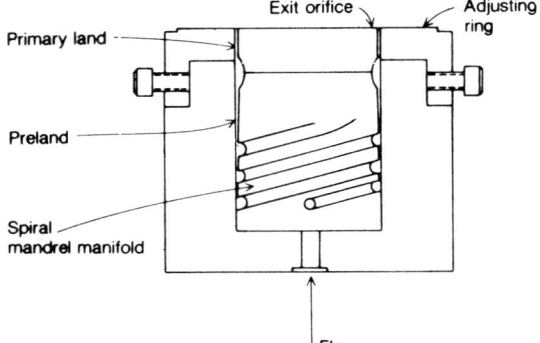

(3) *SPIRAL FEED ROLE*
Advantages:
1. No weld line in film
2. Positive die opening
3. Easy to clean
4. Can be rotated
5. Improved Film Optics
Disadvantages:
1. High head pressure
2. Will not handle low flow
   resins without modification

Fig. 3–12b. Blown film dies, with the spiral groove method used as the best method to distribute melt flow evenly; distribution can be improved by lengthening the spirals and/or increasing the number of distribution points.

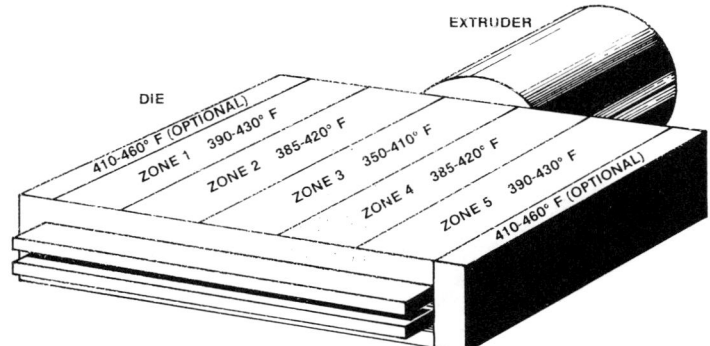

End sections of melt die are higher than center area equalizing the cooling rate of melt as it passes through the die, thus equalizing the flow rate and stock temperature of the sheet across the width of die.

Fig. 3–12c. Schematic for sheet die using temperature control pattern.

the throughput. Also, melt fracture, or sharkskinning, may occur, resulting in a rough surface finish. With its extensional rheology, processors of LLDPE can overcome these problems with wide die gaps in the 0.090 in. (2.3 mm) range. With an increase in the die gap, the head pressure decreases, making possible significant increase in output.

LLDPE can be drawn or stretched in the melt with low induced orientation (see Chapter 1), so a wide die gap does not add undesirable film stresses for LLDPE as with LDPE. The optimum die gap for each application will vary according to resin grade, melt heat, and output rates. Only the die mandrel requires modification for conversion to LLDPE. If only LLDPE is to be run, then the existing mandrel can be machined down. If both resins are to be run, two separate mandrels are required. As cast film is processed at considerably higher heats than blown film, shear stresses are minimized, and no die modifications are required.

A good PS sheet die usually can run some other resins. The gauge control capability of the die for PS or other resins is determined to a great extent by the flow adjustments. Heavy gauge dies might have a lip land length of 3 to 4 in. and a relatively coarse method of adjusting the massive lips. In contrast, a film die would have a lip land length of 0.75 in. and a lip adjustment capable of extremely fine adjustments. Many film dies and thin gauge sheet dies utilize a flexible lip for extremely close gauge control.

Any analysis of die efficiency must include a careful examination of the compatibility of the die with the products to be extruded. If a die is designed for sheet thicknesses of 0.150 to 0.375 in., it is extremely difficult to extrude

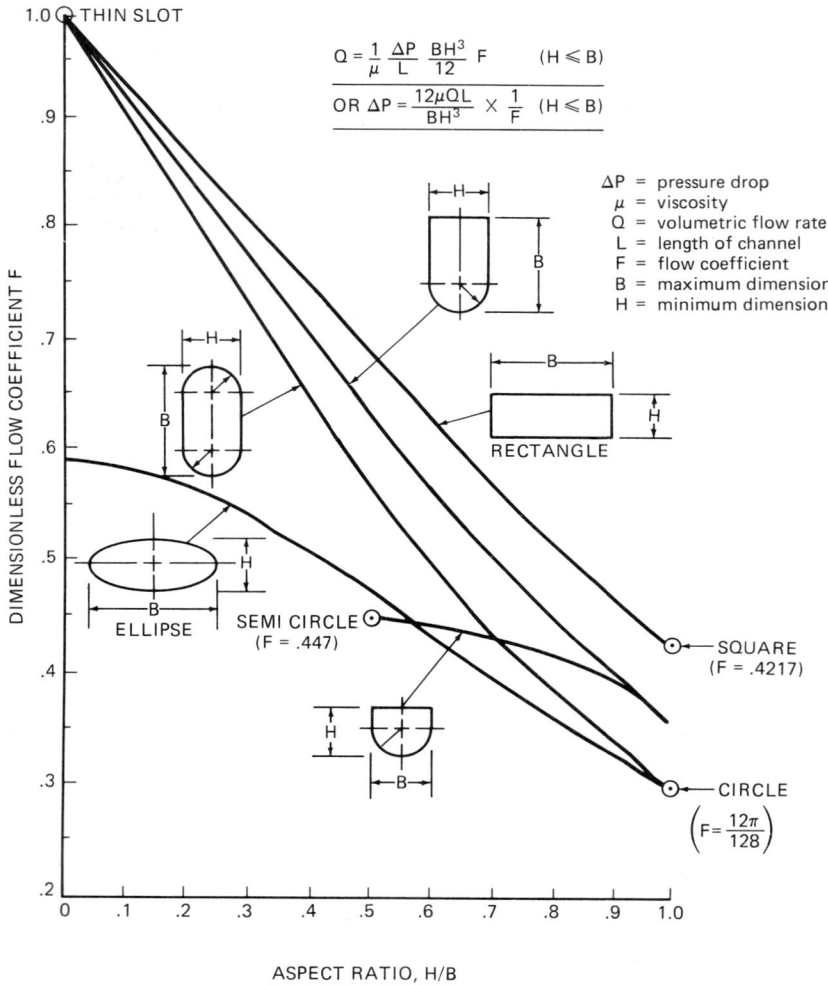

Fig. 3–12d. Flow coefficients calculated at different aspect ratios for various shapes using the same equation.

5 mil (0.005 in.) film. As there is no die design that could be called a "universal" die, it is very inefficient to expect an operator to run a die beyond its capabilities. The result would be poor gauge control, and so forth. If the flow channel geometry is optimized for a resin under a particular set of conditions (heat, flow rate, etc.), a simple change in flow rate or in heat can make the geometry very inefficient. Except for circular dies, it is es-

sentially impossible to obtain a flow channel geometry that can be used for a relatively wide range of resins and a wide range of operating conditions, such as those reviewed for LLDPE and LDPE.

For this reason, adjustment capabilities are provided in the die that basically permit heat and pressure changes. Some dies require that the heat profile be across one direction or in different directions, using individual heating pads, and so on, with appropriate controls. As a guide to film and sheet selection for a die, the following general classification may be helpful: (1) film dies are generally applicable for thicknesses of 0.010 in. or less; (2) thin-gauge sheet dies normally are designed for thicknesses up to 0.060 in.; (3) intermediate sheet dies may cover a thickness range of 0.040 to 0.250 in.; and (4) heavy-gauge sheet dies extrude thicknesses of 0.080 to 0.500 in.

To simplify the processing operation, the die design should consider certain factors if possible. The goals are: to have extrudate (product) of uniform wall thickness (otherwise the heat transfer problem is magnified); to minimize the use of hollow sections; to minimize narrow or small channels; and to use generous radii on all corners, such as a minimum of 0.02 in. (0.5 mm). An "impossible" or difficult process can be designed, but it requires extensive experience (both practical and theoretical), with trial-and-error runs, to make it practical (Figs. 3–13 through 3–17).

Basics of Flow

The non-Newtonian behavior of a plastic (Chapter 1) makes its flow through a die somewhat complicated. One plastic characteristic is that when a melt is extruded from the die, there is some swelling (Fig. 3–14 and Table 3–5). After exiting the die, it is usually stretched or drawn down to a size equal to or smaller than the die opening. The dimensions are reduced proportionally so that in an ideal resin the drawn-down section is the same as the original section but smaller proportionally in each dimension. Because of melt-elasticity effects of the material, it does not draw down in a simple proportional manner; thus the drawdown process is a source of errors in the profile. The errors are significantly reduced in a circular extrudate, such as wire coating (Fig. 3–18). These errors must be corrected by modifying the die and takeoff equipment (Fig. 3–14).

There are substantial influences on the material due to the flow orientation of the molecules; so there are different properties in the flow direction and perpendicular to the flow. These differences have a significant effect on the performance of the part (see Chapter 1).

Another important characteristic of melts is that they are affected by the orifice shape (Fig. 3–15). The effect it produces is related to the melt condition and the die design (land length, etc.), but a slow cooling rate can have a significant influence, especially with thick parts. Cooling is more

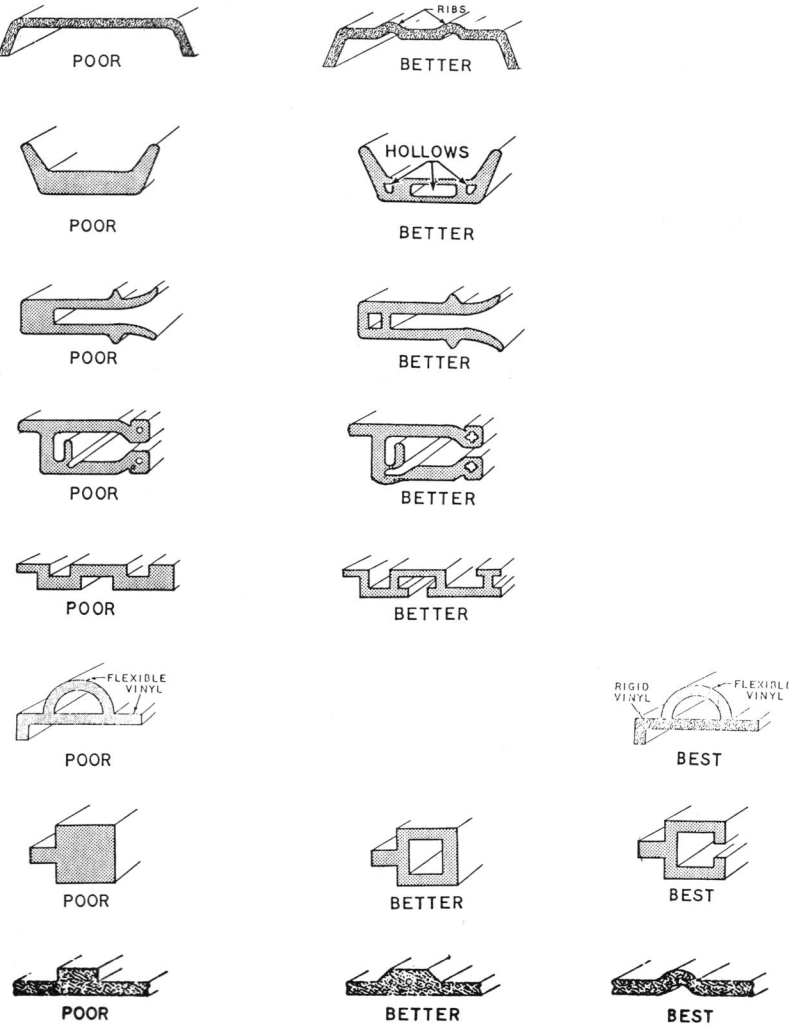

Fig. 3–13. Influence of part design on reducing extrusion process variables.

rapid at the corners; in fact, a hot center section could cause a part to "blow" outward and/or include visible or invisible vacuum bubbles. The popular coat hanger die, used for flat sheet and similar products, illustrates an important principle in die design. The melt at the edges of the sheet must travel farther through the die than the melt that goes through the center of

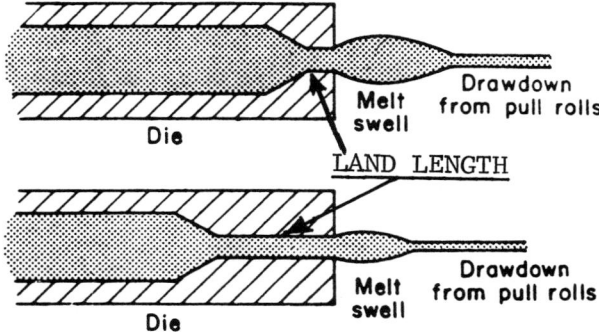

Fig. 3–14. Effect of land length on swell.

the sheet. Thus, a diagonal melt channel with a triangular dam in the center is used to restrict the direct flow to some degree. The principle of built-in restrictions is used to adjust the flow in many dies (Fig. 3–19). With blow molding dies (see Chapter 4) and profile dies, the openings require special attention to provide the proper product shape (Figs. 3–16 and 3–17).

Special Dies

Some special dies are shown in Fig. 3–20; they produce interesting flow patterns and products such as tubular to flat netting dies. For a circular output, a counter-rotating mandrel and orifice have semicircular-shaped slits through which the melt flow emerges. If one part is held stationary, then a rhomboid or elongated pattern is formed; if both parts rotate, then a true rhombic mesh is formed. When the slits overlap, a crossing point is formed

**Table 3–5. General Effect of Shear Rate on Die Swell
of Various Thermoplastics.**

SHEAR RATE, S^{-1} / PLASTIC	DIE SWELL RATIO AT 392°F (200°C)			
	10	100	400	700
PMMA-HI	1.17	1.27	1.35	—
LDPE	1.45	1.58	1.71	1.90
HDPE	1.49	1.92	2.15	—
PP—copolymer	1.52	1.84	2.1	—
PP—homopolymer	1.61	1.9	2.05	—
HIPS	1.22	1.4	—	—
HIPVC	1.35	1.5	1.52	1.53

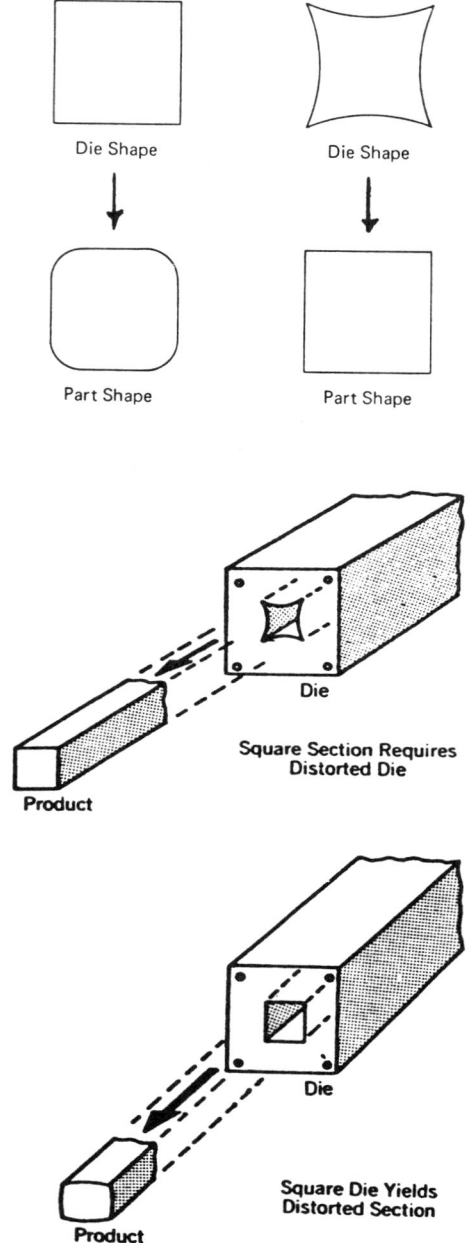

Die Shape

Die Shape

Part Shape

Part Shape

Die

Square Section Requires Distorted Die

Product

Die

Square Die Yields Distorted Section

Product

Fig. 3–15. Effect of die orifice shape on square extrudate.

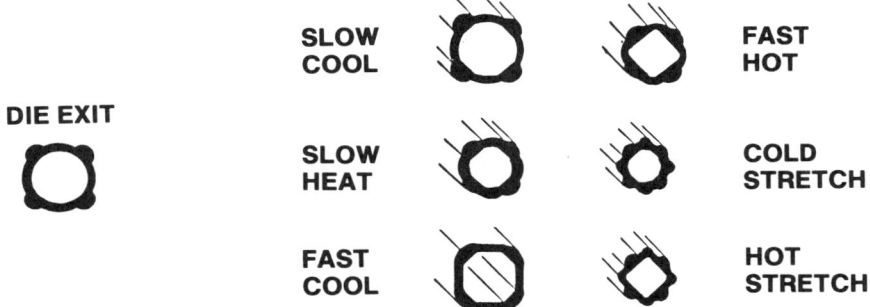

SLOW COOL **FAST HOT**

DIE EXIT

SLOW HEAT **COLD STRETCH**

FAST COOL **HOT STRETCH**

Fig. 3–16. Examples of temperature, pressure, and takeoff speed (time) variations that can potentially influence the shape of the extrudate.

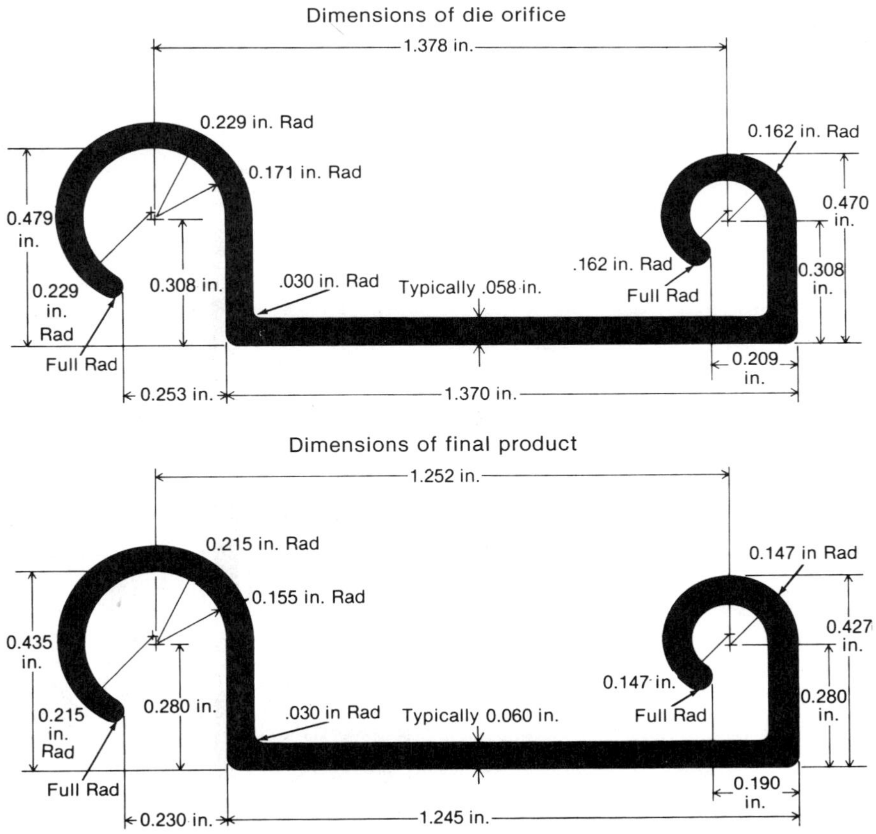

Dimensions of die orifice

1.378 in.

0.229 in. Rad

0.171 in. Rad

0.162 in. Rad

0.479 in.

0.470 in.

0.229 in. Rad

0.308 in.

.030 in. Rad   Typically .058 in.

.162 in. Rad
Full Rad

0.308 in.

Full Rad

0.253 in.

1.370 in.

0.209 in.

Dimensions of final product

1.252 in.

0.215 in. Rad

0.155 in. Rad

0.147 in Rad

0.435 in.

0.427 in.

0.215 in. Rad

0.280 in.

.030 in Rad   Typically 0.060 in.

0.147 in.
Full Rad

0.280 in.

Full Rad

0.230 in.

1.245 in.

0.190 in.

Fig. 3–17. Examples of changes in dimensions of a PVC profile shape from the die orifice to the product.

127

128

(1) DDR in a circular die is the ratio of the cross sectional area of the die orifice/opening to the final extruded shape.

$$DDR = \frac{D_D{}^2 - D_T{}^2}{d_{cw}{}^2 - d_{bw}{}^2}$$

where :

D_T = Diameter of Guider Tip
D_D = Diameter of Die Opening
d_{bw} = Diameter of Bare Wire
d_{cw} = Diameter of Coated Wire

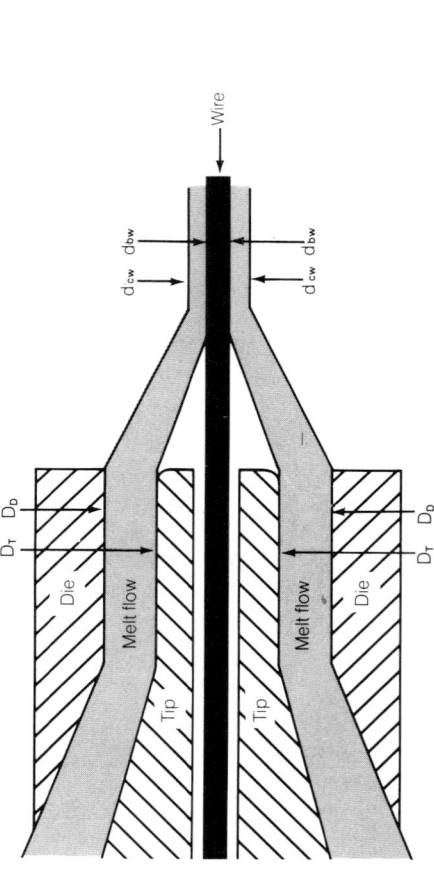

(2) DRB aids in determining minimum and maximum values that can be used for different plastics. Outside these limits can cause at least out of round and melt degradations.

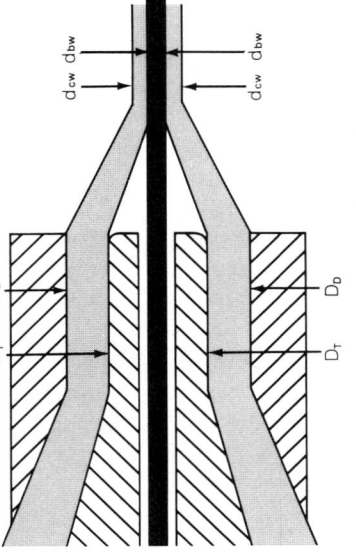

$$DRB = \frac{\dfrac{D_D}{d_{cw}}}{\dfrac{D_T}{d_{bw}}} \approx 1$$

Definition of draw ratio balance = 1

Fig. 3-18. Drawdown ratio (DDR) and draw ratio balance (DRB); applicable to different products, as shown here for wire coating. Plastics have different DDRs and DRBs, which can be used as guides to processability and help establish limits for the various melt characteristics.

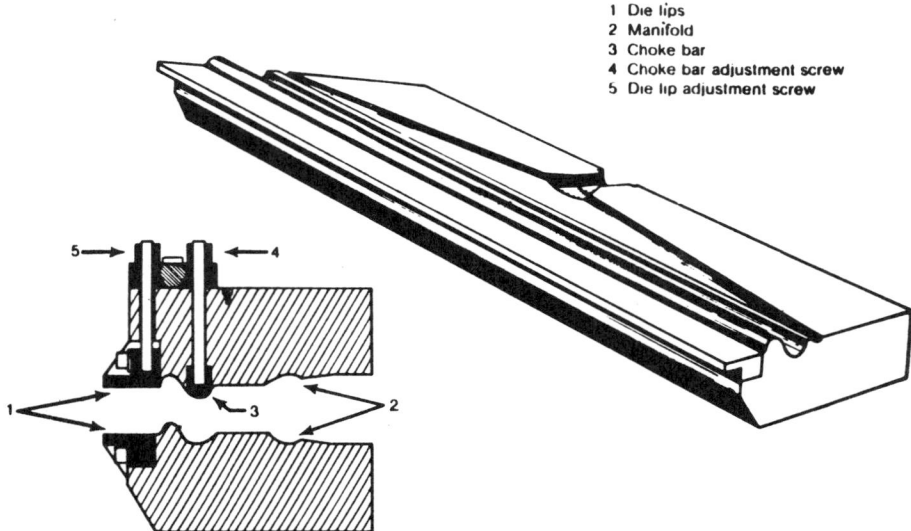

1 Die lips
2 Manifold
3 Choke bar
4 Choke bar adjustment screw
5 Die lip adjustment screw

Fig. 3-19. Example of a coat hanger die, showing lower half and cross section.

where the emerging threads are "welded." For flat netting, the slide is in opposite directions.

Materials of Construction

Usually flat film and sheet dies are constructed of medium carbon alloy steels. Die flow surfaces are chrome-plated to provide corrosion resistance. The exterior of the die is usually flash chrome-plated to prevent rusting. Where chemical attack can be a severe problem (with PVC, etc.), various grades of stainless steel are used (Tables 2-4 and 2-8).

Profile, pipe, blown film, and wire coating dies generally are constructed of hot rolled steel for low pressure melt applications. When high pressure dies are required, 4140 steel is utilized. Chrome plating generally is also applied to the flow surfaces, particularly with EVA. Stainless steel is used for any die subject to corrosion (1-3, 154-157).

Maintenance

The die is an expensive and delicate portion of any extrusion line. Great care should be taken in the disassembly and cleaning of components. Disassembly should be attempted only when the die has had sufficient time to

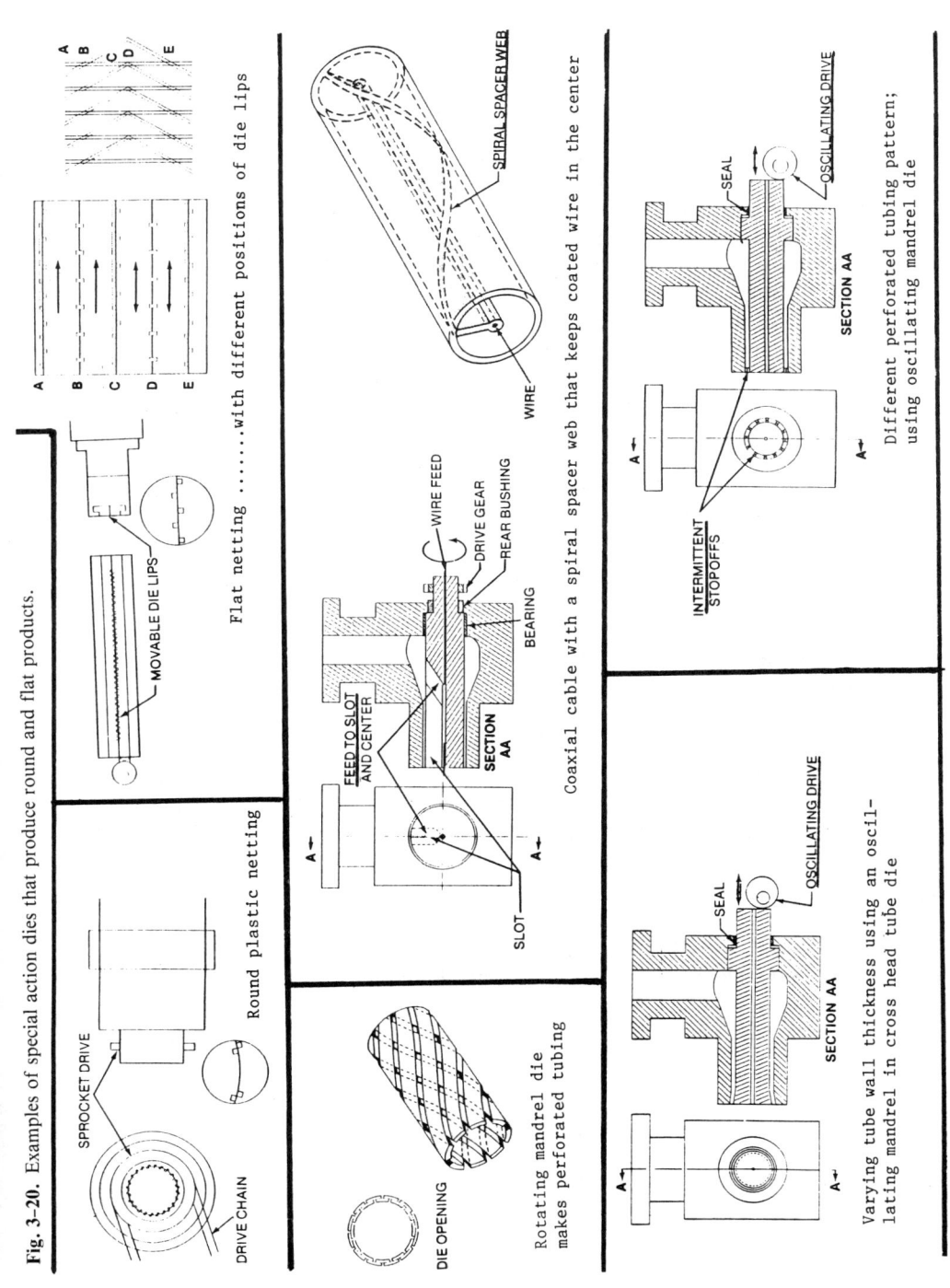

Fig. 3-20. Examples of special action dies that produce round and flat products.

Flat netting ......with different positions of die lips

MOVABLE DIE LIPS

Round plastic netting

SPROCKET DRIVE

DRIVE CHAIN

SLOT

FEED TO SLOT AND CENTER

SECTION AA

WIRE FEED

DRIVE GEAR

REAR BUSHING

BEARING

Coaxial cable with a spiral spacer web that keeps coated wire in the center

SPIRAL SPACER WEB

WIRE

DIE OPENING

Rotating mandrel die makes perforated tubing

Varying tube wall thickness using an oscil-lating mandrel in cross head tube die

SEAL

OSCILLATING DRIVE

SECTION AA

INTERMITTENT STOPOFFS

SEAL

OSCILLATING DRIVE

SECTION AA

Different perforated tubing pattern; using oscillating mandrel die

131

heat-soak or at the end of a run. Experience has shown a temperature of 450°F (232°C) to be adequate for cleaning up most nondegradable resins. For degradables, cleanup should begin immediately after shutdown to prevent corrosive action on the flow surfaces. While the heat is left on, all die bolts should be broken loose. The heat should then be turned off, and all electrical and thermocouple connections removed—carefully; then while it is still hot, the equipment is disassembled and thoroughly cleaned with "soft" brass and copper tools.

If the extruded materials tend to cling to the flow surfaces, it is usually best to purge the die prior to cleanup, with a purging compound (Chapter 2). During assembly, the die bolts should be just snugged tight until the die heat is in the normal operating range. Once this heat is reached and a sufficient heat soak has been allowed (which could take at least 15–30 min.), all bolts should be tightened to the manufacturer's recommended sequence and torque levels. If the die is stored disassembled, care should be taken in its handling to prevent damage to individual components and to flow surfaces, which can include storage in a vacuum sealed container (1–2).

COEXTRUSION

Coextrusion provides multiple molten layers—usually using one or more extruders with melts going through one die—that are bonded together. This technique permits using melt heat to bond the various plastics (Table 3–6), or using the center layer as an adhesive. Coextrusion is an economical competitor to conventional laminating processes by virtue of reduced materials handling costs, raw materials costs, and machine-time cost. Pinholing is also reduced with coextrusion, even when it uses one extruder and divides the melt into at least a two-layer structure. Other gains include elimination

Table 3–6. Examples of Compatibility Between Plastics for Coextrusion.

	LDPE	HDPE	PP	IONOMER	NYLON	EVA
LDPE	3	3	2	3	1	3
HDPE	3	3	2	3	1	3
PP	2	2	3	2	1	3
Ionomer	3	3	2	3	3	3
Nylon	1	1	1	3	3	1
EVA	3	3	3	3	1	3

Code: 1. Layers easy to separate.
2. Layers can be separated with moderate effort.
3. Layers difficult to separate.

or reduction of delamination and air entrapment (1-12, 33-36, 100, 124, 153-161).

In the past, a processor desiring to enter the field had little choice of equipment, but the increased interest in coextrusion has produced a proliferation of equipment. With rapidly changing market conditions and the endless introduction of useful materials, the design of machines has become much more involved. It is important that the processor have flexibility in making selections, but not at the expense of performance, dependability, or ease of operation. One should provide for the material or layer thickness necessary in product changeover without high scrap rates. The target is to incorporate scrap regrind in the layered construction.

It is very important to be able to control the individual layer distribution across the width of the die. It is normal, as the viscosity ratio, or the thickness ratio, of the polymers being combined increases, for the individual layer distribution(s) of the composite film to become displaced. Viscosity differences influence reduction and saving of materials.

A number of techniques are available for coextrusion, some of them patented and available only under license. Basically three types exist: feedblock, multiple manifold, and a combination of these two (Table 3-7).

With the feedblock die, different melts are combined just upstream of the die, prior to entering the die via a special adapter. Laminar flow keeps the layers from mixing together so that the layup exits as an integral construction from the die.

A multiple manifold die involves the combination of melts within the die. Each inlet port leads to a separate manifold for the individual layers involved. The layers are combined at or close to the final land of the die, and they exist as an integral construction through a single lip. Although the multi-manifold die can be more costly than the feedblock type, it has the advantage of more precise control of individual layer thickness.

A third approach combines the feedblock and multi-manifold types, and provides further processing alternatives as the complexities of coextrusion increase. This approach has been used successfully in barrier sheet coextrusions where requirements preclude other alternatives. The feedblock is placed on the manifold or manifolds, permitting the combination of materials of similar flow characteristics in the feedblock while feeding dissimilar materials directly into the die.

Assuming both types have a good manifold design, the multi-manifold can process a broader flow range of resins than the feedblock. No matter which method is used, it is important to maintain melt heat of each layer above the "freezing" temperature of all layers, or poor adhesion results. The heat has to be the most important operating requirement for all resins used.

Table 3–7. Comparison of Feedblock and Multi-manifold Coextrusion Dies.

CHARACTERISTIC	FEEDBLOCK	MULTI-MANIFOLD
Basic difference	Melt streams brought together outside die body (between extruder and die) and flow through the die as a composite.	Each melt stream has a separate manifold; each polymer spreads independent of others; they meet at die pre-land to die exit.
Cost	Lower.	Higher.
Operation	Simplest.	—
Number of layers	Not restricted; seven- and eight-layer systems are commercial.	Generally restricted to three or four layers.
Complexity	Simpler construction; no adjustments basically.	More complex.
Control flow	Contains adjustable matching inserts, no restrictor bar.	Has restrictor bar or flow dividers in each polymer channel; but with blown film dies, control is by individual extruder speed or gear boxes.
Layer uniformity	Individual layer thickness correction of ± 10 percent.	Restrictors and manifold can meet ± 5 percent.
Thin skins	Better on dies > 40 in.	Better on dies < 40 in.
Viscosity range	Usually limited to 2/1 or 3/1 viscosity range of materials.	Range usually much greater than 3/1.
Degradable core material	Usually better.	—
Heat Sensitivity	More.	Less.
Bonding	Potentially better; layers are in contact longer in die.	—

It is also desirable to be able to vary and control the thickness of the plies. As there are two or more extruders feeding the separate inlet channels, the first step is to calibrate them in terms of output rate vs. screw speed. As previously discussed, gear pumps can be used. The next step is to establish the required width to thickness and takeoff speed, and convert this value into an output rate for each resin. From such output rates and calibration curves, screw speeds are established. When a feedblock is not designed properly, problems such as excessive pressure drops, shear rates, or residence time can affect the quality of the extrudate. The tear-drop shape manifold has proved beneficial for coextrusion (see Fig. 3–12a, part 4).

All extrusion processes require some form of melt transfer from the extruder to the feedblock or die. The transfer device can be short, such as an extruder adapter, or longer, as required in coextrusion piping. The pipe design is integral to coextrusion success; in general, the inside diameter should be large enough to avoid excessive pressure drops, and not so large as to cause extended residence time. Its heavy wall provides maximum thermal stability and heat distribution efficiency. The heat should be uniform to avoid hot and cold spots. Low voltage heaters are desirable, but heater tapes are dangerous because of the possibility of nonuniform heat distribution.

Safety consideration should be the determining factor in the design of piping. Pipes have been known to cold-pack, depending on shutdown procedures. The location of the control thermometer is important. Overheating of a zone, particularly on startup, can cause degradation of the plastic into a gas, creating extremely high pressures.

Tie-Layers

Choosing an adhesive layer is by no means a simple operation; there are many different types, each with specific capabilities, with EVAs forming the bulk. Selection of a material is based on its providing good adhesion and surviving the process. For example, high melt strength in a blown film improves bubble stability. At temperatures above 460°F (238°C), EVAs could suffer from gel formation and decomposition. High melt strength also can help in cast extrusion and thermoforming processes, and the melt draw is important in coextrusion of cast film/sheet or a coating. Good melt draw is required to run higher takeup speeds and/or thinner adhesive layers without causing flow-distribution or edge-weave problems. Effects such as "neck-in" and "edge bead" are also minimized by choosing adhesives with a good draw.

Various processing conditions can require the tie resin to fall into a particular melt index (MI) classification. MI is inversely related to molecular weight (MW) (see Chapter 1); a high-MW adhesive will have a low MI. Most adhesive are available in a range of MIs to meet different requirements.

The melt stability or flow is easily influenced by regrind. It is important that the regrind be compatible with the adhesive.

ORIENTATION

Orientation consists of a controlled system of stretching plastic molecules to improve their strength, stiffness, optical, electrical, and other properties. This process has been used for almost a century, and became very prominent during the 1930s for stretching fibers up to ten times; later it was adapted to stretching film/sheet and more recently blow molded containers. Many other products take advantage of its benefits (tape, pipe, profile, thermoformed parts, etc.). Practically all plastics can undergo orientation, although certain types find it particularly advantageous (PET, PP, PVC, PE, PS, PVDC, PVA, and PC). Of the 12 million tons of plastic film sales worldwide, about 13 percent are sales of oriented material (see Chapter 1) (1-8, 161-164).

In extrusion the most important orienting processes are used for blown film, flat film and sheet, and (as reviewed in Chapter 4) blow molding. During blown film processing the blow-up ratio determines the degree of circumferential orientation, and the pull rate of the bubble determines longitudinal orientation (Fig. 3-21). The optimum stretching heat for amorphous resins (PVC, etc.) is just above the glass transition temperature; for

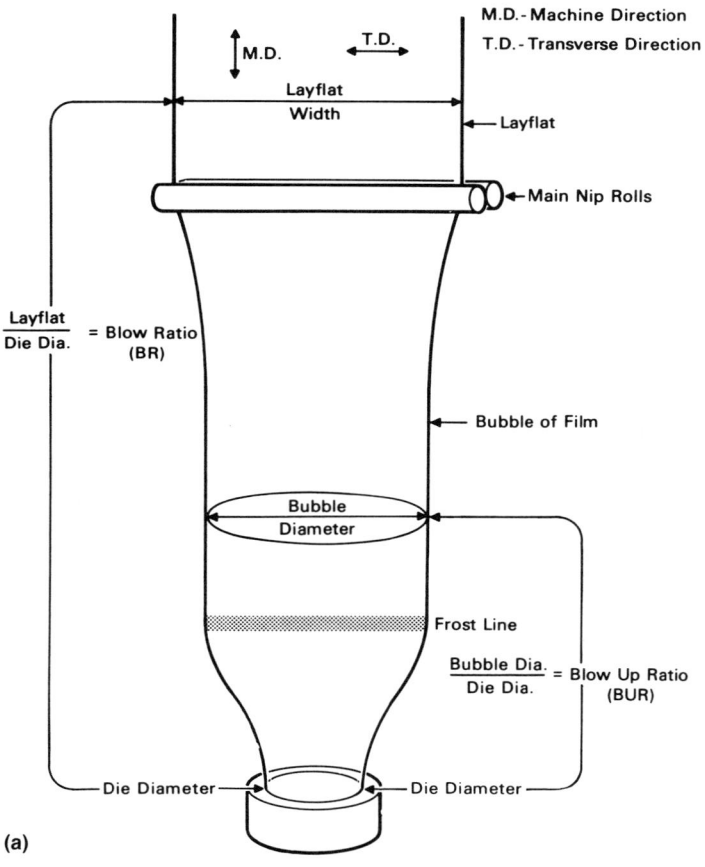

(a)

Fig. 3-21. (a) Terminology for blown film. (b) Orientation of blown film.

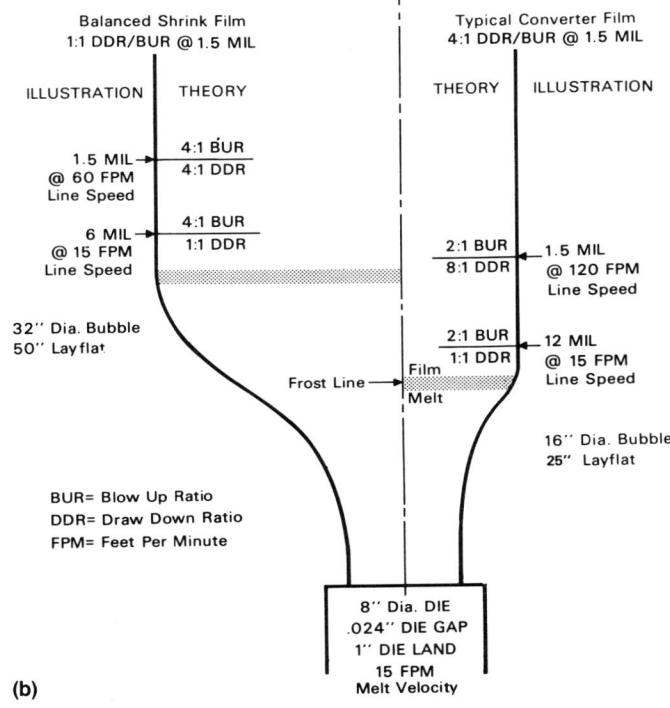

(b)

BUR= Blow Up Ratio
DDR= Draw Down Ratio
FPM= Feet Per Minute

Fig. 3-21. Continued.

crystalline types (PET, PE, etc.) it is just below the melting point (Table 3-8). During the stretching process, the structure changes because of crystallization, thus usually necessitating an increase in heat if further deformation is planned. Afterward, the orientation is "frozen-in" by lowering the heat or, with crystalline types, set by increasing the crystalline portion.

With orientation, the thickness is reduced and the surface enlarged. If film is longitudinally stretched in the elastic state, the thickness and the width are reduced in the same ratio. If lateral contraction is prevented, stretching reduces the thickness only.

The direct injection of liquid additives, such as polyisobutylene (PIB), to produce stretched film prevents difficulties in extruding and offers a processor a wider range of materials from which to select. It provides cost reductions due to the use of more economical formulations. This method is suitable for the injection of cross-linking agents, liquid colors, and the like, via the extruder or a gear pump.

In orienting film or sheet the processor uses a tentering frame (typically

Table 3–8. Examples of Orienting Conditions for Plastics.

| PLASTIC | MELTING TEMPERATURE (T_m) IN °C | GLASS TRANSITION TEMPERATURE (T_g) IN °C | MODULUS OF ELASTICITY, MPa $\times 10^3$ | | | | | DENSITY G/CM^3 |
| | | | BIAXIALLY ORIENTED FILM | | UNORIENTED FILM | ORIENTED FIBER | | |
			LONGITUDINAL	TRANSVERSE		MELT SPUN	GEL SPUN	
PVC	—	70 to 90	2.5 to 2.7	3 to 3.5	2.2	5.5	—	1.35
HDPE	138	−70 to −110	3 to 4	3 to 4	1.2	5	170	0.96
PP	134	−5 to −20	3 to 4	2 to 3	0.9	5	18	0.90
PA	260	50 to 75	2 to 2.5	2 to 2.5	0.5	4.5	19	1.13
PET	250	70 to 110	4 to 5.7	4.5 to 8.5	1.5	15	28	1.35

used for many decades in textile weaving), which is enclosed in a heat-controlled oven, with a very accruate and gentle air flow used to hold the oven at the required orienting heat (Fig. 3–22). The frame has continuous speed control and diverging tracks with holding clamps. As the clamps move apart at prescribed diverging angles, the hot plastic is stretched in the transverse direction resulting in single orientation (O). To obtain bidirectional orientation (BO) an inline series of heat-controlled rolls are located between the extruder and tenter frame. The rotation of each succeeding roll is increased, based on the longitudental stretched properties desired.

PROCESSING LINES

Each line has interrelating operations, as well as specific line operations, to simplify processability (1–5, 8–11, 85, 122, 124, 163, 165–184). Usually the extruder is followed by some kind of cooling system to remove heat at a

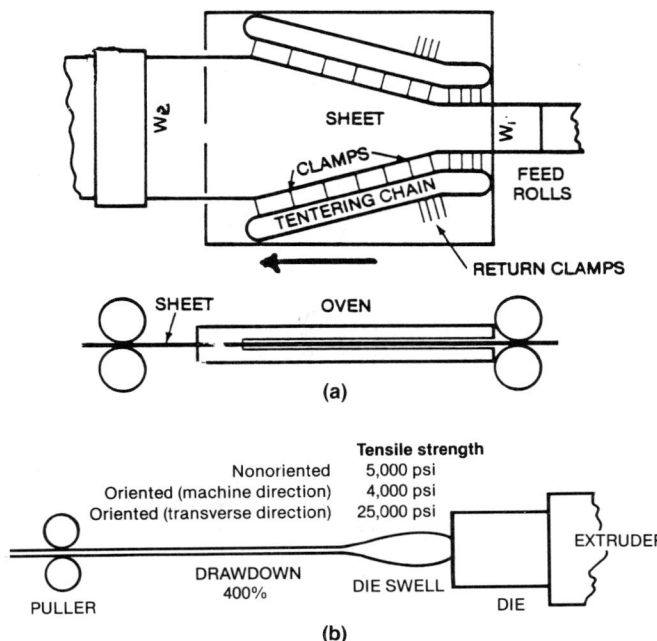

Fig. 3-22. Use of tenter frame to biorient film or sheet. In (a), the feeder-roll speed to puller-roll speed ratio is 1:4 (the ratio of width W_2 to width W_1). Part (b) is a schematic of the drawdown phenomenon with die swell to produce orientation in the longitudinal (machine) direction.

controlled rate, to cause plastic solidification. It can be as simple a system as air and/or water cooling, or a cooled roll contact can be used to accelerate the cooling process. Some type of takeoff at the end of the line usually requires an accurate speed control to ensure product precision and/or save on material costs by tightening thickness tolerances. The simplest device might be a pair of pinch rolls or a pair of opposed belts (caterpillar takeoff). A variable speed drive is usually desired to give the required precision.

Blown Film

More plastics go through blown film lines than other extrusion lines. The process can vary in direction (up, down, or horizontal) and in the method of flattening the film prior to wind-up (Fig. 3–23). Developments in these lines relate to the extruder, dies, takeoff systems, and automation components. The development of new high-speed extruders with a grooved feed zone and barrier screws makes it possible to increase output while providing greater processing flexibility—which, particularly in coextrusion, renders changes of screws unnecessary. Blown film dies have been developed with the goals of low pressure consumption, good self-cleaning, material changes, and ease of maintenance. The automation of blown film plants to reduce

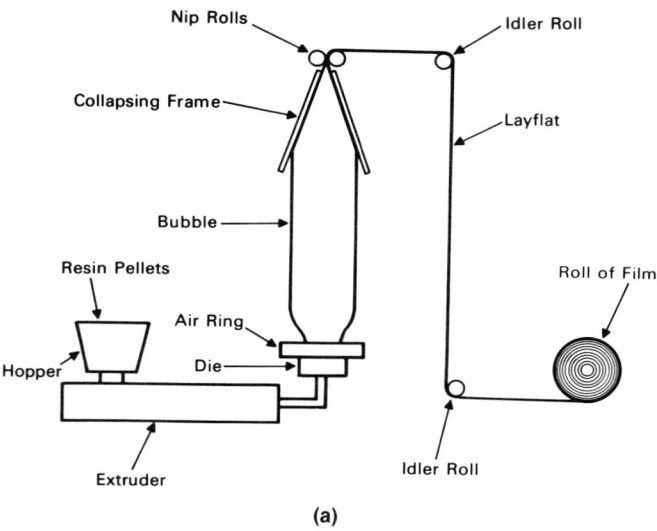

(a)

Fig. 3-23. (a) Schematic of a basic (vertical-up) blown film line. (b) Geometry of collapsing bubble.

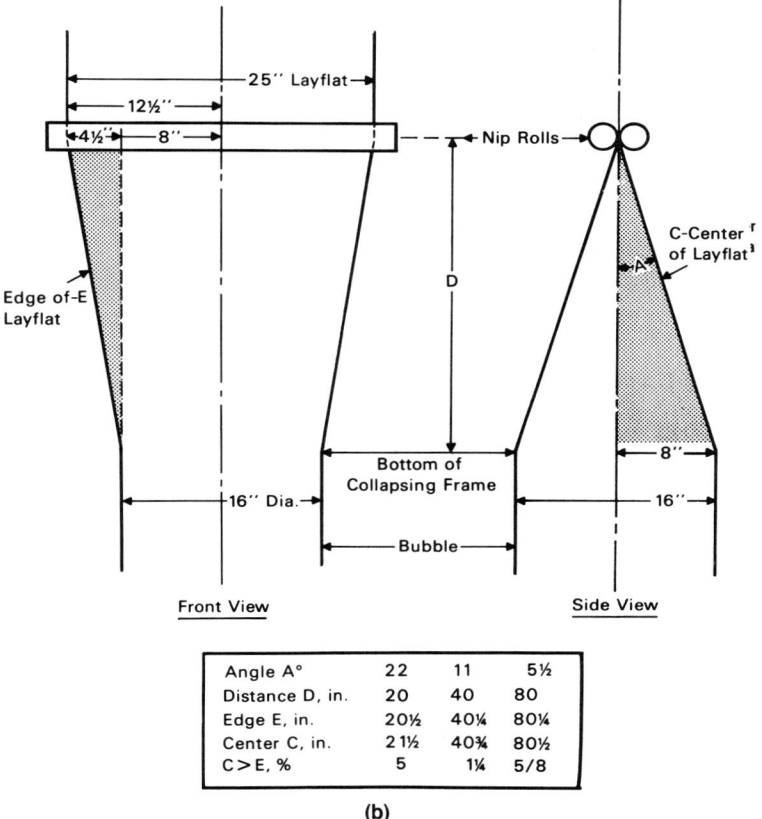

Fig. 3-23. Continued.

film thickness tolerances involves the increased use of linear weight control systems (upstream and downstream), as well as greater opportunities to influence profile thickness via suitable control elements on the die and cooling systems.

Regarding the film direction, horizontal operation entails no overhead installation and a low building height, but requires a larger floor space with probable adverse effects of gravity and uneven cooling. Vertical-down operation has the advantage of start-up without flooding of the annular die gap by exiting hot melt. However, vertical-up operation is the usual method, provided sufficient melt strength exists for an upward start-up, and so on. Special die blow heads are designed, with (usually) a multiple threaded helical mandrel discharging into an expansion space. The tubular melt assumes its final shape in a "smoothing-out" zone, which in all heads is a cylindrical land in a parallel position between the mandrel and the orifice. Its length is about 10 to 15 times the annular gap width (the lower value applies to thin film and the higher to thick film). The gap width is generally 0.5 to 2.0 mm.

Different methods of bubble cooling exist, each with advantages and disadvantages. For example, because of their different extensional rheologies (flow), LLDPE bubbles are less stable than those of LDPE. Proper cooling is very important in obtaining good gauge uniformity. Gentle, very cold air has a better cooling effect than high velocity cool air; the gentle air helps to minimize bubble instability (usual pressure is 150–600 mm water column). Although single-lip air rings have proved adequate for some applications, dual-orifice designs provide enhanced cooling, which effectively stabilizes the bubble and speeds up the line (Fig. 3–24). Internal bubble cooling (IBC) with a dual-lip air ring is also effective in increasing bubble stability at high production rates. However, improperly arranged IBC configurations can cause melt fracture due to chilling of the die lip.

Heat between the die and the pinch rolls influences the haul-off rate. LDPE, for example, leaves the die at 150 to 170°C. On its arrival at the pinch rolls, the temperature should have fallen to 40°C. The film should be wound up at as low a heat as possible in order to prevent excessive shrinkage on the roll, which causes blocking. Thin-walled film can be taken off at speeds of at least 20 to 50 m/min. With film of 150 to 300 μm thickness, rates of at least 10 to 20 m/min are achieved.

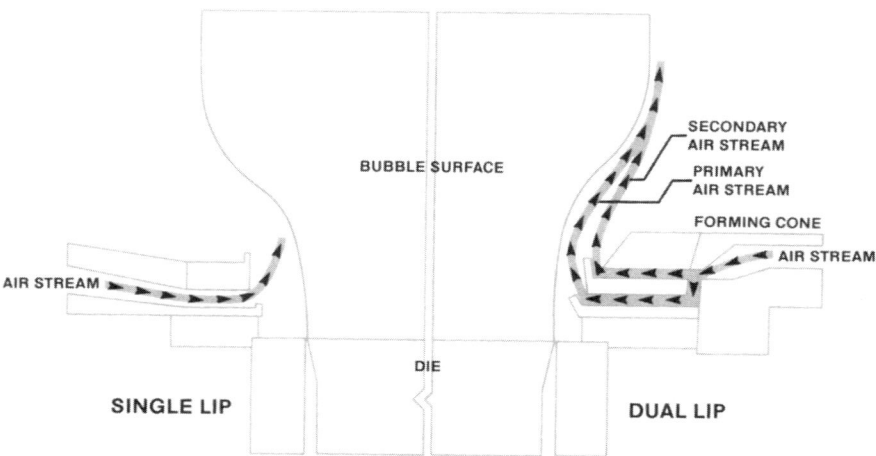

Fig. 3–24. Comparison of conventional single-lip and dual-lip air rings for cooling blown film.

The blow ratio (Fig. 3–21a) is usually between 1.5:1 and 4.0:1, depending on the material being processed and the thickness required. With crystalline types, melt leaving the die changes from a transparent (amorphous) condition to hazy. The level at which this transition occurs is called the frost line (Fig. 3–21b). The visual appearance (whether it is level/straight or shows a varying line and height) of the melt exiting the die can be related to processing conditions (uneven melt flow, heat variations, degree of orientation, etc.). For polyolefins (PE, PP, etc.) to be printed, usually a corona discharge pretreatment is given following the layflat operation. This has the effect of oxidizing and activating the surface. The weldability may be impaired and the blocking tendency increased if the treatment is too intensive.

A certain degree of variation in thickness is unavoidable. When caused by the tubular film die, these variations always occur at the same position. A local film excess usually appears as a line. This can be countered if the die head, the haul-off and wind-up gear, or, with a vertical extruder, the extruder barrel is rotated or moved from side to side at regular intervals (Fig. 3–25).

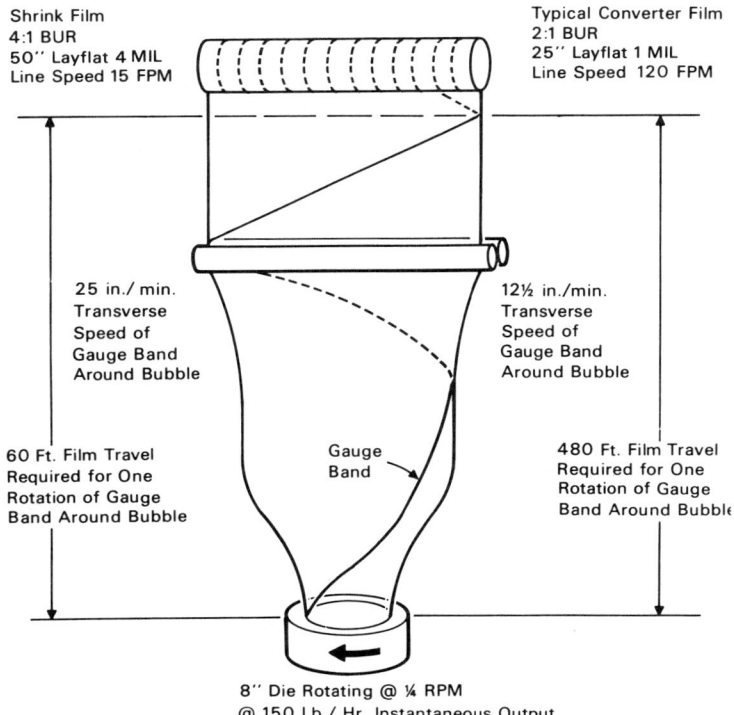

Shrink Film
4:1 BUR
50″ Layflat 4 MIL
Line Speed 15 FPM

Typical Converter Film
2:1 BUR
25″ Layflat 1 MIL
Line Speed 120 FPM

25 in./min.
Transverse
Speed of
Gauge Band
Around Bubble

12½ in./min.
Transverse
Speed of
Gauge Band
Around Bubble

60 Ft. Film Travel
Required for One
Rotation of Gauge
Band Around Bubble

Gauge
Band

480 Ft. Film Travel
Required for One
Rotation of Gauge
Band Around Bubble

8″ Die Rotating @ ¼ RPM
@ 150 Lb./ Hr. Instantaneous Output

Fig. 3–25. Example of averaging out thickness changes in blown film by rotating die head.

Flat Film

Flat films processed through slit dies are cooled principally by using chilled rolls. Many different resins are used, with thicknesses ranging from 15 to 200 μm (Figs. 3–26 and 3–27). Alternatively, certain plastics go directly into a water tank, but that creates many technical difficulties in production. Thus, the chill roll process is preferred; and film up to 3 m in width will have output rates of at least 120 m/min.

In this process, the melt film contacts (as quickly as possible, vertically or at an angle) the first water-cooled highly polished (to 1 μm) chrome-plated roll. An air knife can be used; its placement parallel to the die makes it possible to press the film smoothly onto the first cooling roll by means of a cold air stream. Lubricant plate-out on the cooling rolls is avoided by operation with contact rolls. At haul-off rates of up to 60 m/min, reel change is carried out by hand. At higher rates, automatic changeover equipment is required.

Advantages of the chill roll process (vs. blown film) include: preparing almost transparent film from crystalline resins (the frost line forms about 50 mm above the contact line with the chill roll); no risk of blocking; a simple crease-free wind-up; continuous film thickness control; high output; relatively small space requirement; and the fact that pretreatment for printing can be applied simultaneously to both sides of the film. Disadvantages are: the limitation on maximum width of about 3 m (blown film layflat is at least up to 12 m); loss through edge trimming; and basically only uniaxial orientation.

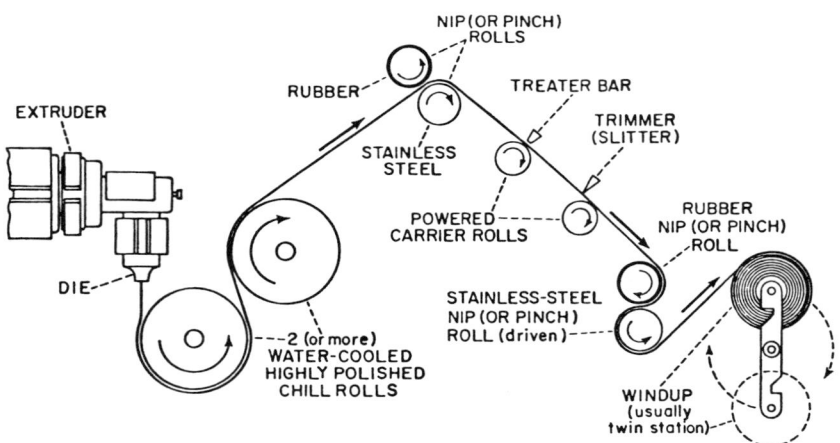

Fig. 3-26. Schematic of chill roll system for flat-film extrusion line.

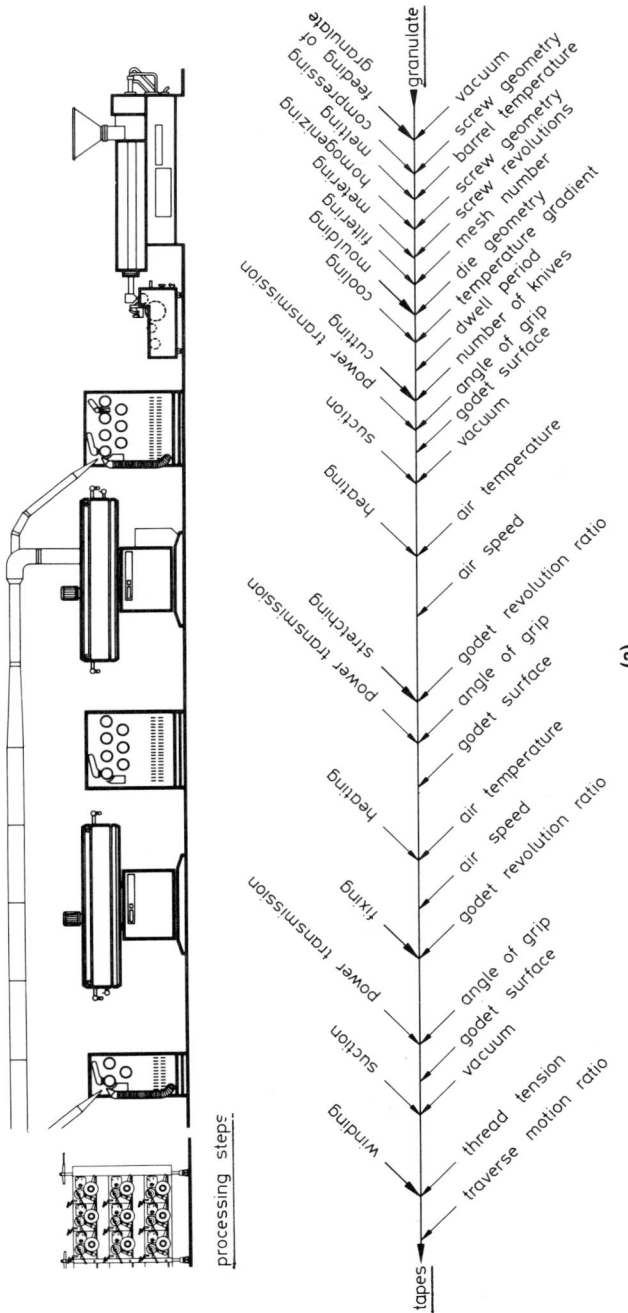

Fig. 3-27. (a) Chill roll process used in oriented film tape line. (b) Example of performance of oriented PP based on orienting heat and stretch ratio.

processing steps

granulate

feeding of granulate — vacuum
compressing — screw geometry
melting — barrel temperature
homogenizing — screw geometry
metering — screw revolutions
filtering — mesh number
moulding — die geometry
— temperature gradient
cooling — dwell period
cutting — number of knives
power transmission — angle of grip
— godet surface
suction — vacuum

heating — air temperature
— air speed
— godet revolution ratio
power transmission — angle of grip
stretching — godet surface

heating — air temperature
— air speed
fixing — godet revolution ratio
power transmission — angle of grip
— godet surface
suction — vacuum
winding — thread tension
— traverse motion ratio

tapes

(a)

145

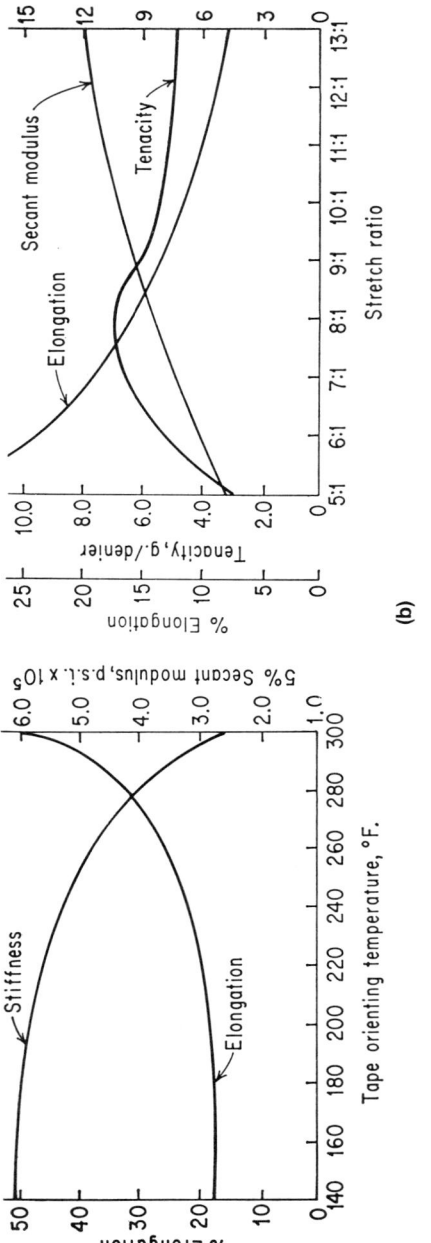

Fig. 3-27. Continued.

Sheet

Sheet is usually defined as being thicker than film, or thicker than 1 to 4 mm ($\approx$ 0.003–0.010 in). Sheet thickness can be at least 2 mm (0.5 in.), and widths can be up to 30 m (10 ft). Basically, hot melt from a slit die is directed to a combination of an air knife with two cooling rolls, or, a more popular choice, to a three-cooling-roll stand (Fig. 3–28), which cools, calibrates, and produces a smooth sheet. To aid the chill rolls, end sections of the die are operated at a higher heat than the center (Fig. 3–12c). Cooling rolls require this type of heat control from their ends to the center.

The operation (as well as design) of a slit die, particularly for wide sheets, requires extensive experience. Its rather high melt pressure can deform the die.

Pipe

A typical pipe line consists of a single or a twin screw extruder, a die, equipment for inside and outside calibration, a cooling tank, a wall thickness measuring device, marking equipment, haul-off and automatic cutting and pallet equipment, or a windup unit for self-supporting pipe coils or lengths that are coiled on a drum (Fig. 3–29). Single screw extruders are generally used when processing PVC compound in granule form; twins handle powders of PVC. The adjustment and control of back pressure are very critical. PVC pipe is a big and very competitive market, so quality and profitability have been the most important requirements for years. Improving the equipment is almost of secondary importance because the equipment for good-quality products is already available.

The consumption of material, the main cost factor, can be gradually min-

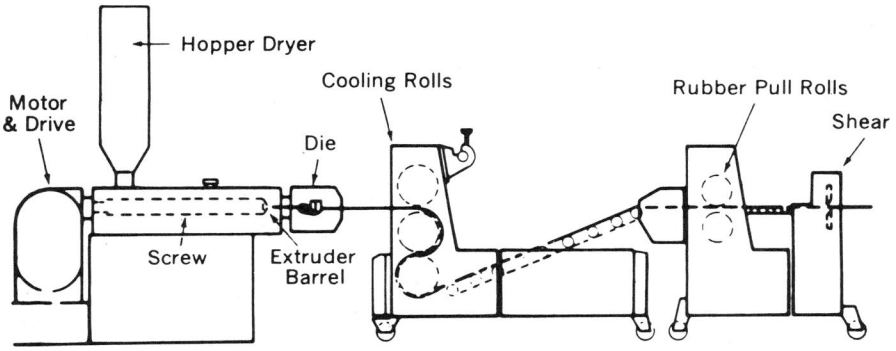

Fig. 3–28. Example of sheet line using three-cooling-roll stand.

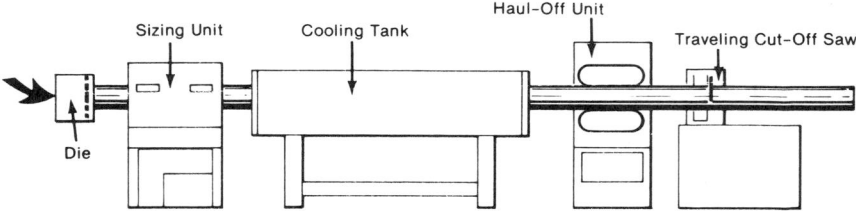

Fig. 3-29. Example of the important downstream equipment used in pipe and profile extrusion.

imized by the use of current measuring and control systems. The goal is always tighter tolerance control to save material. Calibrating discs and pressure calibration methods of many different designs are used to meet various requirements (Fig. 3-30). The operator's expertise in using these calibration systems is as important as controlling the complete line. A system of feedback and control of the extruder by a microprocessor is used to control wall thickness, combining ultrasonic gauging with gravimetric proportioning. Such new technologies are available but have to be debugged. Perhaps it can be said that any equipment from the "past" is definitely noncompetitive, based on all the new equipment that has been made available from upstream, through the extruder, and downstream. In the last few years, all the equipment has been significantly altered to increase profits.

Profiles

As with pipe, the profile market is largely dominated by PVC and is highly competitive. Automation at the processor's level has reached a very advanced stage. As previously reviewed (in discussing dies, etc.) the operator's capability is needed to ensure maximum product efficiency, in regard to controlling die swell, rate of pull, and so on. High performance lines operate at over 2 m/min (Fig. 3-29).

As changes have been made in PVC compounding to optimize processing, there has been a considerable change in the type of impact modifier used. The use of acrylic, with butyl acrylate, has almost replaced modification with EVA. When modification is carried out with acrylate, the elastomer phase is embedded in the form of beads in a continuous PVC matrix. During processing it is retained to the decomposition range. This wide processing latitude has at least made it easier to achieve the present high outputs in profile extrusion. It also has provided low shrinkage, high heat distortion, and good weather resistance.

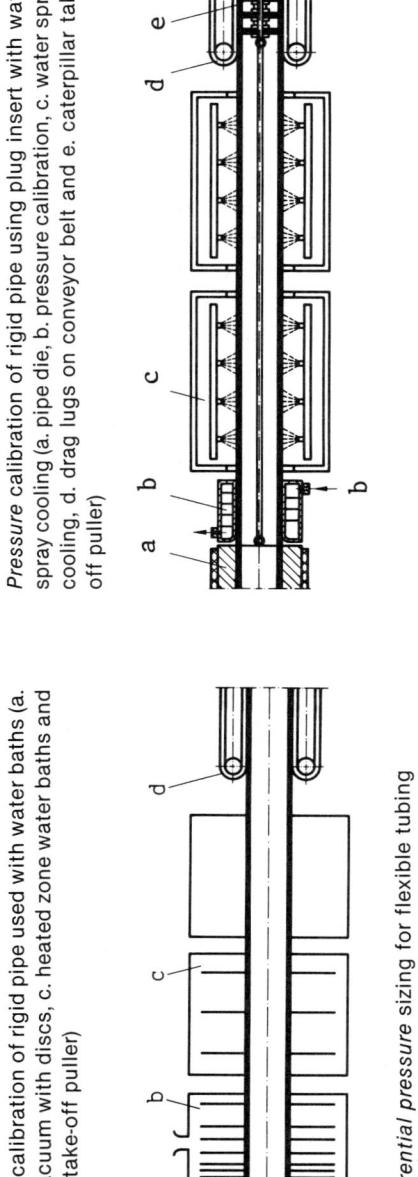

Vacuum tank calibration of rigid pipe used with water baths (a. pipe die, b. vacuum with discs, c. heated zone water baths and d. caterpillar take-off puller)

Pressure calibration of rigid pipe using plug insert with water spray cooling (a. pipe die, b. pressure calibration, c. water spray cooling, d. drag lugs on conveyor belt and e. caterpillar take-off puller)

Differential pressure sizing for flexible tubing

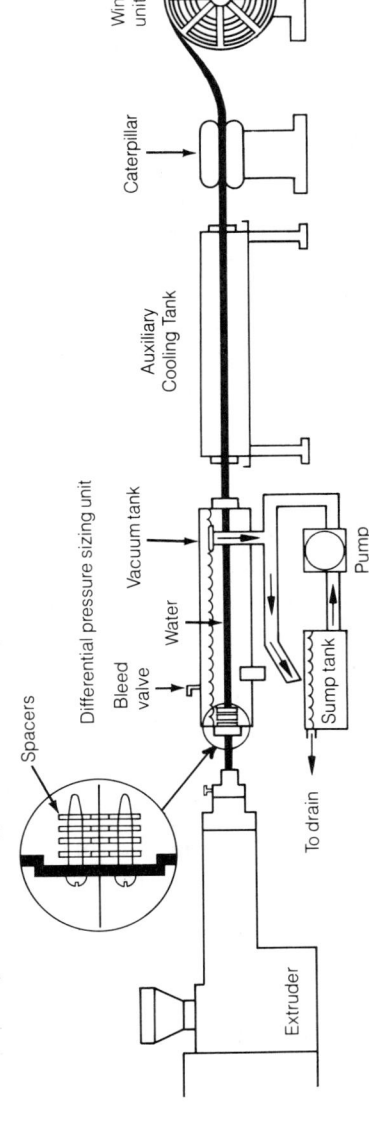

Fig. 3–30. Examples of calibration systems for pipe/tube extrusion lines.

Coatings

Plastic coatings are applied in different forms and shapes on many different products, such as wire, cable, profiles (plastics, wood, aluminum, etc.), films/foils (plastics, aluminum, steel, paper, etc.), rope, and so on. Certain coatings only require snug fits, whereas others require excellent adhesion, usually necessitating cleaning, priming, and/or heating substrates.

Wire Coating. Wire coating is performed by extruding plastic around a wire. This may be accomplished by feeding the wire directly through a hole in the center of the feed screw, or, by far the more popular method, by using a crosshead die (similar to Fig. 3–12a, part 2) through which the wire is fed (Fig. 3–31). Hot melt is extruded over a preheated wire to improve adhesion and reduce shrinkage stresses. Before the wire enters the die, preheating is done by an electric current, radiant heaters, and so on; thick wires or cables can be heated by a gas flame or hot gas. Wire travels at rates up to at least 1,300 m/min (4,000 ft/min). Regardless of the speed, the rate of movement has to be held extremely uniform (or "perfect"). In order to achieve uniformity, all peripheral (expensive) equipment is carefully controlled and monitored, from the wire input drum to the output windup drum.

There are two basic types of dies used, called high and low pressure coaters. With high pressure, the melt meets and coats the wire between the die lips prior to exiting the die. The result is good contact of plastic to wire, tight control of the plastic OD, and the ability to handle plastics that require tight melt control, particularly with operation at "peak" heats and pressures. In the low pressure type, the melt makes contact with the wire after they both exit the die. Plastic hugs the wire, with formation of a loose jacket that facilitates removal of plastic insulation. If spiders are used to support the central mandrel, they are usually thin and streamlined to minimize disruption of the velocity. Adjustment of the wall thickness distribution and concentricity via die centering bolts can be manual or automatic (Fig. 3–12a part 2). Automatic control can be achieved with in-line wall thickness measurement probes, and so forth. With high pressure dies, a vacuum can be used in the die just before the wire goes through its snug central support, to obtain an air-free and better bond. With a low pressure die, low air pressure is applied through the center of the mandrel tube to prevent collapse of the melt tubing on exiting (eliminating melt adherence on the front of the die, which requires down time and cleanup) and to aid in maintaining the plastic ID.

Cooling of the thermoplastic coated wire usually occurs as soon as it leaves the die through water cooling troughs, which may have cascading heated sections. The troughs are usually 20 to 100 ft long. With thermoset

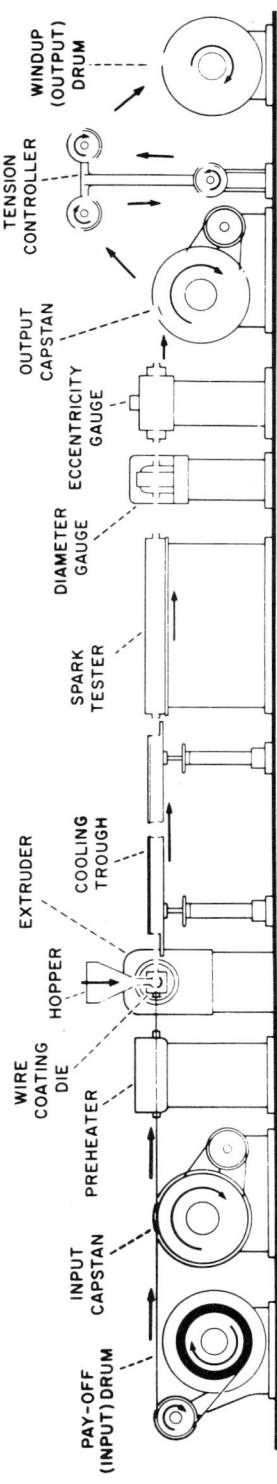

Fig. 3–31. Example of the general layout of a wire coating extrusion line using a corsshead die.

151

plastics or elastomers and natural rubber, the required higher heat of melt solidification is added via hot gas systems, vulcanization cures, and so forth.

Coating Wood Profiles. Wood, as well as other materials (plastic, aluminum, steel, etc.), can be easily coated in profile shape. Procedures used would be similar to wire coating, including possible preheating and cleaning. For noncontinuous profiles, special equipment is available; wood (or other material) is fed automatically from a storage–feeding magazine (when required preheated and/or cleaned), through a crosshead die, and finally through a cooling medium of air, water, gas, and so on (Fig. 3–32).

Coating Films/Foils. Extrusion coatings, using many different plastics, are applied to film, foil, or sheet substrates of plastic, wood, aluminum, steel, paper, cardboard, and so on. Basically a "curtain" of very hot melt is extruded downward from a slit die (similar to flat film slit dies). As mentioned earlier, preheating and/or cleaning operations may be required. How melt contacts the substrate, which is supported by a large, highly polished chill roll with a small rubber or metal nip roll. The nip roll applies the required pressure to ensure proper air-free adhesion. As the melt is usually at its maximum heat and pressure, "delicate" operation is required in the equipment and the surrounding area. Changes in air currents and moisture can cause immediate down time; tighter controls may not be necessary if all that is required is not to open a "door"—particularly a large garage door.

With coating (as well as other flat film or sheet processes), upon exiting the die the hot film will shrink across the width. This shrinkage, or "neck-in," is the amount of shrinkage from the die face to the coating width. With neck-in there is also "beading," which is a thickening at both edges of the film. The neck-in and beading influence the performance of a coating (as well as a film/sheet), including wrinkling, sagging, coating breaks near the bead, induced unwanted stresses, and so on. The quality of coated edges

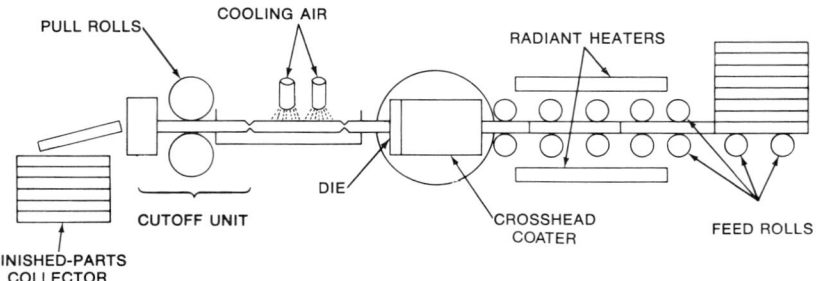

Fig. 3–32. Schematic of noncontinuous profiles being automatically coated with plastics.

is unpredictable and may require more than the usual trimming widths. The amount of neck-in and beading varies for different resins (one should get manufacturers' inputs on minimum amounts, etc.). The processing conditions definitely influence these undesirable effects. (Certain die designs practically eliminate edge bead, but the designs are specific for given resins.) One should determine the minimum neck-in and bead size based on processing conditions, and use those observations as control parameters. Sensors of neck-in and beads can be used in automatic process control. An example of typical surface coverage using PE coating resin is given in Table 3-9. With 3½ in. extruders, coating widths range from 600 to 1200 mm; with 4½ in., from 900 to 2500 mm; with 6 in., from 1,000 to 4,000 mm; and with 8 in., from 3,000 to 5,000 mm.

CONTROLS

Controlling extrusion processes takes two types of systems, one on the extruder and another on the finished extruded product. In turn, these controls have to be interfaced. Extruder controls have been reviewed (heats, pressures, hopper feed rate, screw RPM, back pressures, etc.), as have product controls (wall thicknesses, rate of travel, width indicators, weight, etc.). Extensive efforts are continually being made to achieve the maximum in speed and precision.

For products with specific requirements, typically the best approach has always been to control the weight per unit area or the length of output that is directly related to a constant melt throughput, by weight. As reviewed in this chapter (and others), many machine controls and designs are required, such as feed rate of loading, screw design, and so on. In regard to products such as film, thickness gauges were once very popular for adjusting takeoff speeds. However, better control is achieved now with weight control systems. Depending on the given weights per unit length or area and the speed of the line, the melt throughput rates of the extruder are precalculated ac-

Table 3-9. Example of a Surface Coverage of Polyethylene Coating (Average Density 0.920 g/cm²).

	ONE M² OF SUBSTRATE	
THICKNESS (MM)	REQUIRES (G)	0.45 KG PE COVERS (M)
0.001	5.8	175
0.002	11.6	85
0.004	23.2	42.5
0.008	46.4	21

curately enough to ensure, via a melt throughput control system, that the weight can be kept constant within a tolerance range of at least $\pm$ 0.5 percent. Whereas measurement of takeoff speed using a wheel pulse counter is relatively easy, measurement of melt throughput involves considerable expense.

From a practical standpoint, extrusion lines appear to be relatively simple and compact, with a long operating life; their controls should provide ease of startup and shutdown, and must be easy to service. Of course, the real world is not this simple; the various materials and types of equipment in the lines are not stable, and they all work within rather tight limits. Although all types of very tightly controllable equipment are available, the operator or processor (and others) must recognize that all equipment and all materials have limits, optimum operating conditions, and so on. So to benefit from setting up or operating a line one has to determine what is required, determine the limits of operating materials and equipment, and establish quality control and other requirements as summarized in Fig. 1–1.

Acquiring "all this knowledge" takes time and experimentation. One must be aware of potential problems or limitations. As new ones develop, one should accommodate them logically in processes. This approach provides a means of determining which materials and equipment to purchase or upgrade. One could thus set up an ideal line with so-called interchangeable and controllable features. Given a specification that must be met on a complete line, one must meet the target (or no payment is made). Few processors desire tight specifications, but such responsibility can be managed, one step at a time. The processor should determine as well as is possible what is required to meet product performance (color, dimensions, strength, etc.), and relate the requirements to material and equipment currently in use. It would be unfortunate if one were to purchase the "best" material and equipment available at a particular time and then determine that it did not perform as needed and learn that another piece of equipment was required. This situation happens all too often, as people hastily, or without full knowledge of an operation, make foolish mistakes.

A processor does not have control of all the steps from basic material to finished product. Material processing capability is limited by what is received and by when inspection is made or required; so selection tests are important and must be subject to change (see Chapter 10). In turn, the process line has many variables that must be coordinated. A practical procedure today, with all closed-loop process control systems, is to subdivide the controls into distinct subsystems (Fig. 3–33). They can then be controlled within single control loops or by simple intermeshed circuits. A single-loop feedback circuit has one input and one output signal; disturbances that affect the process are registered by the controller directly, by an ad-

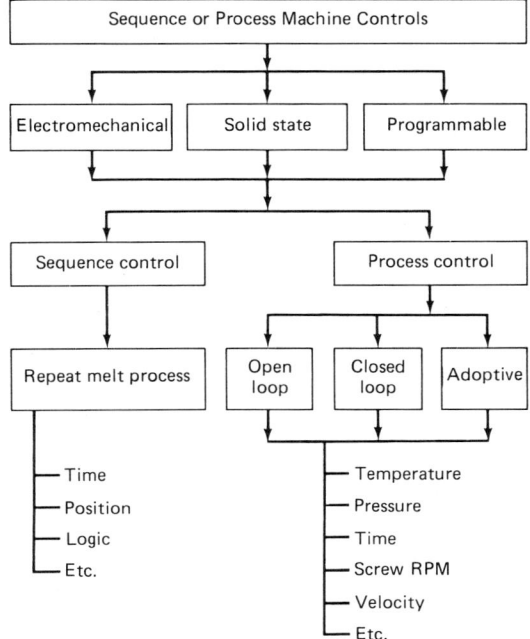

Fig. 3-33. Simplified overview of process controls for extrusion machine.

ditive term in the value of the controlled parameter. A disturbance is registered by its own sensor and interacts with the control signal with an open-loop controller. When the effect of a disturbing factor on a process is well understood, disturbance control provides a rapid reaction to unpredictable influences.

Additional process control signals can be achieved in a single closed-loop system by having it cascade the signal (see Chapter 1 on heat cascading devices). Cascade control offers the advantages of faster reaction time, reduced susceptibility to disturbance, and less effect from incorrect settings. As this chapter's review of all the variables that can exist and are controllable would suggest, interrelating them with multi-performing controllers is desirable. In some cases, computer integrated microprocessor control systems can be used (see Chapter 2 on CIIM); they offer some benefits when properly installed with matching hardware, but cannot do the complete job. A complete package that would properly include all parameters is desirable but at present not very realistic. Computerized integrated controllers are available and used, each for its unique capability. Better complete multiparameter controllers still need to be designed. One must appreciate that use of a controller, whether simple or highly computerized, necessitates that the operator recognize what parameters require control and the degree of control needed, and correlate these factors. For example, if the output rate has to be increased, is a screw RPM increase satisfactory, or should there be a change in back pressure, resin feed rate, and so on, or combinations of certain controls, to provide a product that is acceptable (according to various criteria)? The possibilities of measuring back pressure in the melt and the melt heat at the screw tip are of particular value in processing. In this manner, it is possible to control the reproducibility of the two most important parameters in every processing method: pressure and heat.

From the processor's point of view, the extruder has the following objectives: (1) high throughput proportional to screw speed and basically independent of back pressure; (2) uniform, pulsation-free delivery, with optimum melt heat (with respect to power consumption and quality of extrudate); (3) uniform melt heat locally and also throughout the run; (4) delivery of an orientation-free, relaxed melt; (5) homogeneous mixing of the resin with all its additives; and (6) a pore-free extrudate, free of volatiles, at high output rates for both granular and powder feeds.

With all the varieties of materials available, and their individual grades with many different formulations, it has not yet been possible to design a screw in advance, based on the melt or the rheological/flow physical relationships involved in plastification and conveying. Trial and error, with an observant processor using reliable and reproducible controls, makes the screw perform to its maximum efficiency.

Downstream Controls

Downstream product controls can be interconnected with process controllers so that any variation in the product is immediately reflected as a corrective change. A decision is derived from the feedback signal about the kind and extent of adjustment that should be made to a control variable; in other words, the system has to be matched to the process. A few systems permit the properties of the extrudate to be controlled: (1) pipe thickness distribution is controlled by centering of the die either mechanically or thermally, and wall thickness by the haul-off speed; (2) blown and flat film averaging thickness are controlled by extruder speed and haul-off speed, thickness profile by the die gap restricter bar height (thermal or piezoelectric), film width by the diameter of the calibrator, transverse thickness tolerances by cooling air, and edge shift by side-gusseted triangles; (3) a profile's dimensional stability (weight per unit volume) is controlled by haul-off speed, screw speed, and recorded melt throughput or a product dimension; and (4) all products' throughput is controlled by screw speed, and the running length by haul-off speed and screw speed (gravimetric metering).

Transverse direction (TD) gauge control of cast film and sheet has been successfully used for many years worldwide, by automatic dies employing thermal bolts to locally adjust the die gap. TD control of blown film is now being used in a system that takes a different approach; it cools segments of the die lip, in turn cooling the melt in contact with it, reducing the melt's drawdown capability and thus increasing the thickness. The TD gauge profile can be stabilized to a tolerance of ± 1.7 percent. In addition to minimizing resin consumption, the system also reduces film curvature, which increases as TD tolerance increases (180).

To monitor this gravimetric control system in the machine direction, an electrical sensor is used. The transverse profile of the bubble is read by a capacitance gauge that makes a complete circuit of the bubble in 2 to 3 min. The system's microprocessor takes the data and displays the profile on a color monitor. An additional display indicates the centering status of the die prior to startup, guiding the material adjustment. The system is then activated. Valves arrayed around the die at a spacing of about 0.75 in. direct cold compressed air through cooling channels onto segments of the die lip, with the microprocessor controlling the operation. In running 53 μm LDPE film, the thickness tolerance is reduced from 3.3 to 2.0 percent after 15 min. After another 22 min. the tolerance is reduced to 1.7 percent, which is maintained as long as the control is on.

There are different pulling and cutting units for film, sheet, pipe, and so forth. Tube pullers are driven by various methods. Electric drives are available with tachometer feedback accuracy ± 0.5 to 1.0 percent. Digital drives can hold to ± 0.01 percent.

Table 3-10 compares cutting equipment capabilities. Automatic slitter/ rewinders, for handling blown or cast films such as rigid PVC, PS, and nonuniform-thickness laminates, operate at speeds of 300 to 500 m/min, producing reels up to 3,000 mm in diameter and slit reels up to 1,000 mm in diameter. Film is drawn from an unwind to the slitting/rewinding stations through a translation bridge equipped with idler rollers that facilitate web feed and prevent the deposition of dust due to static. Rewinding/slitting machines are available that can handle highly tension-sensitive materials such as PP and PET, video film, magnetic tape, and heavy gauge laminates, with working widths of 2,500 to 3,000 mm and finished-reel diameters of up to 600 mm for PP and PET. They operate at 600 m/min and produce slit widths of 100 mm and above. Rewind units are also available that operate on single and twin spindle roll-slitting principles. Units

Table 3-10. Examples of Cutting Equipment Capabilities.

CUTTER	LINE SPEED, M/MIN	CUTS/MIN., MAX.	ACCURACY, ± MM	ADVANTAGES	DISADVANTAGES
Saws	150	30	0.015	Easy setup and large capacity.	Requires cleanup via air systems, etc. and uses clamping/travel table units.
Guillotine	90	50	0.015	Large capacity and angle cuts.	Slow blade speed; rigids need profiled bushing and/or blade; high air consumption and few cuts/min.
Flywheel	4,500	12,000	0.004	High cut rates and high accuracy.	Not good under 300 cuts/min.; must adjust blade RPM for speed changes; profile bushings may be needed for rigids; limited to small angle cuts.
Die-set stationary traveling	—	90	0.00	In-line finishing and high accuracy.	Price, slow line speed, long setup time, and long runs only.

incorporate high-accuracy scissor slitting, crush cutting for nonbrittle materials, and razor-blade slitting for materials such as cellophane, PE, PP, PVC, and thin aluminum foils.

Machines for automatic winding of blown and cast film, which includes winding and slitting, are available; they handle all types, widths, and thicknesses, and different winding characteristics, produce different reel diameters on various core sizes, and divide film webs into different lane widths during production. Reel changes are automatic, and reel production can continue for one or two shifts without operator intervention, because of the incorporation of a reel-bar storage system.

Lines produce reels in widths of 900, 1,200, and 1,600 mm and permit winding of all film types on various cores with reel diameters of up to 600 mm. They also provide automatic shaft changes at a predetermined film length and gap winding that is steplessly adjustable from 1 to 10 mm. Jumbo reels run up to 2,000 mm in diameter and weigh up to 3 tons in other systems.

Developments in web tension control systems are providing increased capability and function to eliminate problems (Fig. 3–34). They include ultrasonic roll diameter sensors, pneumatic pressure gauge tension monitors, capston-Mt. Hope tension systems, and so on. As an example, replacement of a web-tensioning system's conventional electromechanical drive with an ordinary ac motor enables processors to lower system cost and improve web consistency, as has been done for many years. A vector control system uses a belt and pulley arrangement to remotely couple an encoder to the shaft of the ac induction motor. This approach provides closed-loop feedback, without requiring that one modify the relatively inexpensive motor by installing a special feedback device on it.

EXTRUDER OPERATION

This section summarizes what has been reviewed in this chapter. Machine operation takes place in three stages. The first stage covers the running of a machine and its peripheral equipment. The next involves setting processing conditions to a prescribed number of parameters for a specific material, with a specific die, and in a specific line. The final stage is devoted to problem solving and fine-tuning of the operation, which will lead to meeting product performance requirements at the lowest cost of operation. A successful operation requires close attention to many details, such as the quality and flow of feed material, a heat profile adequate to melt but not degrade the material, and a startup and shutdown that will not degrade the plastic. Processors must become familiar with a troubleshooting guide.

Care should be taken to prevent conditions that promote surface con-

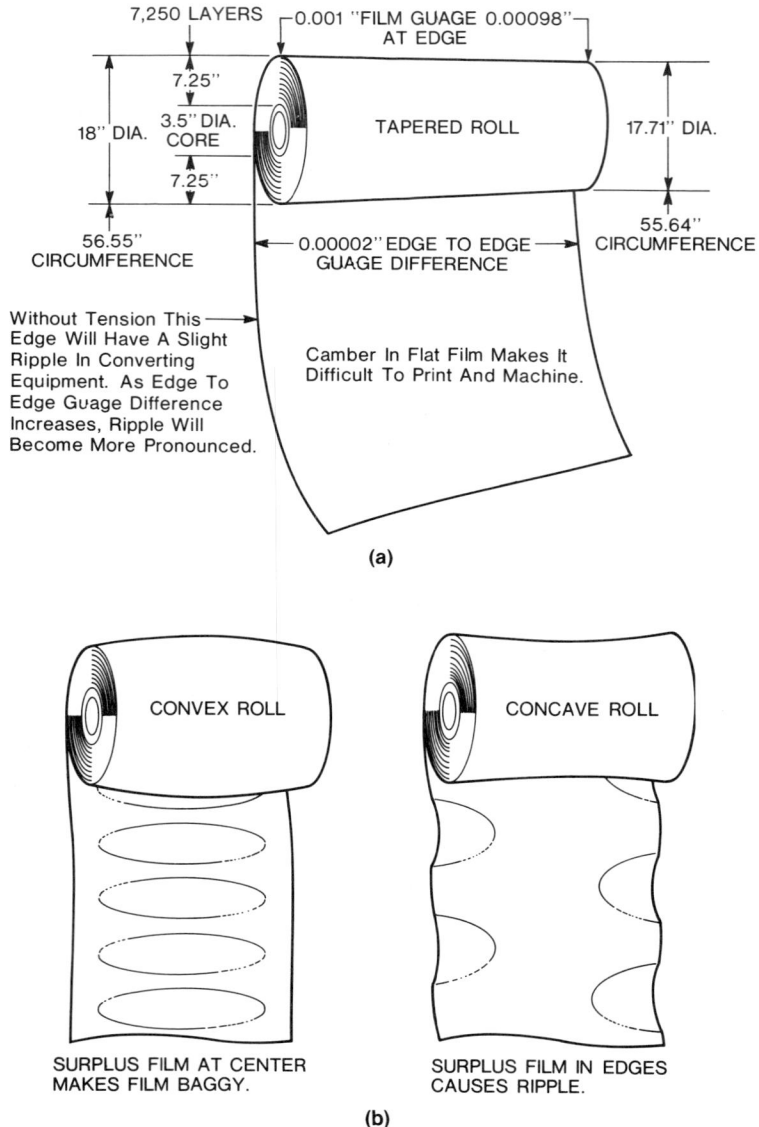

Fig. 3-34. Influences that affect performance of film during windup. (a) Effect of tension. (b) Effect of uneven thickness.

densation of moisture on the resin and moisture absorption by the pigments in color concentrates. Processors must avoid contamination from other plastics, dust, paper clips, and so on, and take special care in cleaning feed hoppers, hopper dryers, blenders, scrap granulators, and other material handling equipment. Resin silos, containers, and hoppers should be kept covered to prevent contamination. Also, certain established procedures for startup should be followed to prevent contamination, overheating, and excessive pressures.

It is important to provide safe operating conditions for personnel and equipment. One must realize that high production rates cannot be achieved until all parts of the extrusion system reach the optimum operating conditions. This condition is best achieved by gradually increasing heat and rates until desired optimum conditions are reached.

Prior to startup one must take certain precautions. (1) Unless the same resin is already in the machine from a previous run, the entire machine should be cleaned or purged, including the hopper, barrel, screw, breaker plate, die, and downstream equipment. If a resin was left in the barrel for a while, with heats off, the processor must determine if the material is subject to shrink and could cause moisture entrapment from the surrounding area, resulting in contamination that would require cleanup (this situation could also be a source of corrosion in the barrel/screw). One must (2) check heater bands and electrical connections, handling electrical connections carefully, and (3) check thermocouples, pressure transducers, and their connections carefully. Also, the processor should: (4) be sure the flow path through the extruder is not blocked; (5) have a bucket or drum, half filled with water, to catch extrudate whenever purging or processing plastics that have contaminating gaseous byproducts; and (6) inspect all machine ventilation systems to ensure adequate air flow.

Startup procedures also involve precautions. (1) Starting with the front and rear zones (die end and feed section), one should set heat controllers slightly above the resin melt point and turn on the heaters. Heat-up should be gradual from the ends into the center of the barrel to prevent pressure buildup from possible melt degradation. One should: (2) increase all heaters gradually, checking for deviations that might indicate burned-out or runaway heaters by slightly raising and lowering the controller set point to check if power goes on and off; and (3) after the controllers show that all heaters are slightly above the melt point, adjust the settings to the desired operating heats, based on experience and/or the resin manufacturer's recommendation, checking to ensure that the heat increase is gradual, particularly in the front/crosshead. (4) The time required to reach temperature equilibrium may be $\frac{1}{2}$ to 2 hours, depending on the size of the extruder. Overshooting is usually observed with on/off controllers. (5) Hot melts can behave many different ways, so no one should stand in front of the extruder during startup, and one should never look into the feed hopper because of the potential for blowback due to previous melt degrading, and so on. (6) After set heats have been reached, one puts the resin in the hopper and starts the screw at a low speed such as 2 to 5 RPM. (7) The processor should observe the amperage required to turn the screw, stop the screw if the amperage is too high, and wait a few minutes prior to restart. (8) In working with a melt requiring high pressure, the extruder barrel pressure should not exceed 7 MPa (1,000 psi) during the startup period. (9) One should let the machine run for a few minutes, and purge until a good-quality extrudate is attained visually (experience teaches what it should look like; a certain size and amount of bubble or fumes may be optimum for a particular melt, based on one's experience after setting up all controls). If resin was left in the extruder, a longer purging time may be required to remove any slightly degraded resin. (10) For uniform output, it is essential that all of the resin be melted prior to entering its metering zone, and that this section run full. (When time permits, after running for a while, the processor should stop the machine, let it start cooling, and remove the screw to evaluate how the resin performed from the start of feeding to end of metering. Thus one can see if the melt is progressive and can relate it to screw performance and product performance. (11) One then turns up the screw to the required RPM, checking to see that maximum pressure and amps are not exceeded, and (12) adjusts the die with the controls it contains, if required, at the desired running speed. Once the extruder is running at maximum performance, the processor sets up controls for takeoff equipment, which may require more precision settings, if required. One may get into a balancing act of interrelating extruder and downstream equipment.

In regard to shutdown, the procedures vary slightly, depending on whether or not the machine is to be cleaned out or just stopped for a brief time period. If the same type of resin is to be run again, cleanout generally is not required. The goal is to avoid degradation by reducing exposure of the resin to high heat. If cleanout is required, because a different type of resin is to be processed, it is necessary to disassemble the equipment at a heat high enough to allow cleanout prior to material solidification.

The procedure for shutdown without cleanout is as follows: (1) one should empty the hopper as well as possible; (2) if coating, one should take the action required to remove the substrate (with wire coating, one removes the wire from the crosshead die; in coating paper, film, etc., usually the extruder is on a track and can be withdrawn from the substrate; etc.); (3) one reduces all heat settings to the melt heat; (4) one reduces the screw speed to 2 to 5 RPM, purging the resin, if required, into a water bucket or drum prior to reducing the heat to melt heat; (5) when the screw appears to be empty, one stops the screw, and shuts off the heaters and the main power switch (however, steps 4 and 5 must be completed before the melt heat drops significantly, or a premature shutdown will occur with material remaining in the barrel); and (6) if a screen pack with breaker plate is used, one disconnects the crosshead from the extruder and removes the breaker plate and screen. If necessary, appropriate action is taken to clean them.

For cleanout of the extruder at shutdown, the first three steps are the same as steps 1, 2, and 4 in the preceding paragraph. The procedure then is: (4) When the extruder appears to be empty, one stops the screw, and (5) shuts off and disconnects the crosshead heaters. One then reduces other heaters to about 170 to 330°C (400 to 625°F), depending on the resin temperature at the melt point. (6) The processor disassembles the crosshead and cleans it while still hot. One removes the die, and the gear pump if used, and removes as much resin as possible by scraping with a copper spatula or brushing with a copper wire brush. One removes all heaters, thermocouples, pressure transducers, and so on. One should consider using an exhaust duct system (elephant trunk) above the disassembly and cleaning area, even if the resin is not a contaminating type; it keeps the area clean and safe. (7) One pushes the screw out gradually while cleaning with a copper wire brush and copper wool. Care should be exercised if a torch is used to burn and remove resin; tempered steel, such as Hastelloy, may be altered, and the screw distorted or weakened, and subjected to excessive wear, corrosion, or even failure (broken). (8) After screw removal, one continus the cleaning, if necessary, and (9) turns off heaters and the main power switch.

Final cleaning of parts, particularly disassembled parts, is best done manually, or, much better, in ventilated burn-out ovens, if available, at 540°C (1,000°F) for about 1½ h. For certain parts, with certain resins, the useful life could be shortened by corrosion. One should check with the part manufacturer. After burn-out, one removes any grit that is present with a soft, clean cloth. If water is used, one can air-blast to dry. With precision machined parts, water cleaning could be damaging, because of potential corrosion when certain metals are used.

TROUBLESHOOTING

Throughout this chapter and the book there are discussions of why problems develop and how they can be eliminated or kept to a controllable minimum. To do the best job of eliminating or reducing problems, one must understand the complete process, and there is no substitute for experience. Only through hands-on problem solving can an operator, or even management, really begin to understand the complexity of extrusion and know how to handle the wide variety of problems that invariably occur. (See Chapter

2, under "Troubleshooting," for an approach that includes screw wear and inspection, and is applicable to extrusion.)

Much useful information on problems and solutions in extrusion is available from material and equipment suppliers. Trade journals publish special issues on this subject with very useful tables, and industrial literature includes practical guides with helpful details (see reference 2 and Chapter 18). The size of this book does not permit such treatment, but an excellent summary of troubleshooting is given in Table 3–11, from *Plastics World* (185).

Table 3–11. Troubleshooting Guide: Common Extrusion Problems and How to Solve Them.

PROBLEM	CAUSES(S)	SOLUTION(S)
GENERAL CONSIDERATIONS		
Surging	Resin bridging in hopper	Eliminate bridging
	Incorrect melt temperature	Correct melt temperature
	Improper screw design	Check design
	Rear barrel temperature too low or too high	Increase or decrease rear temperature
	Low back pressure	Increase screen pack
	Improper metering length	Use proper screw design
Gels	Melt temperature too high	Lower melt temperature
(Contaminants that look like small	Not enough progression in screw	Use new screw
specks or bubbles)	Bad resin	Check resin quality
	Melt temperature too high	Check melt temperature
Melt fracture	Melt temperature too low	Increase melt temperature
(Rough surface finish. Also called "shark-skin.")	Die gaps too narrow	Heat die lips
		Increase die gaps
		Use processing aids
Bad color	Color concentrate incompatible with resin	Ensure melt index of concentrate base material is close to melt index of resin
Bubbles	Wet material	Dry thoroughly
	Overheating	Decrease temperature; check thermocouples
	Shallow metering section	Use proper compression ratio screw
Overheating	Improper screw design	Use lower compression screw
	Restriction to flow	Check die for restrictions
	Barrel temperature too low	Increase temperature

Table 3–11. Continued.

PROBLEM	CAUSE(S)	SOLUTION(S)
Die Lines	Scratched die	Refinish die surface
	Contamination	Clean head and die
	Cold polymer	Check for dead spots in head; adjust barrel and head temperature to prevent freezing
Flow lines	Overheated material	Decrease temperature
	Poor mixing	Use correct screw design
	Contamination	Clean system
	Improper temperature profile	Adjust profile

BLOWN FILM

Wrinkles	Dirty collapsing frame	Clean frame
	Too much web tension	Adjust tension
	Improperly designed air ring	Use new air ring
	Gauge variations	See gauge variations
	Insufficient cooling	Use refrigerated air
		Increase flow of air
		Reduce output
	Misalignment between nip rolls and die	Check alignment
Folds, creases	Excessive stretching between nip and roller	Reduce winding speed
	Nip assembly drive not constant	Adjust or replace drive
Blocking	Inadequate cooling	Use better cooling method
	Excessive winding tension	Adjust tension
	Excessive pressure on nip rolls	Adjust pressure
	Bad resin	Check resin
Port lines	Melt temperature too low	Increase melt temperature
	Die too cold or too hot in relation to melt temperature	Adjust die temperature
Splitting	Excessive orientation in machine direction	Increase blow-up ratio
	Die lines	See die lines
	Degraded resin	Reduce melt temperature
	Poor resin choice	Ensure resin is suitable
Die lines	Nick on die lip	Change die
	Dirty die	Clean die
	Inadequate purging	Increase purging time between resin changes

Table 3–11. Continued.

PROBLEM	CAUSE(S)	SOLUTION(S)
Gauge variations (Machine direction)	Surging	Check temperature
		Check hopper for bridging
	Inconsistent take-up speed	Check take-up speeds
Gauge variations (Transverse direction)	Non-uniform die gap	Adjust gap
		Center air ring on gap
Printing problems	Insufficient treatment	Use properly treated film
	Additives interfering with ink	Use resins with no interfering additives
		Erratic treatment
		Reduce slip levels to about 600 ppm for water-based inks

SHEET

Note: Most of the problems covered under blown film also relevant to sheet extrusion.

Poor gauge uniformity	Melt flow is not stable	Use gear pump to stabilize flow
Viscosity not stable	Poor mixing	Use static mixer
Streaks	Contaminated system	Clean hopper
		Check screw and die; clean if necessary
Total discoloration	Excessive regrind	Check amount of regrind used
Discontinuous lines	Too much moisture	Increase resin drying
		Use hot regrind

PIPE AND TUBING

Note: This data pertains to extrusion lines using water-filled vacuum sizing tanks.

Poor output	Improper die or screw design	Ensure die and screw are designed for desired output
Inside diameter:		
Blisters	Insufficient vacuum	Increase vacuum
	Excessive moisture	Maintain normal % of moisture in compound
	Gases entrapped	Reduce temperature
	Water inside pipe	Stop water access
Burn streaks	Mandrel heat too high	Check mandrel heat
	Stock temperature too high	Reduce temperature slowly
Grooves	Mandrel is coated with material	Clean mandrel
Wavy surface	Screw clearance set improperly	Adjust clearance
	Puller drive slipping	Adjust or replace puller drive

Table 3-11. Continued.

PROBLEM	CAUSE(S)	SOLUTION(S)
Outside diameter:		
Burn streaks	Material hung up on die	Clean die
	Temperatures too high	Reduce temperatures slowly
Uneven circumference	Too much air pressure on puller	Reduce air pressure
	Insufficient air pressure	Check air pressure and all connections
Discolored	Stabilizer level too low	Check stabilizer level
Pock marks	Air bubbles adhering to pipe in flotation tank	Install wiper in tank
	Improper adjustment of spray rings that surround water tank	Readjust spray rings
Oversized	Air supply too high	Adjust air supply
	Insufficient water supply	Increase water supply
	Pipe hot when measured	Allow pipe to cool before measuring
Wall too thick	Misadjusted die bushing	Adjust die bushing to achieve uniform thickness
	Wrong die set-up	Use correct set-up

Table 3-12 presents a troubleshooting guide for plastic foam film. One must be aware that causes of problems are not always obvious; for example, Fig. 3-35 shows contamination in the die presenting a problem that is almost impossible to resolve by equipment controls.

To determine the source of a problem, it is necessary to understand the basics of a process and apply them to problem solution. For example, with film and sheet dies, tip adjustments only control the uniformity of the transverse thickness. To control the average thickness a proper relationship between the extruder pumping rate and the speed of roll-up is required. Closing the die lip opening has very little effect on the extrusion rate and does not make the entire sheet thinner.

Regarding surges, any variation of the extrusion rate with time originates in the extruder and thus must be corrected in the extruder. This could be the most common and most difficult problem to correct, as it has many interrelated sources, none of which are in the die. A surge can be due to inconsistent feed from the hopper to the barrel, varying frictional forces in the barrel feed zone, improper screw design, and other causes.

Table 3-12. Troubleshooting Guide for Plastic Foam Film.

PROBLEM	PROBABLE CAUSE(S)	CORRECTION
Random poor cell structure	Low melt pressure.	Increase screw speed.
		Reduce die-lip temp.
		Decrease gauge of screen packs.
		Use resin of lower melt index.
		Reduce die gap.
	Hangup in die.	Reduce land length.
		Clean die.
	Stagnant low-pressure areas in head.	Increase screw speed.
	Irregular cells in spider area or opposite die-ring feed.	Use bottom-fed spiral die.
Poor skin formation	Too much blowing agent.	Increase head pressure.
	Linear skin speed too low.	Reduce blowing-agent level.
		Increase screw speed.
	Loss of melt pressure in die land.	Reduce land length.
		Increase L/D ratio.
		Increase screw speed.
	Die-block temperature too low.	Increase temperature.
Pinholes in film or bubble burst on surface	Too much blowing agent.	Reduce blowing-agent level.
	Die temperature too high.	Reduce die temperature.
	Resin melt index too high.	Decrease resin melt index.
		Reduce processing temperature.
	Blowing agent decomposing too soon.	Reduce processing temperature.
		Increase screw speed.
		Reduce blowing-agent level level.
	Poor flow within polymer skin.	Improve flow in head and die.
Cells collapsing	Resin melt index too high.	Decrease melt index.
		Reduce processing temperature.
	Cooling too fast.	Reduce cooling rate.

Melts basically move in laminar flow in the die, where no mixing action takes place; so poor mixing and incomplete thermal homogenization must be corrected prior to melt flow in the die. In regard to melt heat, because they are generally very efficient heat insulators, melts are in the dies for only a few seconds. It is possible to alter frictional effects on the melt skin

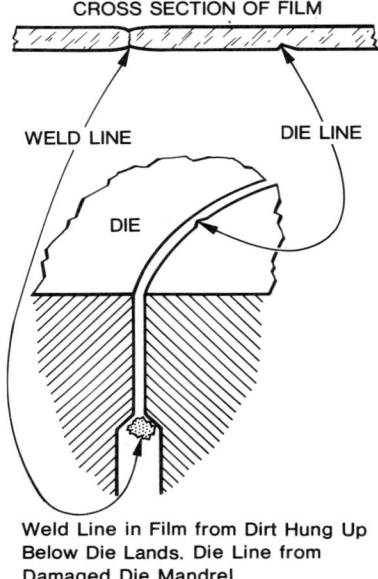

CROSS SECTION OF FILM

WELD LINE   DIE LINE

DIE

Weld Line in Film from Dirt Hung Up
Below Die Lands. Die Line from
Damaged Die Mandrel

Fig. 3-35. Example of a contaminated die and a troubleshooting problem.

through changes in the die heat, but there is usually a better chance to produce an even heat in the extruder.

Die adjustments do not significantly influence the extrusion rate; the lip opening, restrictor bar clearance, die heat, and overall die design have small to moderate effects on the back pressure. However, the extrusion rate is only slightly affected by substantial changes in back pressure. Thus attempts to open the lip or increase the die heat to increase the extrusion output rate can be exercises in futility, as well as the source of many additional problems.

Chapter 4

BLOW MOLDING

INTRODUCTION

Blow molding (BM), the third most popular method of plastics processing, offers the advantage of manufacturing molded parts economically, in unlimited quantities, with little or virtually no finishing required. It is principally a mass production method. The surfaces of the moldings are as smooth and bright, or as grained and engraved, as the surfaces of the mold cavity in which they were processed. Among the special techniques available are stretch blow molding and coextrusion. One can improve the cooling efficiency and reduce cycle time with gases (CO_2, etc.). Other developments include shuttle postcooling, insertion of printed film in the mold to avoid the need for subsequent decorating, and so on.

Blow molded parts demonstrate that, from technical and cost standpoints, BM offers a promising alternative to other processes, particularly injection molding (IM) and thermoforming. The technical evolution of BM, plus accompanying improvements and new developments in plastics, has led to new BM parts. With the coextrusion technology now established and the hardware in place, the variety of achievable properties can readily be extended by the correct combination of different materials (see Chapter 3). The potential for BM products includes much more than the simple bottles that have been made for many decades. Now the expertise and economics of the method are such that many ideas once deemed futuristic are much closer to realization (2, 96, 186–212).

BM offers a number of processing advantages, such as molding extremely irregular (reentrant) curves, low stresses, the possibility of variable wall thicknesses, the use of polymers with high chemical resistance, and favorable processing costs. Reentrant curves are the most prominent features— so much so that it is difficult to find examples that do not incorporate them (2). They combine aesthetics with strength and cost benefits.

170

A very significant difference exists between BM and IM. BM usually only requires 25 to 150 psi pressures, with possibilities for certain resins of up to 200 to 300 psi. For IM, the pressure is usually 2,000 to 20,000 psi, and in some cases up to 30,000 psi. The lower pressures generally result in lower internal stresses in the solidified plastics and a more proportional stress distribution. The result is improved resistance to all types of strain (tensile, impact, bending, environmental, etc.) (see Chapters 1 and 10).

As the final mold equipment for BM consists of female molds only, it is possible simply by changing machine parts or melt conditions to vary the wall thickness and the weight of the finished part. If the exact thickness required in the finished product cannot be accurately calculated in advance, this flexibility is a great advantage from the standpoint of both time and cost. With BM, it is possible to produce walls that are almost paper-thin. Such thicknesses cannot be achieved by conventional IM, but, with certain limitations, can be produced by thermoforming. Both BM and IM can be succesfully used for very thick walls. The final choice of process for a specific wall section is strongly influenced by such factors as tolerances, reentrant curved shapes, and costs.

BM can be used with plastics such as PE that have a much higher molecular weight than is permissible in IM (see Chapter 1 on MW). For this reason, items can be blown that utilize the higher permeability, oxidation resistance, UV resistance, and so on, of the high-MW plastics. This feature is very important in providing resistance to environmental stress cracking. This extra resistance is necessary for plastic bottles used in contact with the many industrial chemicals that promote stress cracking.

With BM, the tight tolerances achievable with IM are not obtainable. However, in order to produce reentrant curved or irregularly shaped IM products, different parts can be molded and in turn assembled (snap-fit, solvent-bonded, ultrasonically bonded, etc.). In BM of a complete irregular/complex product, even though IM tolerances cannot be equaled, the cost of the container is usually less. No secondary operations such as assembly (adhesives, etc.) are required. Other advantages also are achieved, such as significantly reducing (if not eliminating) leaks, reducing total production time, and so forth.

BM can be divided into three major processing categories: (1) extrusion BM (EBM), which principally uses an unsupported parison; (2) injection BM (IBM), which principally uses a preform supported by a metal core pin; and (3) stretch BM, for either EBM or IBM, to obtain bioriented products, providing significantly improved cost-to-performance advantages. Almost 75 percent of processes are EBM, almost 25 percent are IBM, and about one percent use other techniques such as dip BM (2). About 75 percent of all IBM products are bioriented. These BM processes offer different advantages in producing different types of products, based on the materials to be used, performance requirements, production quantity, and costs.

BM requires an understanding of every element of the process, starting with the basic "extruder" used in conventional extrusion and IM machines. (For information on the machines used to plasticate/melt materials for BM, see Chapters 2 and 3.)

With EBM, the advantages include high rates of production, low tooling costs, incorporation of blown handleware, a wide selection of machine builders, etc. Disadvantages are a usually higher scrap rate, the use of recycled scrap, and limited wall thickness control or resin distribution. Trimming can be accomplished in the mold for certain type molds, or secondary trimming operations have to be included in the production lines, and so forth.

With IBM, the major advantages are that no flash or scrap occurs during processing, it gives the best of all thickness and material distribution control, critical neck finishes are molded to a high accuracy, it provides the best surface finish, low-volume quantitites are economically feasible, and so on. Disadvantages are its high tooling costs, the lack (to date) of blown handleware (there is only solid handleware), its being relatively limited to relatively smaller blown parts (whereas EBM can easily blow extremely large parts), and so forth. Similar comparisons exist with biaxially orienting EBM or IBM. With respect to coextrusion, the two methods also have similar advantages and disadvantages, but mainly major advantages. With IBM, for example, PET can be processed (mono- or multilayer) and stretched into the popular 2- and 3-liter carbonated beverage bottles. Table 4–1 provides a cost comparison of the different BM techniques for PVC and PET, the plastics predominantly processed in BM.

EXTRUSION BLOW MOLDING

In EBM, a parison (tubular type of hot melt) is formed by the extruder melt output (Figs. 4–1 and 4–2). Turning continuously, the screw feeds the melt through the die head, generally as an endless parison directly through a die. A die head can have one or more openings; so one or more parisons can be extruded. The size of the part and the amount of material to produce a part (shot size) dictate whether or not an accumulator is required. The basic nonaccumulator machine offers a continuous flow of plastic melt. With an accumulator, the flow of the parison through the die is cyclic (Figs. 4–3 and 4–4). The connecting channels between the extruder and the accumulator, as well as the accumulator itself, are designed to prevent restrictions that might impede flow or cause the melt to hang up (see Chapter 3). Flow

Table 4–1. Manufacturing Cost Comparison of 16-oz Blow Molded Bottles.

	STANDARD EXTRUSION BLOWMOLDING 2-PARISON HEAD 4-FOLD	STRETCH BLOW MOLDING PVC (2) SINGLE PARISON HEADS 4-FOLD	STRETCH BLOW MOLDING PET
1.0 Machine cost incl. head, molds, ancillaries (lic. fee, stretch PVC and PET)	$270,000	$450,000	$850,000
2.0 Hourly machine costs			
Depre'n, 5 yr, 30 K hr, $/hr	$ 9.00	$ 14.85	$ 28.33
Financing cost, 5 yr. 12.5%	2.80	4.65	10.20
Labor, 1 man	13.00	13.00	13.00
Energy at $.06 per kWh	2.50	5.35	11.00
Floor space	1.50	2.00	4.00
Maintenance and consumable mtl.	2.25	3.75	4.50
Total hourly mc costs	$ 31.05/hr	$ 43.60/hr	$ 71.03/hr
3.0 Bottle specs. hourly/Annual prod.			
3.1 16 oz finish wt. (454 g) —regular 37 g (1.3 oz) —stretch PVC 20 g (0.7 oz) —stretch PET 20 g (0.7 oz)			
Cycle time/bottles per hour	8.4 sec/1,714	7.5 sec/1,920	4,000
bottles per yr., millions	10,286	11,520	24,000
4.0 Annual costs			
4.1 16 oz (454 g)			
Resin: —37 g $.70/lb ($ 70/ 0.45 kg or $ 1.54 kg)	$585,200		
—20 g $.66/lb ($ 1.46 kg)		$334,950	
—20 g $.60/lb ($ 1.32 kg)			$634,360
Machine costs	186,300	261,600	426,180
total p.a.	$771,500	$596,550	$1,060,540
Royalty (PET)-DuPont-per year			30,000
Cost per thousand	$ 75.00	$ 51.78	$ 45.44

Notes: 1. Figures are not be to considered as absolute costs, but rather reflect comparisons between various machine options.

2. All calculations are based upon 100% efficiency.

3. All bottle weights are finish weights (flash being considered as 100% reusable).

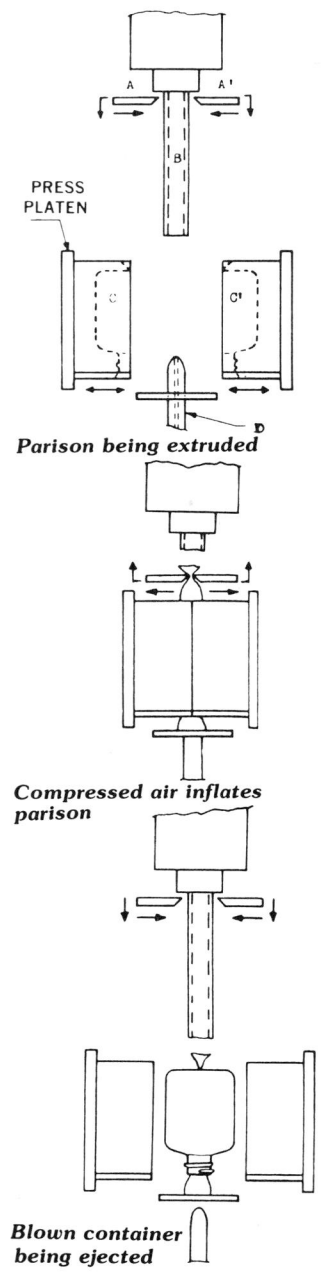

Fig. 4–1. Basic extrusion blow molding process. A = Parison cutter, B = Parison, C = blow mold cavity, D = blow pin.

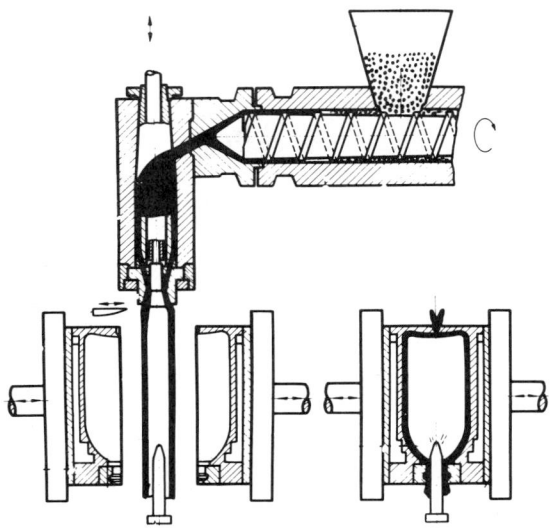

Fig. 4–2. Typical phases in blow molding (not to scale).

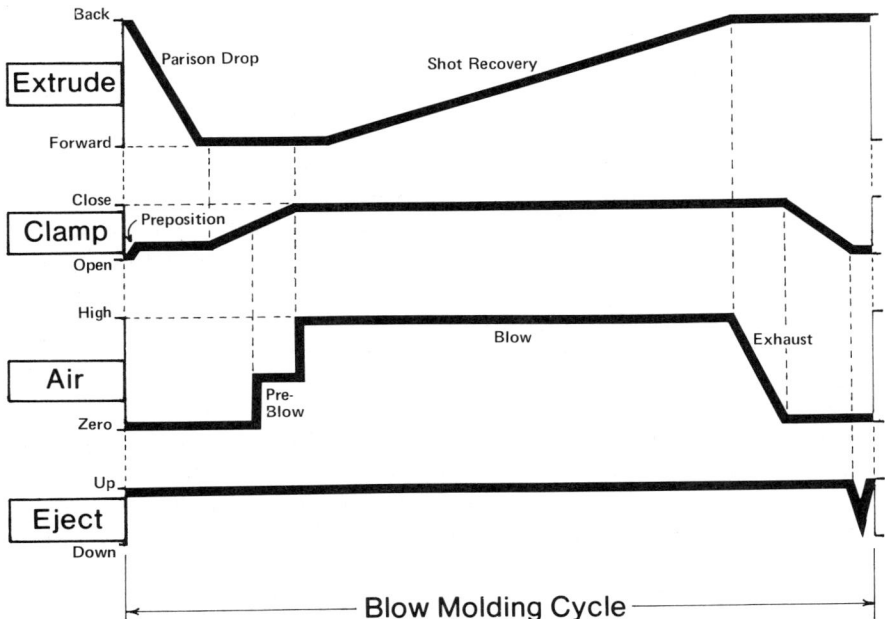

Fig. 4–3. Schematic of blow molder using an accumulator head.

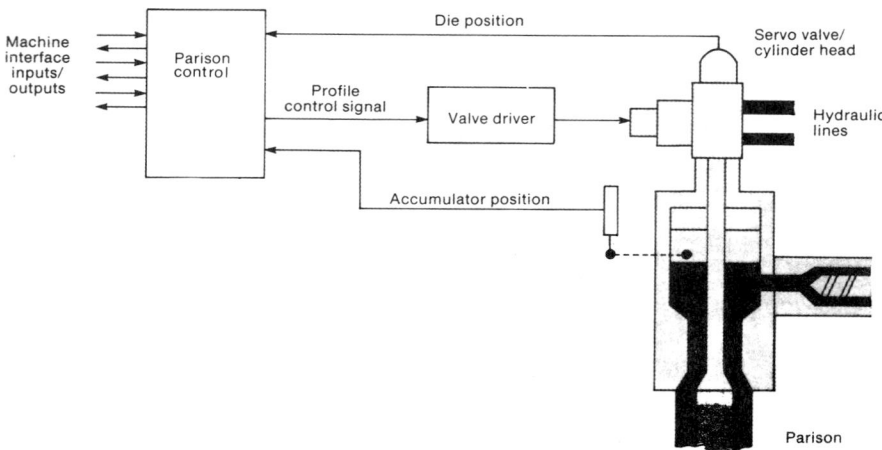

Fig. 4-4. Schematic of an accumulator head with programmable process controller; controls melt characteristics (interrelates with extruder performance), rate of melt flow to form parison, and profiling thickness of parison as it extrudes from die.

paths should have low resistance to melt flow to avoid placing an unnecessary load on the extruder.

To ensure that the least heat history (residence time) is developed during processing, the design of the accumulator should provide that the first melt in is the first to leave when the ram empties the chamber; the goal is to have the chamber totally emptied on each stroke.

When the parison exits the die and reaches a preset length, a split cavity mold closes around it and pinches one end of it. Usually a blow pin is located opposite the pinched end of the "tube." Compressed air inflates the parison against the female cavity of the mold surfaces. Upon contact with the relatively cool mold surface, the blown parison cools and solidifies to the part shape. Next the mold opens, ejects the part, and then repeats the cycle by again closing around the parison, shaping it, and so on.

Various techniques are used to introduce air. It can enter through the extrusion die mandrel (as with most pipe lines, Chapter 3), through a blow pin over which the end of a parison has dropped (Fig. 4-1), or through blowing needles that pierce the parison. The wall distribution and thickness of the blown part are usually controlled by parison programming, the blow ratio, and part configuration.

The mold clamping methods are hydraulic and/or toggle, similar to, but less sophisticated than, those used with IM. Sufficient daylight is needed in the mold platen area to accommodate parison systems, ejection of blown parts, unscrewing equipment, and/or other special equipment. Clamping systems vary, depending on part configuration and the location of parting lines (2).

There appears to be an unending series of new developments in BM to improve productivity and reduce costs. For example, with cooling there are reliable CO_2 systems, air chillers that reduce the temperature of the blown air to around $-95°F$ ($-70°C$), and blow pins that permit heated air in the blown part to exit, so that a continuous flow of fresh/cool air enters the part. With such systems, the output can be increased 10 to 30 percent.

The control and monitoring functions range from extremely simple ones to expensive complete microprocessor systems. Some machines use electrical relays that permit molding with inexpensive machines that do not require a very sophisticated maintenance shop or skilled operators. However, to produce good-quality parts at the lowest cost, machine systems generally must be more sophisticated, to meet any degree of performance (see Chapter 3).

Melt

Melt properties are of critical importance to BM—more so than for conventional extrusion. To a large extent they determine the quality achieved. The melt viscosity decides whether sagging or lengthening of a parison during extrusion can be compensated, particularly in noncircular parisons (Fig. 4-5). Because engineering resins have so far been used mainly with IM, most processors attempt to use easy-flowing, low-molecular-weight IM-grade resins. But in BM, particularly EBM, the objective is very different; the melt should be viscous and of high molecular weight (high melt strength) (Chapter 1). This requirement generally also ensures another important feature—better impact strength. The melt viscosity should be nearly independent of the shear rate and the processing heat.

In raw material data sheets there are usually no characteristics that indicate whether the resin is suitable for BM, or up to what limit an application is sensible. With increasing markets for these materials, more useful data sheets for all types of resins are becoming available.

Another important melt condition, extensibility, relates to how successfully large blowup ratios can be met, and whether edges and corners can be properly blown without thin spots. For ease of processing, there should be a large difference between the processing heat and solidification heat. This characteristic has a particularly advantageous effect on the quality of the blown surface. The greater the difference is, the longer the time that the melt has to be shaped, resulting in better surface definition. With a large difference, the contact of the parison with the surface of the mold is not

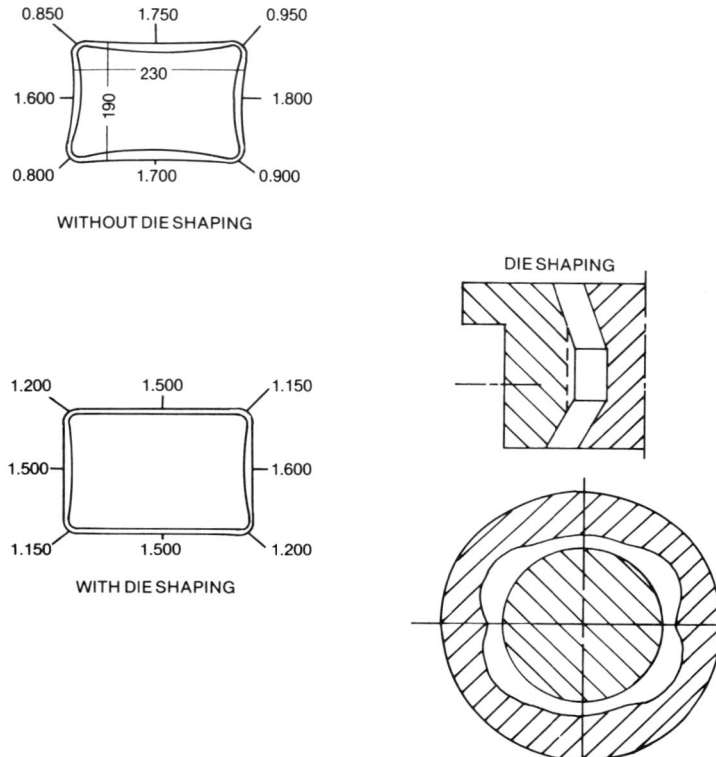

Fig. 4–5. Noncircular blow mold die, with and without wall thickness die shape (dimension in millimeters).

influenced (or is affected very little). With little difference in the heating profile, uneven contacting of the mold can cause an uneven surface appearance, uneven stress, and so on. Increasing the mold heat can offset some of the negative effects. Unfortunately a large difference in the heat profile involves increasing the cycle time, but this increase can be minimized to meet quality production requirements with no rejects.

Parison Thickness Control

Electronic parison programming is an effective way to control material usage and improve both quality and productivity. The most common method used is orifice modulation (Fig. 4–4). The die is fitted with a hydraulic positioner that allows positioning of the inside die diameter during the parison drop. The OD to ID relationship of the tapered die orifice opening is varied in a programmed manner to increase or decrease the parison wall thickness.

Electronic parison programming utilizes an electronic unit, commonly called a programmer, that uses a closed-loop servo system supplying proper signals to control the amount, direction, and velocity of the movement of the hydraulic positioner. Programmers are designated by the number of program points available, which usually ranges from 5 to 100 points.

In extrusion of a parison, especially a large parison, the wall thickness will vary as the weight of the resin increases, and it sags. Parison control may be helpful, but another method of minimizing wall thickness variations is to increase the pressure of the melt in the die, either by regulation of the barrel back pressure or, possibly, by pressure variations via a ram when an accumulator is used. An even wall thickness distribution on the parison circumference and wall thickness control in the longitudinal direction often are no longer sufficient to meet quality specifications. Increasing requirements and ever more geometric shapes necessitate a partial wall thickness adjustment, which is realized by shaping the flow channels in the die. The wall thickness along the length can be controlled by the conventional programmers reviewed. Circumferential distribution is controlled by patented predeformation of the die ring.

In regard to parison control, a compromise is necessary between the desired net weight and the need to maintain a sufficient safety margin over a set of minimum specifications, which include: minimum wall thickness, drop speed, drop strength, dimensional stability, and fluctuations in net weight. Most of these parameters can be directly affected by the molder's ability to control the parison wall thickness. The most common and practical way of doing this has been to adjust the gap between the die and mandrel (Table 4-2).

For some resins, it is advantageous to maintain a constant shear rate at the die orifice. Constant shear ensures that the melt is uniformly stressed, and that a uniform surface texture is produced. To provide constant shear, an accumulator ram speed is controlled according to a predetermined profile. As the parison drop increases, the ram stroke increases. Electronic and mechanical systems are being used with adjustments that are reproducible at speeds of 20 mm/s. This action can be applied only when a torpedo head is used. An adjustable core "plug" at the center of the torpedo moves longitudinally and controls the orifice opening without changing the volume of melt in the accumulator (Fig. 4-4). In a central feed head with the outer die ring moving, the volume of the melt in the flow channel varies, making it difficult to control the speed of the flow.

Table 4–2. Examples of Differently Performing Extrusion Blow Molding Dies.

TYPE DIE	FEATURE	ADVANTAGE/DISADVANTAGE
Simple die	Fixed die gap	Simple; inexpensive; no adjustment facility
Die profiling	Premanently profiled; preferred in die land area	Fixed circumferential wall thickness change; time-consuming; complex
Die centering	Can be permanently shifted laterally to correct parison drop path	Compromise between required drop path and equal wall thickness
Open-loop axial die gap control	Can be axially shifted during extrusion	Equal circumferential wall thickness change possible; no feedback
Servohydraulic closed-loop axial die gap control	As above, with greater speed, accuracy, and flexibility	Equal circumferential wall thickness change possible, with feedback
Stroke-dependent die profiling	Permanently ovalized die gap	Fixed, unequal circumferential wall thickness change possible; affects entire parison length
Die/mandrel adjustable profiling	Settable adjustment of die gap profile	Settable, unequal circumferential wall thickness change possible; rapid optimization
Servohydraulic closed-loop radial die gap control	Programmable ovalization and shifting of die gap	Programmable circumferential wall thickness change possible, independent of parison length

Accumulator

Processing capability developments in the past involved commodity resins, as they were predominantly used. Now there is more focus on engineering resins. Overlapping the melt streams in accumulator heads to obtain uniform wall thickness distribution is now a more fundamental requirement. The accumulator heads that are available differ in their feed channel designs, and they are frequently protected by patents; so one must be careful when purchasing them. In a parison head, the melt is divided into separate streams by the mandrel or spiders. Weld lines form where the flow fronts reunite. As the parison is deformed differentially in BM, these weld lines are potential weak sections in areas of extreme deformation.

If it is necessary to minimize streaks, another factor must be considered. No oxygen should be permitted to reach the melt and cause oxidation during downtime. A special die with an automatic shutoff prevents this action.

During downtime, the barrel heat should not drop lower than 20°C above the softening point, in order to prevent excessive volume contraction of the melt.

Accumulators can provide specialty products, an example being clear stripping systems. Two separately extruded melts enter the head and fuse together prior to exiting. Like overlapping layer formation of the parison, this feature also provides a positive way to strengthen weld lines.

Parison Swell

A very important factor in BM is the effective diameter swell of the parison. Ideally, the diameter swell is directly related to the weight swell and would require no further consideration. In actual practice, the existence of gravity, the finite parison drop time, and the anisotropic aspects (the parison has directional properties) of the BM operation prevent reliable prediction of parison diameter swell directly with the weight swell. After leaving the die, the melt—which has been under shear pressure—undergoes relaxation that causes cross-sectional deformation or swell.

Parison swell tends to be the most difficult property to control in efforts to produce low-cost, lightweight products. Diameter swell is easy to see; with the parison dropping, one may be able to see it actually shrink even after it stretches. If it is shrinking in length, then the wall must be thickening, and the parison is heavier per unit length, a behavior known as weight swell. Table 4-3 gives swell action for some common plastics.

For a given die cross section, the weight varies in proportion to the deformation. The time dependence and the increasing effect of weight with increasing parison length cause a narrowing of the diameter and a reduction in wall thickness from the lower end toward the die. Excessive length and prolonged hanging may lead to collapse, even with the usual procedure followed of providing some air pressure flow during parison formation. With this section, unlike the other processing methods, the characteristics of the

Table 4-3. Average Parison Swell for Some Commonly Used Plastics.

PLASTICS	SWELL, PRESENT
HDPE (Phillips)	15–40
HDPE (Ziegler)	25–65
LDPE	30–65
PVC (rigid)	30–35
PS	10–20
PC	5–10

melt within and after it exits the die are extremely critical in designing the die to suit the blow mold. As previously reviewed, a high melt strength is desirable to minimize this problem.

The diameter swell is an important processing parameter that must be controlled for BM of unsymmetrical containers and particularly with blown side-handle products. Resins with an excessively high diameter swell could experience curtaining and stripping (adherence to mold cavity) problems. An inadequate diameter swell contributes to an increase in rejects for defective handles.

Pinch-off

The pinch-off is a very critical part of the EBM mold, where the parison is squeezed and welded together, requiring good thermal conductivity for rapid cooling and good toughness to ensure long production runs. The pinch-off must have structural soundness to withstand the resin pressure and repeated closing cycles of the mold. It usually must push a small amount of resin into the interior of the part to slightly thicken the weld area. It also provides a cut through the parison to provide a clean break later when flash is removed.

Most molds use a double-angle pinch-off with 45° angles and a 0.010 in. (0.25 mm) land; see view (1) in Fig. 4–6. When the blown part is large relative to the parison diameter, the plastic will thin down and even leave holes on the weld line, requiring pinch-off (2). Using shallow angles (15°) has a tendency to force the plastic to the inside of the blown part, thereby increasing the thickness at the weld line. A pinch-off with a dam (3) also helps to solve problems. The flash pocket depth is related to the pinch-off and is very important for proper molding and automatic trimming.

A gross miscalculation of pocket depth (which must be learned through experience) can cause severe problems. For example, if the pocket depth is too shallow, the flash will be squeezed with too much pressure, putting undue strain on the mold, mold pinch-off areas, and machine clamp press section. The molds will be held open, leaving a relatively thick pinch-off, which is difficult to trim properly. If the pocket is too deep, the flash will not contact the mold surface for proper cooling. In fact, between molding and automatic trimming, heat from the uncooled flash will migrate into the cool pinch-off and cause it to heat up and cause unwanted problems like sticking to the trimmer. Or, during trimming, it can stretch instead of breaking free.

The knife-edge-cutter width of the pinch-off depends on the resin used, the wall thickness of the blown part, the size of the relief angle, the closing speed, and the time when blowing starts. As a general guide for small parts

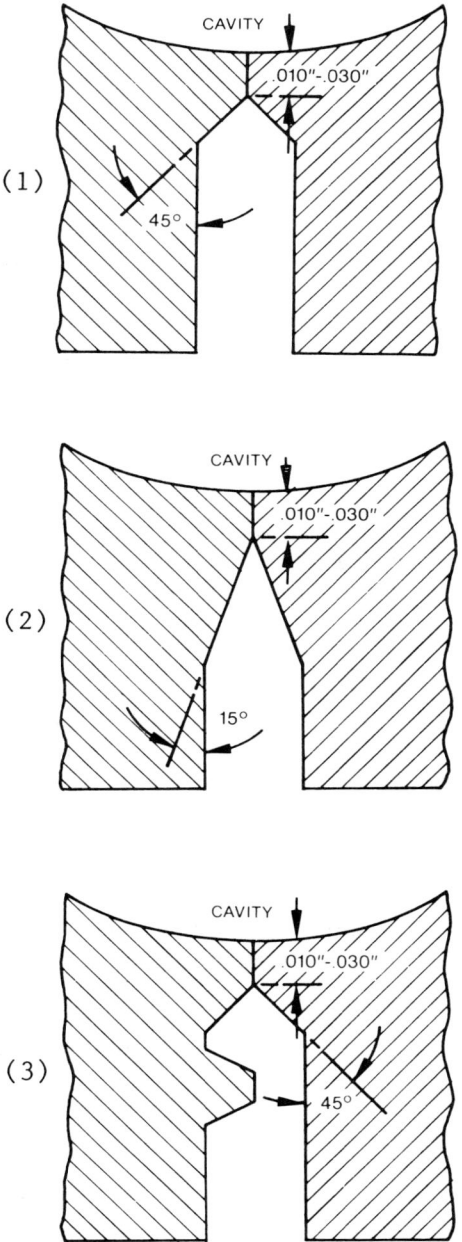

Fig. 4–6. Typical pinch-off double-angle designs.

(up to 10 mil), the width is 0.004 to 0.012 in. (0.1–0.3 mm). When processing LDPE, one uses the narrowest edge. Edge tapering, starting from the center, can help. Its internal edge should be machined smooth and possibly broken slightly to prevent surface damage of the parison or the molded part during closing and opening. Such surface damage could include dull patches next to the weld line or scratches and scoring during ejection.

Deflashing is not as easy with engineering resin products as with the PEs. Even if the processor maximizes the clamping force and the design of pinch-off edges, conventional methods such as knocking-off, punching, or cutting can be a problem for most production operations. For these plastics, other cutting methods generally are used, such as mechanical separation, water jet cutting, or laser beam cutting. Robots are used to handle these types of cutters.

Blowing the Parison

The air used for blowing serves to expand the parison tube against the mold cavity walls, forcing the melt to assume the shape of the mold and forcing it into the surface details, such as raised letters, surface designs, and so on (Table 4–4). The air performs three functions: it expands the parison against the mold, exerts pressure on the expanded parison to produce surface details, and aids in cooling the blown parison. During the expansion phase of the blowing process, it is desirable to use as high a volume of air as is available so that expansion of the parison against the mold walls is accomplished in a minimum time. A maximum volumetric flow rate into the cavity at a low linear velocity can be achieved by making the air inlet orifice as large as possible.

Blowing inside the neck sometimes is difficult. Small orifices may create

Table 4–4. Guide for Air Blowing Pressure.

PLASTIC	PRESSURE, PSI
Acetal	100–150
PMMA	50–80
PC	70–150
LDPE	20–60
HDPE	60–100
PP	75–100
PS	40–100
PVC (rigid)	75–100
ABS	50–150

a venturi effect, producing a partial vacuum in the tube and causing it to collapse. If the linear velocity of the incoming blow air is too high, the force of this air can actually draw the parison away from the extrusion head end of the mold, resulting in an unblown parison. The air velocity must be carefully regulated by control valves placed as close as possible to the outlet of the blow tube. Normally the gauge pressure of the air used to inflate commodity and engineering resin parisons is from 30 to possibly 300 psi (0.21–2.1 MPa). Often too high a blow pressure will "blow out" the parison. Too little pressure will yield end products lacking adequate surface details. As high a blowing pressure as possible is desirable to give both minimum blow time (resulting in higher production rates) and finished parts that reproduce the mold surface. The optimum blowing pressure generally is found by experimentation on the machine that will be used in production. The blow pin should not be so long that air is blown against the hot plastic opposite the air outlet. That action can result in freeze-off and stresses in the container at that point.

Air is a fluid, just as the parison is, and as such it has a limited ability to blow through an orifice. If the air entrance channel is too small, the required blow time will be excessive, or the pressure exerted on the parison will not be adequate to reproduce the surface details in the mold. General guidelines for determining the optimum diameter of the air-entrance orifice during blowing are: (1) up to one quart (0.95 L) use $\frac{1}{16}$ in. (1.6 mm); (2) for one quart to one gallon (0.95–3.8 L), use 0.25 in. (6.4 mm); and (3) for one to 54 gallons (3.8–205 L) use 0.5 in. (12.7 mm).

The pressure of the blowing air will cause variations in the surface detail of the molded items. Some PEs with heavy walls can be blown with air pressure as low as 30 to 40 psi. Low pressure can be used because items with heavy walls cool slowly, giving the resin more time, at a lowered viscosity, to flow into the indentations of the mold surface. Thin walls cool rapidly; so the plastic reaching the mold surface will have a high melt viscosity, and higher pressures will be required, of 50 to 100 psi (0.3–0.7 MPa). Larger items such as one-gallon containers require a higher air pressure, of 100 to 150 psi. The plastic has to expand farther and takes longer to get to the mold surface in larger items. During this time the melt heat will drop slightly, producing a more viscous mass that in itself requires more air pressure to reproduce the details of the mold.

A high volumetric air flow at a low linear velocity is desired. A high volumetric flow gives the parison a minimum time to cool before coming in contact with the mold, and provides a more uniform rate of expansion. A low linear velocity is desirable to prevent a venturi effect (see above). Volumetric flow is controlled by the line pressure and the orifice diameter. Linear velocity is controlled by flow control valves close to the orifice.

The blowing time differs from the cooling time, being much shorter than the time required to cool the thickest section to prevent distortion on ejection. The blow time for an item may be computed from Table 4-5 and the formula:

$$\text{Blow time, s} = \frac{\text{Mold volume, cu ft}}{\text{cu ft/s}} \times \frac{\text{Final mold psi} - 14.7 \text{ psi}}{14.7 \text{ psi}}$$

This is for free air; but there will be a pressure buildup as the parison is inflated, so the blow rate has to be adjusted. The value of cu ft/s is obtained from Table 4-5, according to the line pressure and the orifice diameter. The final mold pressure is assumed to be the line pressure for purposes of calculation. Actually, the blow air is heated by the mold, raising its pressure. Calculations ignoring this heat effect will be satisfactory when blow times are under one second (for small to medium-size parts); but if blow times are longer, the air will have time to pick up heat, resulting in a more rapid pressure buildup and shorter than calculated blow times (2).

Cooling

Processors often employ methods to speed up cooling. For example, postcooling of blow molded parts can shorten the blow cycle, as previously reviewed. Shuttle machines, which maximize production in continuous EBM, are preferred by many molders because they can produce finished containers in the machine; but trimming cannot proceed until the scrap areas, the thickest areas of the part, have been cooled sufficiently—so the cycle depends on getting parts cool enough to trim. If trimming is not done in the machine, the parts can be demolded when they are cool enough to maintain their shape. Wheel machines are sized to do the job by mounting

Table 4-5. Discharge of Air in cu ft/s at 14.7 psi and 70°F.

GAUGE PRESSURE, PSI	ORIFICE DIAMETER, IN. (MM)			
	$\frac{1}{16}$ (1.6)	$\frac{1}{8}$ (3.2)	$\frac{1}{4}$ (6.4)	$\frac{1}{2}$ (12.7)
5	0.993	3.97	15.9	73.5
15	1.68	6.72	26.9	107
30	2.53	10.1	40.4	162
40	3.10	12.4	49.6	198
50	3.66	14.7	58.8	235
80	5.36	21.4	85.6	342
100	6.49	26.8	107.4	429

a sufficient number of mold stations to provide adequate cooling for each blown container as it takes its turn around the wheel.

Resins vary in cooling requirements. It is not usually necessary to postcool PVC; it gives up its heat much more readily than the polyolefins (and thus is more appropriate for a dedicated operation than for custom blow molding). Also the bigger the part, the more cost-effective its cooling becomes.

Postcooling can be extremely beneficial when directed at the thickest areas, especially at pinch-off, neck finish, and thick handle areas.

Pre-postcooling is very important in developing any "fast" cooling control. It originates with the proper placement of cooling lines in the molds. Unfortunately most blow molds are inefficiently designed for maximizing cooling rates. Gradually improvements are developing, similar to what is available and used in injection molds (1). In fact, a few computer aided design programs have been produced just for blow mold cooling analysis (2).

Efficient mold cooling also depends on accurate venting. Air trapped between the mold cavity and the plastic will significantly slow down cooling and also could cause problems with the part. Certain molds incorporate vacuum lines to ensure proper contact.

Unfortunately, it is not always possible to run the mold as cold as possible because of a high ambient dew point. With high humidity, the mold "sweats," causing the surface of a blown part to be damaged. Mold sweating is easily solved by increasing the mold heat, which in turn increases the cycle time. To eliminate the problem and keep the cycle time short, as is done in other processes, the mold is enclosed in a dry air curtain. Also one should consider dehumidifying the room, or just the molding machine in an enclosure or tent, if practical.

Clamping

The improvements made in clamping units provide a great variety of movement and action in the larger BM machines. Small machines still need certain improvements to ensure good flash removal with low deformation of the clamping units. The most important development has been the use of proportional valves in hydraulic systems. With this technology, a machine runs more smoothly and more exactly to permit a wide variety of action in the mold, as well as accurate control of the closing speed (Fig. 4–7).

The delayed closure action in the final phase of mold closing determines pinch-off weld formation, and the reproducibility of this delayed closure phase ensures the uniformity of BM parts. A disadvantage, still not resolved, is that in many cases it is necessary to work with scale settings and

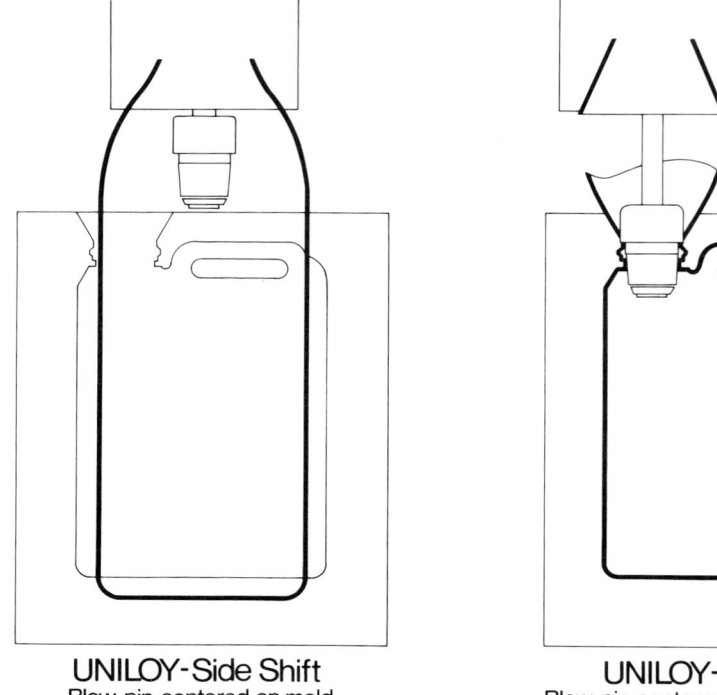

UNILOY-Side Shift
Blow pin centered on mold

UNILOY-Side Shift
Blow pin centered on container neck

Fig. 4-7. Mold movement can locate the blow pin in any required horizontal position.

not actual physical values, in recording and setting the closing speed as well as in other aspects of BM. However, proportional valve technology offers the great advantage that the full clamping force is available within a very short time through suitable control of the valves during mold closure (this is similar to injection molding clamp action; see reference 1). This action facilitates easy pinch-off removal. As reviewed, flash removal is more difficult with engineering resins because of rapid freezing of the resin. Thus to facilitate flash removal with these resins, more care (control) is required in holding down the BM pinch-off edge during mold closing.

The mold closing speed and clamping force have an important influence on product quality. If closing speed is increased, the lengthening of the parison due to low melt viscosity is minimized. Also the formation of the edge layers of the parison, which freeze very rapidly, is better controlled.

For resins with low viscosity and/or for BM when there is a large opening in the delay position, a closing speed of 100 in./s (40 cm/s) helps to resolve problems. Closing speeds such as these can be operated repeatedly only with a speed control system. The required clamping force for the part is determined less by the blowing pressure acting on the part surface than by the length of the flash to be removed. For engineering resins, a clamping pressure of 1,000 to 2,000 lb/in^2 (2,000–4,000 N/cm^2) at the pinch-off edge should be considered. Closing movements in which the flash is separated before the melt has started to solidify are also used. Most of these systems are patented.

Many articles require molds with large thickness and shape variations between the two mold halves because of the particular contours of the part. For such molds, particularly those having very complex (reentrant) shapes, clamping platens and molds can be sectionalized, with each section separately controlled. Thus, sectionalizing can permit blowing complex parts quickly and repeatedly. The hydraulic systems permit sections to operate at different time periods and speeds. When only two mold halves are used, they can meet at the prescribed center point at the same time.

Shrinkage

The shrinkage behavior of different resins and the part geometry must be considered. Generally shrinkage is the difference between the dimension of the mold at room temperature (72°F) and the dimensions of the cold blown part, usually checked 24 hours after production. The elapsed time is necessary to allow the part to shrink. Trial and error determines what time period is required to ensure complete shrinkage. Coefficients of expansion and the different shrinkage behaviors depend on whether plastic materials are crystalline or amorphous (see Chapter 1).

Lengthwise shrinkage tends to be slightly greater than transverse shrinkage. Most of the horizontal shrinkage occurs in the wall thickness rather than a body dimension. With PE, higher shrinkage occurs with the higher-density polymers and thicker walls. Lengthwise shrinkage is due to the greater crystallinity of the more linear type plastics. Transverse shrinkage is due to slower cooling rates, which result in more orderly crystalline growth.

Mold shrinkage is dependent on many factors, such as resin density, melt heat, mold heat, part thickness, and pressure of blown air. Typical PE shrinkages are as follows: LDPE at a thickness up to 0.075 in. has a tolerance of 0.010 to 0.015 in./in., and at a thickness over 0.075 in. has a tolerance of 0.015 to 0.030 in./in.; whereas HDPE at a thickness up to 0.075 in. has a tolerance of 0.020 to 0.035 in./in., and at a thickness over 0.075 in. has a tolerance of 0.035 to 0.055 in./in. Once the operating conditions are established, tolerances of ±5 percent may be expected. When fillers are included in these resins, as well as others, their shrinkage behavior changes, even to the extent of their having very little shrinkage when certain fillers are used.

The cavity defines the shape of a part. In BM, as with other processes, the cavity dimensions are enlarged slightly to compensate for part shrinkage. For polyolefins, particularly in the neck section, slightly higher shrinkage rates are used for extrusion BM than for injection BM. The amount of shrinkage is 2 percent for the body and as high as 3.5 percent for the neck finish. However, rigid resins are relatively unchanged.

The most common special feature of the mold is the "quick change" volume control insert. Rigid volume control is necessary for certain products, such as dairy containers. Here HDPE is used (for many excellent reasons), and the container slowly shrinks and changes in size for many hours after molding. Because of production "volume control" requirements, some dairies must fill containers molded half an hour before fill and then switch to filling containers molded several days previously. Volume-control inserts that displace the difference in volume between the two types are added to the mold, usually as a disc in the side wall, to ensure that the volume and fill levels are the same in both containers at the time of filling. The device works because HDPE shrinkage is reduced virtually to zero for the life of the container when filled with milk or juice and stored at cold temperatures.

Generally, with all BM, shrinkage may be reduced by raising the blowing pressure and lowering the mold heat. Rapid cooling is desired, but cooling too rapidly can cause surface imperfections, distortion, and frozen-in stresses. Raising the stock heat may not appreciably affect the outside dimensions, but it causes more shrinkage to occur in the wall thickness, as higher heats lessen strain recovery and reduce blowing stresses.

Close tolerances on the capacity of the blown part also influence mold construction. Where multiple molds are required, the capacity must be closely regulated from one mold to another. Generally, some of the casting processes have been more successful in reducing capacity variation than duplicating or hobbing. For final capacity adjustment, an insert bottom plate or other adjustable insert is used. By this method, slight changes can be made on each mold insert without noticeable variation in the finished parts.

Plastics will sometimes thin out at an insert or parting line. With resins such as LDPE, this phenomenon is of little importance, but with HDPE and similar resins, it can be a serious problem. When inserts are used, they should be fitted tightly. The parting line miter should be as close as possible, and sufficient clamping force should be provided to prevent partial opening of the mold. Solid molds have a slight technical advantage over molds with inserts with respect to the thinning action.

To aid in controlling shrinkage, vacuum-assisted molding ensures closer contact of the parison with the mold wall. The rate of heat transfer is effectively accelerated, usually with cycle reductions of up to 10 percent. Table 4-6 shows the effective cycle gains made by using a vacuum in conjunction with various wall thicknesses in an HDPE container. The advantage of vacuum assist is further enhanced by utilizing maximum blow pressures. With parts trimmed in the mold, the problem of deflashing is significantly alleviated when reduced blow cycles and vacuum assist are used. This improvement is due to the more rapid rate and degree of part cooling.

With either EBM or IBM, the requirement for a thermally controlled parison at the lowest heat increases the demand on internal pressure/volume to ensure adequate movement or expansion of the parison in the transverse direction. It is possible that vacuum-aided stretch blowing can provide faster expansion rates of the parison and improve performance at the higher stretch–blow ratios. With PVC, PS, and other low shrink alloyed resins, the vacuum assist may need little or no pressure beyond one atmosphere. The function of vacuum (-14.7 psi) has the basic effect of $-(-14.7$ psi)

Table 4-6. Cycle Time Reduction Using Vacuum-Assisted Blow Molding of HDPE.

WALL THICKNESS IN. (MM)	BLOW PRESSURE (PSI)	NORMAL CYCLE TIME (S)	CYCLE TIME WITH VACUUM ASSISTANCE (S)	CYCLE TIME VACUUM ONLY (S)
0.040 (1.0)	40	14	12	12
	60	12	11	
	80	11	10	
	100	10	9.2	
	125	10	9	
0.050 (1.21)	40	16	14	15
	60	15	13	
	80	14.5	11	
	100	14.5	10	
	125	14	10	
0.060 (1.52)	40	20	16	18
	60	18	15	
	80	18	15	
	100	17	14	
	125	16.5	14	

Coolant input kept constant at 28°F (-2°C); coolant flow kept at a constant 31 gpm (gal/min); vacuum set at 27″ Hg gauge; surge tank capacity is 10 cu ft at 27″ Hg.

+ (+14.7 psi) = +29.4 psi. The normal shrink allowances can usually be reduced by one-third. If a vacuum retained for in-mold labeling is used, it effectively provides a simultaneous vacuum assist system to the blowing cycle.

INJECTION BLOW MOLDING

IBM basically has three stages (Figs. 4–8 to 4–10). In the first stage, hot melt is injected through an injection molding machine nozzle into a manifold and into one or more preform cavities. An exact amount of resin is injected around a core pin(s). Hot liquid from a heat control unit is directed by hoses through mold heating channels around the preform cavity; these channels have been predesigned to provide the correct heat control on the melt within the mold cavity. The melt heat is decreased to the required amount.

The two-part mold opens, and the core pin(s) carries the hot plastic to

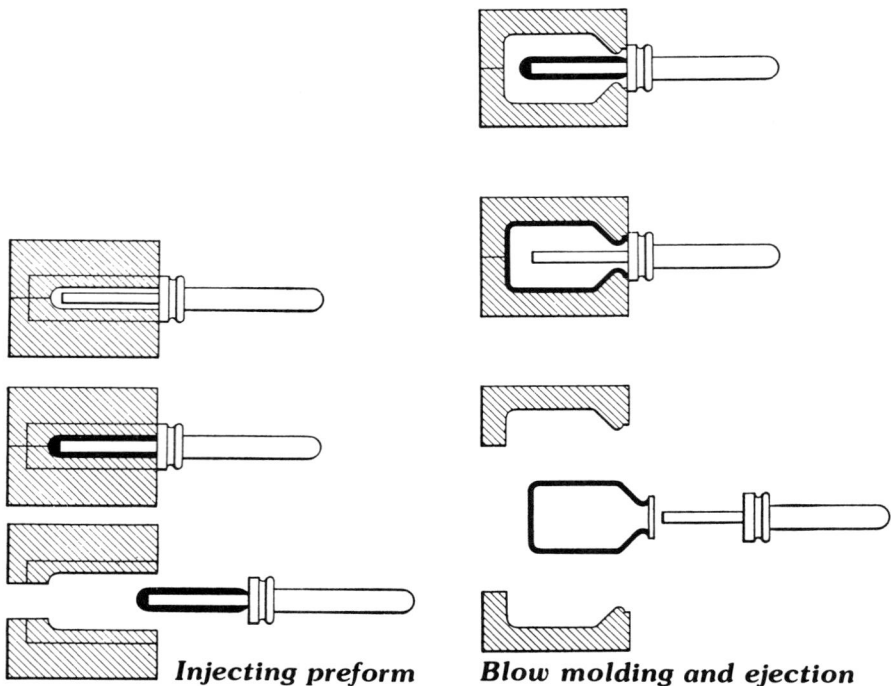

Injecting preform   **Blow molding and ejection**

Fig. 4–8. Basic injection blow molding process.

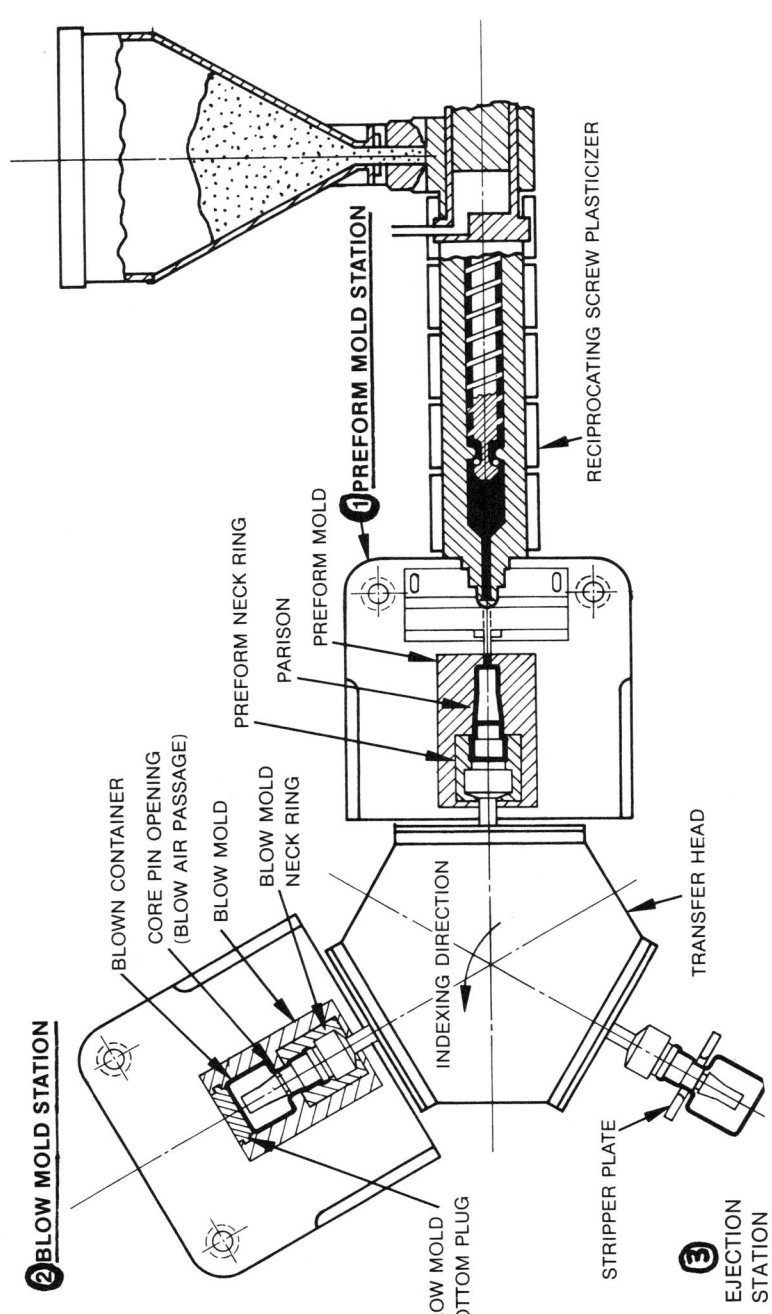

② BLOW MOLD STATION

BLOWN CONTAINER

CORE PIN OPENING
(BLOW AIR PASSAGE)

BLOW MOLD

BLOW MOLD
NECK RING

PREFORM NECK RING

PARISON

PREFORM MOLD

① PREFORM MOLD STATION

RECIPROCATING SCREW PLASTICIZER

INDEXING DIRECTION

TRANSFER HEAD

BLOW MOLD
BOTTOM PLUG

STRIPPER PLATE

③ EJECTION
STATION

Fig. 4-9. Schematic of a three-station injection blow molder.

193

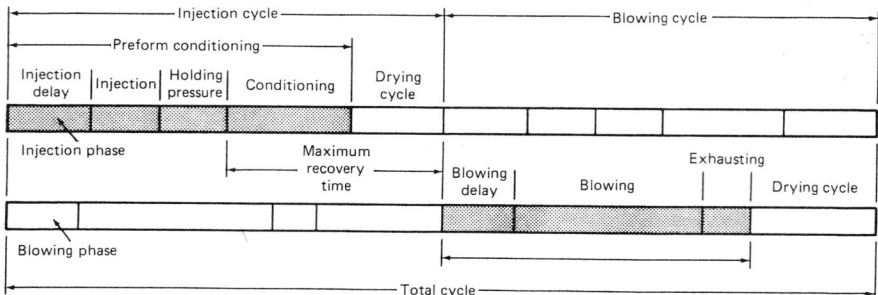

Fig. 4–10. Injection blow molding complete cycle, starting with injection molding the preform followed by the blowing cycle.

the second stage blow mold station (counterclockwise). Upon mold closing, air is introduced via the core pin, and the plastic blows out and contacts the mold cavity surface. Controlled chilled water circulates through predesigned mold channels around the mold cavity (usually 40–50°F), and solidifies the hot plastic.

The two-part blow mold opens, and the core pin carries the complete blown container to the third stage, which ejects the part. Ejection can be done by using a stripper plate, air, a combination of stripper plate and air, robots, and so forth.

IBM can have four or more stations (stages). A station can be located between the preform and blow mold stages to provide extra heat conditioning time. A station between the blow mold and ejection stages can provide additional cooling and/or provide secondary operations, such as hot stamping, labeling, and so on. A station between the ejection and preform stages can be used to detect if ejection has not occurred, to add an insert to the core pin, and so forth.

The process parameters in preform production that determine the quality of the part are the injection speed, injection pressure, hold-on (packing) pressure, heat control of the preform, and melt mix (see Chapter 2). The process permits using resins that cannot be used in EBM, specifically those with no controllable melt strength such as PET, which is used in carbonated beverage bottles (stretched–blown). The information on blowing parisons, cooling, clamping, and shrinkage that was presented for EBM is also applicable to IBM.

There are several different IBM methods available, with different means of transporting the core rods from one station to another. These methods include the shuttle, two-parison rotary, axial movement, and rotary with three or more stations used in conventional IM clamping units. A variation of IBM is displacement BM or dip molding. A premeasured amount of hot melt is deposited into a cupel, the shape of a preform. A core rod is inserted into the cupel, displacing the melt and packing it into the neck finished area. It then moves to a blow station where it basically follows the procedure described above. Advantages of IBM include lower-cost machines and the fact that a nearly stress-free bottle is blown.

STRETCHING/ORIENTING

High-speed EBM and IBM are taken an extra step in stretching or orienting. Figure 4–11 shows stretched IBM; with EBM the stretching action is similar to this, as it occurs during compressed air inflation (Fig. 4–1). In EBM, the

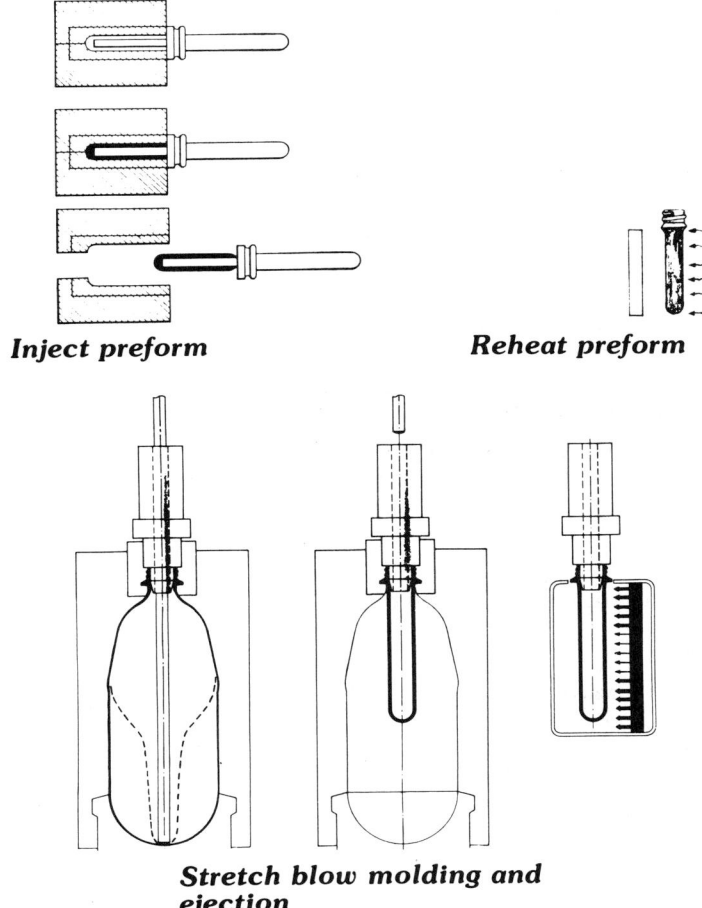

Inject preform   *Reheat preform*

*Stretch blow molding and
ejection*

Fig. 4–11. Schematic of stretch injection blow molding.

parison, which is mechanically held at both ends, is stretched rather than just blown. Stretching can include the use of an expanding rod within the IBM preform or an external gripper with EBM.

By biaxially stretching the extrudate before it is chilled, significant improvements can be obtained in the finished containers, as described in Chapters 1 and 3. As explained, this technique allows the use of lower-material-grade resins or thinner wall thicknesses with no decrease in strength; both approaches reduce material costs. Stretched BM gives many resins (mono- or bioriented) improved physical and barrier properties (Table 4–7). The process allows wall thicknesses to be more accurately controlled and also allows weights to be reduced.

Draw ratios used to achieve the best properties in PET bottles (typical 2- and 3-liter carbonated beverage bottles) are 3.8 in the hoop and 2.8 in the axial direction, and will yield a bottle with a hoop tensile strength of about 29,000 psi (200 MPa) and an axial tensile of 15,000 psi (104 MPa).

Stretch blow is extensively used with PET, PVC, ABS, PS, AN, PP, and acetal, although most TPs can be used. The amorphous types, with a wide range of thermoplasticity, are easier to process than the crystalline types such as PP. If PP crystallizes too rapidly, the bottle is virtually destroyed during the stretching. New grades of PP called "clarified" have virtually zero crystallinity and overcome this problem.

There are in-line and two-stage processes. In-line processing is done on a single machine, whereas two-stage requires either an extrusion or an injection line to produce the solid parisons or preforms. With either type of process, a specific reheat blow machine is used to produce the bottle. With in-line systems, the hot-firm plastic preform or parison passes through conditioning stations that bring it down from the "melt" heat to the proper orientation heat (Table 4–8). A rather tight heat profile is maintained in the axial direction. Advantages of this approach are that the heat history is

Table 4–7a. Volume Shrinkage of Stretch Blow Molded Bottles.*

TYPE OF BOTTLE	PERCENT
Extrusion blow molded PVC	—
Impact-modified PVC (high orientation)	4.2
Impact-modified PVC (medium orientation)	2.4
Impact-modified PVC (low orientation)	1.6
Non-impact-modified PVC (high orientation)	1.9
Non-impact-modified PVC (medium orientation)	1.2
Non-impact-modified PVC (low orientation)	0.9
PET	1.2

*Seven days at 80° F.

Table 4–7b. Gas Barrier Transmission Comparisons for a 24-oz Container Weighing 40 g.*

	RATE, SQUARE METERS PER DAY	
TYPE OF BOTTLE	OXYGEN, CC	WATER VAPOR, G
PET (oriented)	10.2	1.10
Extrusion blow molded PVC	16.4	2.01
Stretch blow molded PVC (impact-modified)	11.9	1.8
Stretch blow molded PVC (non-impact-modified)	8.8	1.3

*At 100°F.

minimized (crucial for heat-sensitive resins), and the preform or parison can be programmed for optimum material distribution.

With the two-stage process, cooled preforms or parisons are conveyed through an oven (usually using quartz lamps) that reheats them to the proper orientation heat profile. Advantages include minimization of scrap (for EBM, there being no scrap with IBM using either the in-line or the two-stage process), higher output rates, and the capability to stockpile preforms or profiles and improve thread finishes (EBM). In-line stretch IBM, which offers more flexibility from a material view, does not give the degree of parison programming available in EBM-stretched. An extruded parison can be heat-stabilized, and then the parison is held externally and pulled to give an axial orientation while the bottle is blown radially. Another technique stretches and preblows the parison before completing the blowing operation. In both of these operations, scrap is produced at the bottle neck and at its base. The processes completely orient the whole bottle, including the threaded end; this neck orientation does not occur with stretch IBM. With EBM, threads are post-mold-finished.

Extrusion processes also can use programmed parisons and two sets of molds. The first mold is used to blow and heat-condition the preform—

Table 4–8. Stretch Blow Molding Processing Characteristics.

	MELT TEMPERATURE		STRETCH ORIENTATION TEMPERATURE		MAXIMUM
PLASTIC	°F	°C	°F	°C	STRETCH RATIO
PET	490	250	190–240	88–116	16/1
PVC	390	199	210–240	99–116	7/1
PAN	410	210	220–260	104–127	9/1
PP	334	168	250–280	121–136	6/1

actually to cool it to the orientation heat. The second mold uses an internal rod to stretch and blow the bottle. With two-stage EBM, the preforms or parisons are usually open-ended. They are reheated, and then stretched by pulling one end while the other end is clamped in the blow mold. Minimum scrap can be produced, and the compression-molded threads provide a good neck finish. This system utilizes a conventional extrusion line and a reheat blow machine whose oven contains quartz lamps to provide the proper heat. (When used to heat the preform or parison, they rotate in a heat-profile-controlled oven to provide uniform heat circumferentially.)

MOLD CONSTRUCTION

As blow molds do not have to withstand high pressure, a wide selection of construction materials is available. The ultimate selection will depend on a balance of the following factors: cost, thermal conductivity, and required service life. The more commonly employed materials for small parts are aluminum and aluminum alloys, steel, beryllium copper (Be/Cu), and cast zinc alloys (Kirksite, etc.). Aluminum molds are excellent heat conductors, are easy to machine, can be cast, and are reasonably durable, particularly when fitted with harder pinch blades and neck inserts (Table 4-9).

Molds with aluminum pinch-off areas could last 1 to 2 million cycles if

Table 4-9. Materials Used in the Construction of Blow Molds.*

MATERIAL	HARDNESS**	TENSILE STRENGTH		THERMAL CONDUCTIVITY BTU/IN./FT² H °F
		PSI	MPA	
Aluminum				
A356	BHN-80	36,975	255	1,047
6061	BHN-95	39,875	275	1,165
7075	BHN-150	66,700	460	905
Beryllium copper				
23				
and	RC-30	134,850	930	728
165	(BHN-285)			
Steel				
0-1				
and	RC 52-60	290,000	2,000	243
A-2	(BHN-530-650)			
P-20	RC-32	145,000	1,000	257
	(BHN-298)			

*BHN = Brinell and RC = Rockwell hardness (C scale).
**Specific gravities (lb/cu in.): Al = 0.097, Be/Cu = 0.129-0.316, steel = 0.24-0.29.

properly set up and maintained. Most molds, however, use Be/Cu and steel pinch-offs, providing more durability, and they can easily be repaired. Be/Cu is preferred because of its high thermal conductivity, but steel is predominantly used because of its wear resistance and toughness.

Although Be/Cu and Kirksite are better conductors of heat, aluminum is by far the most popular material for molds, because of the high cost of Be/Cu and Kirksite's short life (soft zinc alloy). Aluminum is light in weight, a relatively good conductor of heat, very easy to machine, and low in cost. A major disadvantage is that it is easily damaged, particularly if it is abused. Its potential porosity may easily be eliminated by coating the inside cavity with a sealer such as automotive radiator sealant.

Aluminum is used for single molds, molds for prototypes, and large numbers of identical molds, as found on wheel-type blowing equipment or equipment with multiple die arrangements. (Regarding prototyping, it is best to utilize a material that the production mold will use to duplicate heat transfer conditions.) Aluminum may tend to distort after prolonged use. Thin areas, as at pinch-off areas, can wear in aluminum.

Be/Cu molds are corrosion-resistant and very hard when compared to aluminum, making Be/Cu the choice for PVC. However, it is about three times as heavy as aluminum, costs about six times as much per volume, has a lower thermal conductivity, and requires about one-third more time to machine.

In blow molding plastics that produce corrosive volatiles (PVC, nylon, acetal, etc.), it is necessary to employ corrosion-resistant steels or to use plating, even gold plating. However, platings can lead to reductions of strength and hardness, and damage to edges and surfaces can easily occur. With proper care, life spans of 10^5 to 10^6 cycles can be expected. For continuous production and longer runs, steel molds are used.

Hardening is not necessary, and chrome plating is not customary. For large-volume parts with a content of 4 cu yd (3 cu m) and up, welded machined steel plate construction can be considered. Cast metal is of no value for the mold bases. Given the right care, the life of steel molds is the longest, at over 10^8 parts per mold. Inserts subject to wear and tear have to be either refurbished or exchanged at intervals of 10^5 or 10^6 cycles (Table 4-10).

Unfortunately, most BM molds do not yet provide the high level of cooling that has been achieved for decades with IM molds. A greater effort is now being made to properly incorporate cooling channels in BM molds, to provide necessary and controllable heat transfer. CAD systems now are available for proper design of the molds (1, 2). Internal cooling with liquid N_2, CO_2, or refrigeration and ice crystals can improve efficiency by up to 50 percent.

Table 4–10. Guide to Selecting Materials of Construction for BM Molds.*

PROPERTY	STEEL	MACHINED			CAST		
		ALUMINUM	BE/CU		ALUMINUM	KIRKSITE	BE/CU
Pinch life	4	3	2		2	1	3
Cavity life	4	3	4		2	1	3
Surface finish	4	3	4		2	1	3
Heat control	2	4	4		2	1	3
Mold modifications	2	4	2		1	1	2
High volume	4	3	4		2	1	2
Mold lead time	2	3	2		4	4	3
Low cost	2	3	1		4	4	3
Prototype cost	1	3	2		3	4	3
Complex shapes	3	4	3		3	2	2
Moving mold parts	4	3	3		3	1	1

*4 = best to 1 = poorest.

PLASTICS

In the past, practically all BM resins were commodity types (PE, PP, PVC, and PS); and, more recently, engineering types have been used. Typical heats used for BM some resins are given in Table 4–11. The polyolefins (PE and PP) and rigid PVC have proved to be the most suitable materials for BM. In regard to heat control and rheology, PE is processed relatively easily. The thermal sensitivity of PVC and the reprocessing of the flash can cause a number of difficulties if not handled properly (Chapter 1, 3, and 9). In retrospect, it can be said that PVC, in particular, which imposed important and high performance requirements on the processing operation, imparted major impulses to the further development of EBM technology and provided the impetus for a thorough engineering analysis of melt flow through the extruder and blow heads.

The wide diversity of products being manufactured makes it clear that processing plants will have to solve logistical problems such as the separation of material cycles and the drying of materials, as has been done in the standard extrusion and injection molding plants.

Most engineering resins are hygroscopic and therefore must be properly dried before processing. Moisture has a generally adverse effect on the melt viscosity, surface, and physical properties of the product. To avoid these problems, resins must be dried, usually to a residual moisture content of < 0.02 percent, by weight. The optimum drying heat and time are given in the manufacturer's data sheet (see Chapter 9).

Table 4-11. Guide to Processing Temperatures of Plastics for Blow Molding.

PLASTIC	TEMPERATURE, °C
LDPE	130–180
MDPE	150–200
HDPE	160–220
HMWPE	180–230
PVC	190–205
PP	200–220
PS	280–300
PA	240–270
POM	150–280
SB	170–210
ASA	200–230
ABS	180–230
ABS/PC	230–250
PPE	240–250
PBT	245–260
PBT/PC	240–260
PUR	180–190

Blow molds for engineering resins are made from the same materials that usually are used for processing HDPE, principally steel and aluminum. The pinch-off edges should always be made of steel. With commodity resins, such as PE and PP, a sand-blasted cavity surface can be used to aid in venting and also to provide a smooth surface on the blown part (a characteristic of the melt that prevents penetration of the "rough" surface). With engineering resins, the surface of the cavity is reproduced precisely; so sand-blasting does not aid venting. Conventional methods of mold venting through slits in the parting line are of minor importance or could even cause problems with these resins, but appropriate venting measures can be taken (1–7).

It is necessary to provide a heat control system for the mold to obtain the required part finish (Table 4–12). The mold surface heat depends on the resin being processed and usually is about 40 to 50°C below the softening temperature. A higher mold heat means a longer cooling time although engineering resins may require the higher heat to provide their highest-quality performance. The effect of this heat control, however, is not great enough to compensate for extrusion defects (Chapter 3). The shrinkage of engineering resins averages 0.7 percent and, because of their low melt viscosity in comparison with HDPE, is not directionally dependent.

Table 4–12. Examples of Recommended Temperatures for Cavities in Blow Molds.

	TEMPERATURE	
PLASTIC	°C	°F
PE and PVC	15–30	59–85
PC	50–70	122–160
PP	30–60	85–140
PS	40–65	105–150
PMMA	40–60	105–140

Any scrap (flash, rejects, etc.) can be recycled. As reviewed throughout this book, it is vital to granulate the material properly and prevent contamination. Processing of up to 100 percent dry regrind is possible, as well as blending with virgin resins. However, with increasing regrind content, melt viscosity is reduced, parison swell worsens, and the performance properties of the blown container may be reduced or unacceptable (see Chapters 1–3 and 9).

CONTROLS

Different types of microprocessor-based modules control BM machines and melt parameters, ranging from single to multiple functions. The modules interact at high speeds, coordinating process variables such as heat, timing, parison or preform molding speed and pressure, melt wall thickness, and so on. (See Chapters 2 and 3 on controlling injection and extrusion machines.)

Control technology is used to improve machine cycle rates, as in employing proportional hydraulics to safely speed up mold movements. In addition, production monitoring systems have become part of some BM plants, helping managers make effective decisions. These improvements in monitoring and controlling have contributed significantly to the manufacture of products with zero defects and to profits.

BM controls closely regulate many process variables such as viscosity, stretch, orientation, and swell. Additional productivity in a BM operation can arise from the coordination of various functions in a single control system (as reviewed in Chapter 2, regarding computer integrated systems). Controls clearly play an important role, but they are not the answer to many processing problems, as explained in Chapters 1 through 3. They have to be properly integrated with the complete process, as summarized in Fig. 1-1 and Table 4-13.

TROUBLESHOOTING

As described in Chapters 2, 3, and 10, there is a logical approach to setting up a troubleshooting guide. Details reviewed in those chapters concerning the approach to be used are applicable to BM and other processes. Table 4-14 lists some of the common BM problems with information on causes and solutions (192).

In setting up and operating a machine, one should not make too many changes simultaneously. The goal is to make one adjustment/change at a

Table 4-13. Factors to Consider when Reviewing the Complete Blow Molding Process.

Part design	Part configuration (size/shape)—relate shape to flow of melt in mold to meet performance requirements, which should at least include tolerances.
Material	Chemical structure; molecular weight amount and type of fillers/additives; heat history; storage; handling.
Mold design	Number of cavities; layout and size of cavities cooling lines/side actions/knockout pins./ etc.—relate layout to maximize proper performance of melt and cooling flow patterns to meet part performance requirements, preengineer design to minimize wear and deformation of mold (use proper steels, aluminum, etc.), and lay out cooling lines to meet temperature to time cooling rate of plastics (particularly crystalline types).
Machine capability	Accuracy and repeatability of temperature/time/velocity/ pressure controls of extrusion injection unit; accuracy and repeatability of clampling force; flatness and parallelism of platens; even distribution of clamping; repeatability of controlling pressure and temperature of oil; minimizing oil temperature variation; no oil contamination (by the time you see oil contamination, damage to the hydraulic system could have already occurred); machine properly leveled.
Molding cycle	Setting up the complete molding cycle to repeatedly meet performance at the lowest cost by interrelating material/machine/mold controls such as those listed in this chapter.

Table 4–14. Guide to Common Blow Molding Problems

PROBLEM	CAUSE(S)	SOLUTION(S)
Rough parison; orange peel	Melt fracture; melt temperature too low	Polish all tooling Raise melt temperature
Poor gloss	Mold too cold	Increase die surface temperature
Black specks in part	Contamination from degraded material	Purge to clean system Keep materials clean
Gels in parison	Excessive fines in regrind Moisture in resin Screw too deep	Screen out regrind fines Dry material before use Use higher shear screw and lower barrel temperatures
Bubbles in wall	Moisture in trapped air	Increase extrusion pressure If moisture, lower screw speed; reduce feed zone temperature
Uneven wall thickness circumferentially	Pin not centered in die ring	Adjust die pin position
Parison hooking	Head temperature not uniform	Stagger heaterband gaps on head
Incomplete blow	Extrusion rate too high Blow-up air pressure Blow-up time too short Parison is cut at pinch-off	Reduce screw speed Increase blow air pressure Reduce mold closing speed
Holes in parison and/or bottles	Contaiminated or degraded resin Trapped air Moisture in resin	Purge and clean tooling and screw Let extruder run for a few minutes Dry the resin
Parison stretches	Resin melt index too high Melt temperature too high	Use lower melt index Reduce melt temperature Increase screw speed Boost extrusion rate
Parison blow-out	Blow-up too rapid Melt temperature too high Pinch-off too sharp Blow-up ratio too high	Program blow-up start with low air pressure and increase Align molds Use larger parison

Table 4-14. Continued

PROBLEM	CAUSE(S)	SOLUTION(S)
Die, weld, and spider lines in parison	Damaged die ring	Repair or replace die tooling
	Mandrel spider legs cause improper knitting	Streamline spider legs
		Reduce die temperature to increase back pressure
	Contamination from material	Clean die head
Webbing in handle	Parison walls touch when mold closes	Align parison closer to handle side of mold
	Wrong parison diameter	Increase die diameter
		Reduce melt temperature
Rocker bottoms	Blowing air not vented before mold opens	Increase air exhaust time
	Insufficient cooling	Clean mold cooling channels
		Increase blow time
Tails not pulled	Parison is too short	Lengthen the parison by increasing extruder speed
	Plastic or foreign matter holding mold	Clean mold parting surfaces
Bottles thin in various areas	Parison curling	Adjust die ring concentricity
	Parison too long or short	Increase/decrease extruder speed and adjust parison temperature
		Reduce head temperature
Moils not separating from neck finish	Cutting ring is dull	Sharpen or replace cutting sleeve
	Poor contact between cutter ring and striker plate	Increase overstroke and downward pressure of blow pin
Weak shoulders on bottles	Parison sag	Reduce melt temperature and decrease/increase extrusion rate
	Parisons too long or short	
	Container too light	Program increased weight
Slanted neck finish	Blow pin/cutter entry too deep	Raise blow pin until it just cuts
	Parison folding over	Replace dull knife blade
		Adjust knife-cut delay timer

(continued)

Table 4-14. Continued

PROBLEM	CAUSE(S)	SOLUTION(S)
Parts sticking in mold	Mold too hot	Improve mold cooling
	Cycle too short	Lengthen cycle
Mold parting line indented in part	Blow-up air introduced prematurely	Delay blow-up
	Hooking parison	Reduce mold temperature
Handle missing	Insufficient die swell	Position parison closer to handle
		Use larger tooling
Sink marks	Air trapped in mold	Improve venting
		Lower mold temperature
Parison tails	Parison is too long	Reduce extruder speed
	Pinch-off improperly designed	Design pinch-off to compression cool tail
Poor detail definition	Blow air pressure too low	Increase blow air pressure and blow time
	Poor mold venting	Improve venting
	Cold mold	Increase mold temperature

Coextrusion Blow Molding:
Most of the Above Tips Also Apply to Blow Molding Multilayer Containers.

Skips in barrier layer	Temperature of barrier material too high	Reduce barrier material temperature
	Pressure fluctuations at extruder	Maintain constant pressure at extruder screw tip
	Degraded material in head	Purge head and/or extruder
Barrier integrity of handle breached	Too little material in handle	Program more material into handle and pinch-off area
	Poor pinch-off	
Layer separation, blistering or bubbles in container	Adhesive layer too cold, did not flow around structure. Adhesive too hot to stick to adjacent layer	Adjust temperature of adhesive material up or down.
	Adhesive layer cooled too fast	Raise mold temperature to prevent fast cool down
	Moisture in materials	Dry materials

time, and wait for any change to occur—allowing up to at least 15 min. In turn, record the action taken and the result, as well as the machine used, the resin used, other machine/control settings, and the operator. This type of log helps one to study and evaluate operating performance. Basically this approach is not new, except perhaps that there is no formal procedure used to keep records. This procedure helps one to eliminate duplication and unnecessary performances, providing factual information for a trouble-shooting guide. Some processors find that having a specially trained person available on each shift to help the operator to troubleshoot processing problems is a successful way to reduce downtime.

Chapter 5

FORMING

INTRODUCTION

Formed or shaped plastics provide a great variety of marketable products, in a wide size range (see Fig. 5-1). Different techniques are used, with thermoforming being the most productive and the most diversified. Other techniques are basically similar to thermoforming but usually use less heat than it requires and are more limited as to the type of plastic used; these processes include cold forming, stamping or compression forming, flow molding, rubber pad molding, diaphragm forming, coining, forging, and so on. Formed parts are used in many different applications and production lines (form, fill and seal, etc.). Food, electronic devices, medical products, and other parts use continuous thermoforming operations at the end of high-speed production lines to reduce the handling of products, provide hermetically sealed contents, reduce costs, and so forth.

THERMOFORMING

Thermoforming usually consists of heating extruded thermoplastic (TP) sheet, film, and profile to its softening heat and forcing the hot and flexible material against the contours of a mold by pneumatic means (differentials in air pressure are created by pulling a vacuum between the plastic and the mold, or the pressure of compressed air is used to force the material against the mold), mechanical means (plug, matched mold, etc.), or combinations of pneumatic and mechanical means (1-5, 22, 213-229).

The process involves (1) heating the sheet (film, etc.) in a separate oven and then transferring the hot sheet to a forming press, (2) using automatic machinery to combine heating and forming in a single unit, or (3) a continuous operation feeding off a roll of plastic or directly from the exit of an extruder die (postforming). Practically all the materials used are extruded TPs; very small amounts are calendered or cast. To date very few

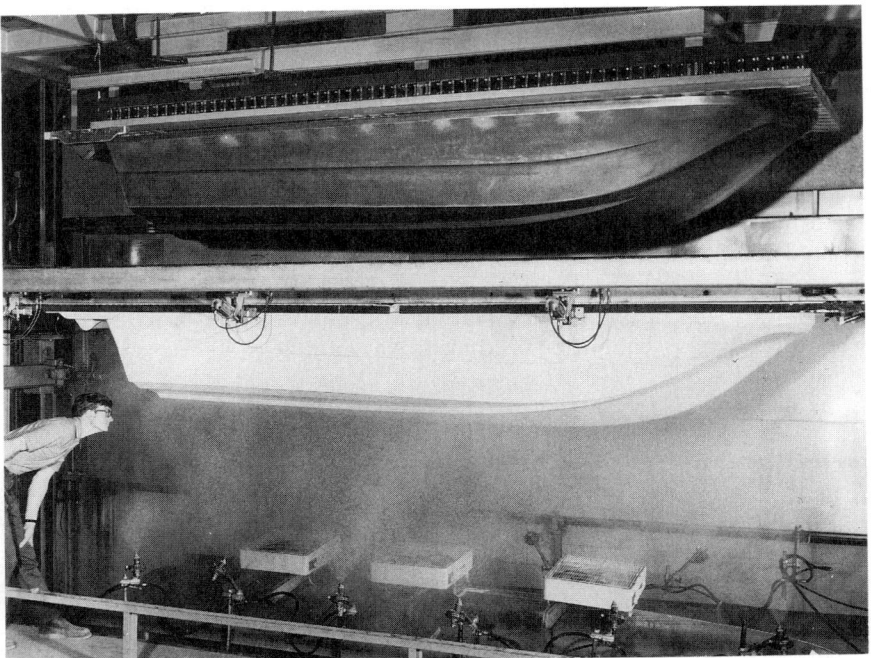

Fig. 5-1. Precisely timed cooling for ABS thermoformed hull for a 15-ft runabout. (Courtesy of the GE-Borg-Warner.)

thermosets (TSs) are used, as the markets have not developed for them. These TSs can be either unreinforced or reinforced (Chapter 7). Practically any TP can be used, but certain types are easier to use, permitting deep draws without tearing or excessive thinning in certain areas such as corners. The ease of forming depends on material characteristics; it is influenced by minimum and maximum thickness, pinholes, ability of the material to retain heat profile gradients across the surface and the thickness, the controllability of applied stress, the rate and depth of draw, the mold geometry, the stabilizing of uniaxial or biaxial deformation, and, most important, minimizing the thickness variation of the sheet.

One of the oldest thermoforming techniques is bending, which is relatively easy to handle. The production of finished parts made by bending often also involves joining (adhesive or welding) or mechanical operations (milling, drilling, polishing). If the sheet is heated only locally in the bending operation, no special forming tools are needed. The width of the heating zone and the thickness of the sheet determine the bending radius. Limitations are related to the softening point of the sheet and the intrinsic rigidity of the heated sheet (sag should be minimized). Extensive use of the method is made in bending transparent plastics (such as PMMA and PC) up to $3\frac{1}{2}$ in. (90 mm) thick, for use in store displays, staircases, partitions in banks, aircraft windows, and so on. With this type of plastic, if restrictions in the bending area are minimized, the thickness at the bend can remain unchanged.

In many applications of conventional thermoforming, low-cost tooling is used compared to that of other processes, particularly in cases of limited production and/or the forming of very large parts (Fig. 5-2). Thermoforming of thin parts has an advantage over most other processes, where very thin "walls" cannot be produced. An example of its use is in skin and blister packaging.

Fig. 5-2. ABS 76 × 230 in. sheets are conveyed to an IR oven in back of a console, where the sheets are individually heated and formed into 15-ft outboard-powered runabouts. The complete automatic process of conveying the sheet, heating, forming, and cooling takes 10 min.

To improve the strength or structural performance of formed plastics, the processor can utilize design features such as corrugations, box shapes, and so on. These features are very easily incorporated with thermoforming. One example of minimizing the thickness of a product and improving its strength is a formed drinking cup with a rolled edge (which also eliminates cutting lips).

With most forming (not including bending) there can be up to 50 percent scrap trim or web. This material could be wasted, but it is usually recycled and blended with virgin resin. Individual sheet stock formed into round shapes could have 50 percent or more scrap. With square forms, there could be up to 25 percent scrap.

The various thermoforming techniques are generally described in terms of the means used to form the sheet, such as bending, vacuum forming, pressure forming, plug-assist forming, matched mold forming, and so on. The different methods enable the processor to form different-shaped products to meet various performance requirements. Most of these techniques are reviewed in Figs. 5-3 to 5-12. An evaluation of these methods shows that simple to very complex shapes can be formed, and the shape as well as the surface condition can be accurately controlled outside, inside, or on both sides.

Compressed air thermoforming is a technique "borrowed" from sheet steel processing; its main use in plastics processing is in plastic cup production. Machines are also coming into use for the production of engineering components, such as computer covers. A flexible-pressure pad is used to press the hot sheet into the mold cavity, with the aid of an air cushion maintained under pressure. Compared with vacuum forming, this

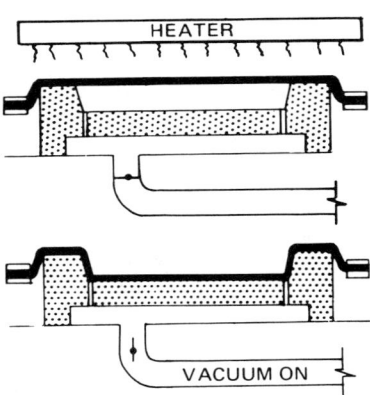

Fig. 5-3. Straight forming: vacuum.

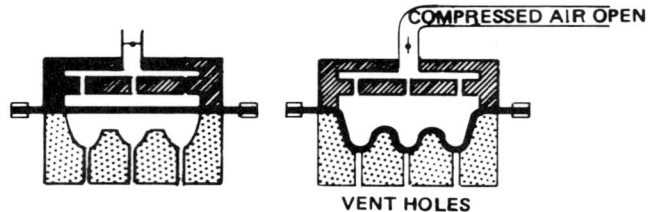

Fig. 5-4. Straight forming: pressure.

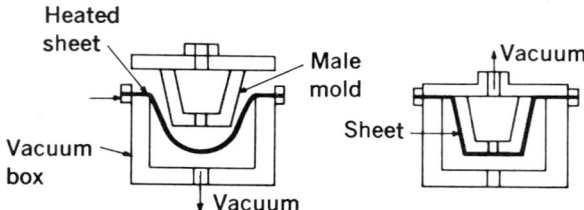

Fig. 5-5. Snapback forming.

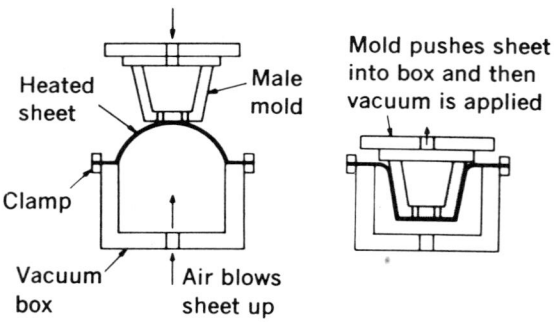

Fig. 5-6. Forming with a billow snapback is recommended for parts requiring a uniform, controllable wall thickness.

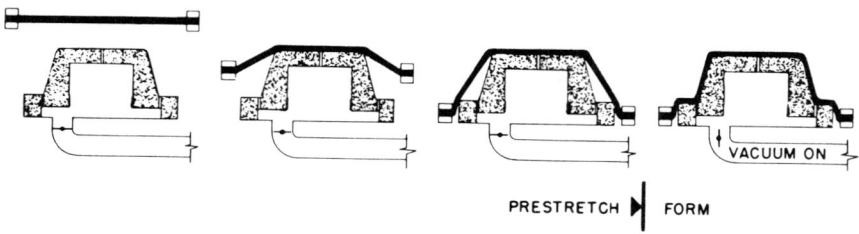

Fig. 5-7. Drape forming.

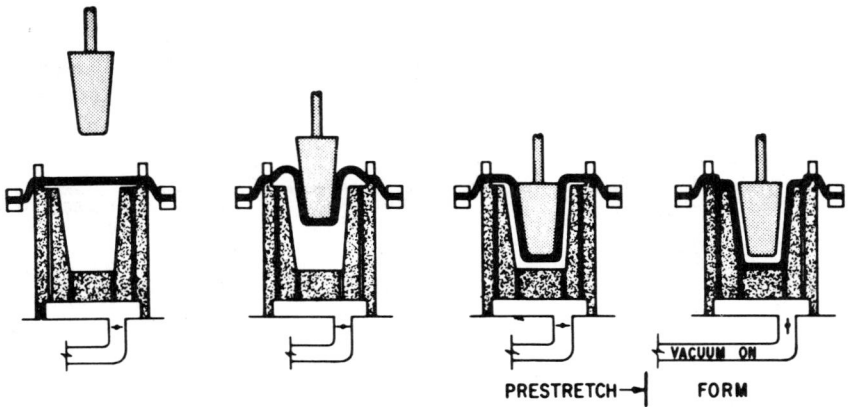

Fig. 5–8. Plug-assist forming.

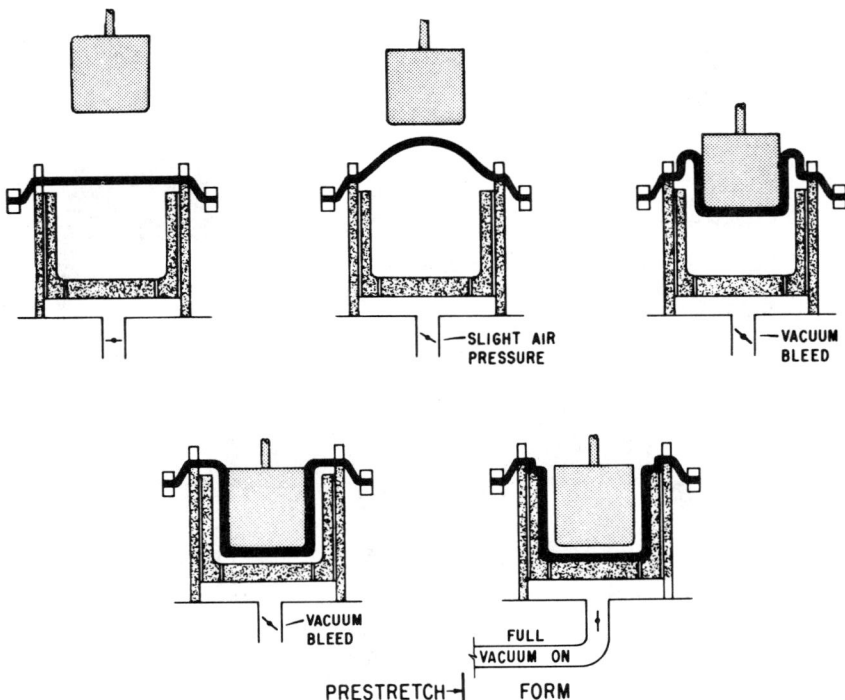

Fig. 5–9. Plug-assist, reverse-draw forming.

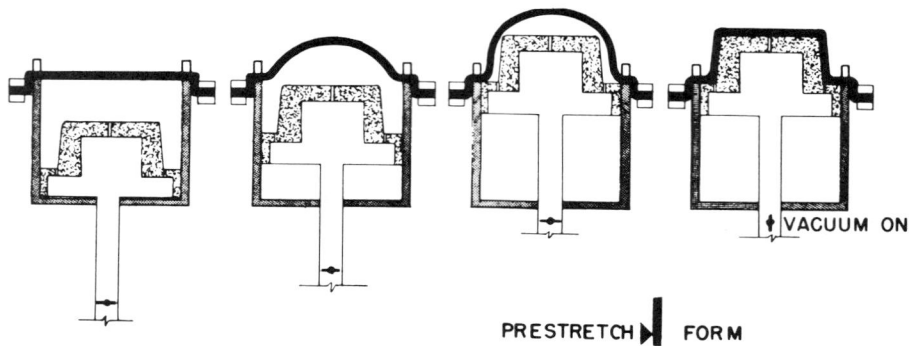

Fig. 5–10. Air-slip forming.

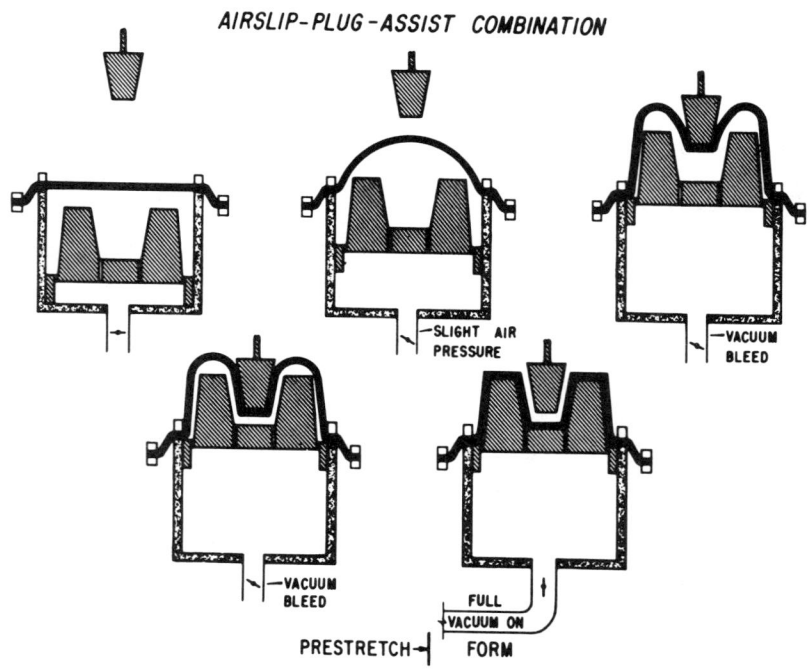

Fig. 5–11. Plug-assist, air-slip forming.

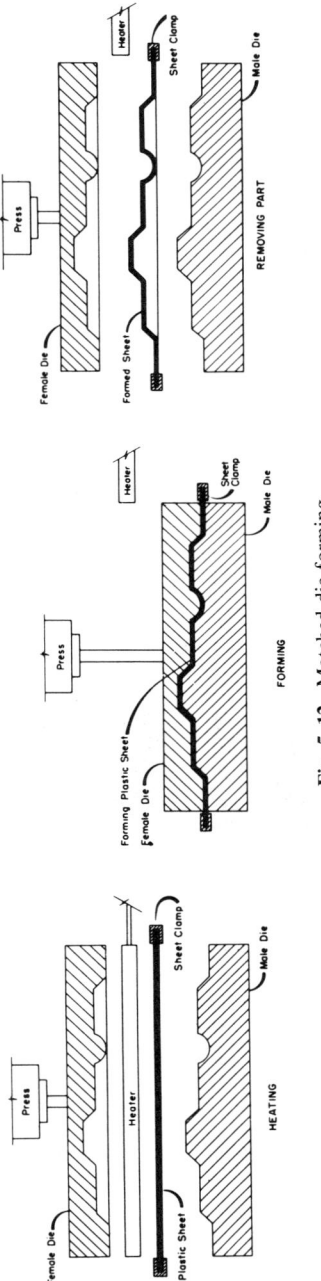

Fig. 5-12. Matched-die forming.

215

method has the considerable advantage that depressions and protrusions of the molding tool are very accurately reproduced. Even undercuts can be molded with very high precision, as well as surface features with different degrees of roughness.

A compressed air thermoforming machine is essentially a deep drawing machine with a controlled heat shield, except that, in place of an upper ram, a pressure bell is used. The heated rubbery material is pressed into the molding tool using a differential pressure of up to 100 psi. After cooling and demolding, the formed part is precisely shaped.

THERMOFORMING MACHINES

The machines range from a "home-made," simple single-stage operation to multistage computerized process-controlled equipment. With single-stage machines, precut sheets are loaded individually into a clamping frame, moved into a heating chamber, and moved back to their original position where the forming takes place. Figure 5-13 shows a single-stage plug-assist former. A two-stage unit consists of two forming stations with one heating chamber.

Another type of machine uses three or more stages. They are usually built on a horizontal circular framed that rotates (Fig. 5-14). The rotary table

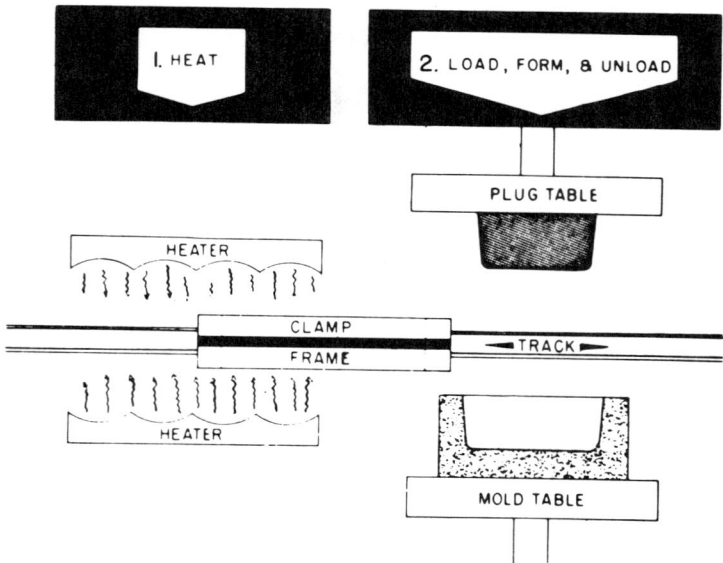

Fig. 5-13. Single-stage thermoforming machine schematic.

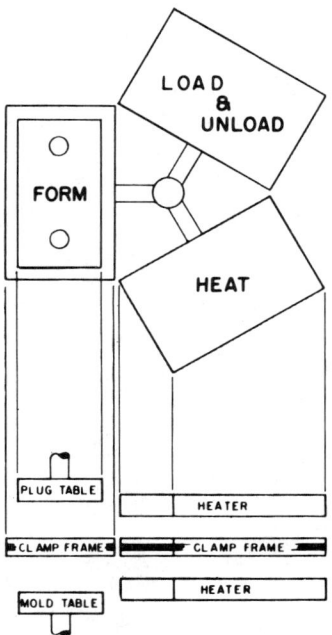

Fig. 5–14. Three-stage thermoforming machine schematic.

operates like a merry-go-round, indexing through the various stations. A three-stage machine would have stations for loading and unloading, heating and forming, and cooling; stations would be indexed 120 degrees apart.

To speed up output, in-line sheet-fed machines are used. Two parallel continually moving tracks hold and move a clamped sheet through the required stations of heat and forming. All movements are indexed so all actions are repeatable. To further increase the output, continuous rolls of sheet or sheet material are fed directly from an extruder (Fig. 5–15). A set of continuously conveying chains/tracks indexes the sheet as it moves accurately through the heating, forming, trimming, and packaging stations.

Other stations can be included, such as decorating. Multicavity molds are used extensively. As is typical of injection molding and extrusion machines (Chapters 2 and 3), they can have sophisticated computer controls to ensure proper operation of all machine and material functions.

Trimming of plastics, if not performed correctly, can be damaging to formed parts and slow down or stop the output. Tools for trimming include shear discs, steel rule dies, and saws. The cutting action can be done with the usual punch press, as well as press brakes and other devices. Punch and

Fig. 5-15. Example of a high speed, continuous thermoformer. (Courtesy of Brown Machine Co.)

die clearances should be held to a minimum. The generally acceptable rules that are applied to metals are not applicable to plastics. Also, what is good for one plastic may not work on another (Chapter 1). Plastics have different cutting habits—some tend to be brittle, rubbery, and so on. Material suppliers and tool manufacturers can provide useful information about trimming.

In-line thermoforming production lines, particularly those being fed directly from an extruder, have to be completely synchronized, or inferior parts result, and/or the cost of the operation goes up. For example, if the trimmer operation has to be slowed down, the extruder output has to be reduced. In fact, a slowdown can result in a shutdown if the extruder cannot operate at the slower speed required. All functions and stations have to be properly interrelated.

OTHER FORMING METHODS

Certain thermoplastics, when formed, require handling normally not available with conventional thermoforming machines; so other processes of formation have evolved. Most of these methods tend to reduce the amount of heat required or even eliminate it entirely. One popular technique is high pressure forming, which is like conventional compression molding (Chapter 6). The techniques that are used modify conventional metal working tools. They can be classified as: (1) cold forming (performed at room temperature with unheated tools), (2) solid-phase forming (plastic is heated below the melting point and formed), and (3) compression molding of reinforced/composite sheets (heat is used). Other methods so used are classified as forging (which includes closed-die forming, open-die forming, cold pressing, etc.), stamping, rubber pad or diaphragm forming, fluid forming, coining, spinning, explosive forming, and so on.

Cold forming and solid-phase forming include the use of ABS/PC, PC, conventional PP, and HMWHDPE. By using solid-phase forming, processors can make more efficient use of ultrahigh molecular weight, high density plastics that are difficult or impossible to process by other methods (Chapters 2, 3, and 4). Forming by these techniques can usually use existing metalworking equipment with minor modifications. Tooling is inexpensive, and production rates can be high. Flash, trim, or weld lines can be eliminated by using some of these processes.

Thermoplastic composites can be stamped to produce high performance parts. Fiber reinforcements can be used, including glass, graphite, aramid, and so on, in different patterns (short fibers, woven, etc.; see Chapter 7). Products are molded in quick, high-productivity processes, using less energy than that needed to manufacture comparable aluminum and steel parts. Tooling costs decrease because of part consolidation. Stamping involves two very different forming processes: solid-state forming and flow molding (or fast compression molding). Each has its advantages and disadvantages.

Solid-state forming basically uses a male metal plug mold that matches a female metal cavity mold and can be used only with crystalline-type resins. Below their glass transition temperatures (T_g) amorphous-type resins are generally too stiff to be rapidly formed into stable products. Crystalline types can be permanently deformed at temperatures between their T_g and melting point (Chapter 1). Molecular orientation, the mechanism that allows this to occur, relates to the draw ratio. Draw ratios can vary from 5:1 for PET and nylon to 10:1 for low molecular weight PP.

The major advantage of solid-state forming is that parts can be produced in very fast cycle times, usually 10 to 20 s. The surface finish of these parts is rather smooth, as the fibers do not surface.

Flow molding is not limited to crystalline types because the resin is melted prior to forming. The forming temperature is usually lower than those of IM or extrusion. Plastics need not be trimmed, as the composite is "compression"-molded to completely fill the mold cavity. Most important, flow molding permits more complex parts to be formed than solid-state forming. The process cycle time is usually about a minute, which is faster than the time needed for most thermoset composites.

The surface is molten during forming, so the surface finish tends to have a fibrous finish. Fiber separation could occur for extremely complex parts. The use of braided woven fabrics and continuous fiber mat reinforcements practically eliminates separation (Chapter 7). Discontinuous fiber reinforced composites, such as those made by the slurry process, can be molded into complex shapes without separation.

PNEUMATIC CONTROL

This review on vacuum thermoforming can be related to most of the other forming processes. With a vacuum system a sheet is subjected to heat to meet its optional processing temperature, or technique that forces it against the shape of a mold. The hot, pliable material is moved rapidly to the mold (for example, by gear drives) and/or is moved by an air pressure differential, which holds it in place as it cools. When the proper set temperature is reached, the formed part can be removed and still retain its shape.

Two important needs in this cycle are to sustain the pressure and to maintain uniform heating of the plastic. Generally, the faster a vacuum is created, the higher the part quality. It is important that the mold be at the proper heat so the fast vacuum will produce a part with no internal stress (or very little). During forming the vacuum gauge should never fall below 20 in. of mercury (Hg), which at sea level is 9.82 psi of atmospheric pressure on the part. As a TP cools, this pressure cannot provide sufficient force to form the part and will not hold the plastic tight against the mold (Table 5-1).

A vacuum under 20 in. Hg is not satisfactory; at least 25 in. Hg is required. For proper pressure regulation, a vacuum storage or surge tank is necesary to retain a minimal even vacuum. For long forming cycles, a surge tank will permit the use of a smaller vacuum pump than would otherwise be required. To determine the vacuum surge tank size in cubic feet, use the following formula (229):

$$V_o \times P_o + V_m \times P_m = V_1 \times P_1$$

where:

V_o = Surge tank volume, including piping to vacuum control valve
V_m = Mold area volume
V_1 = $V_o + V_m$
P_o = Absolute pressure in surge tank (0.5 psi)
P_m = Initial pressure in the mold (at sea level 14.7 psi; or with prestretched forming, use 17.7 psi)
P_1 = Desired atmospheric working pressure

Table 5-1. Vacuum Pressure Measurements.

PRESSURE, PSI		
GAUGE*	ABSOLUTE**	PRESSURE, IN. HG
0.0	14.7	0.0
−1.0	13.7	2.04
−2.0	12.7	4.07
−4.0	10.7	8.14
−6.0	8.7	12.20
−8.0	6.7	16.30
−9.0	5.7	18.32
−9.9	4.9	20.00
−10.0	4.7	20.36
−11.0	3.7	22.40
−12.0	2.7	24.43
−12.3	2.4	25.00
−13.0	2.7	26.47
−13.7	1.0	27.89
−14.0	0.7	28.50
−14.2	0.5	28.91
−14.3	0.4	29.00
−14.6	0.1	29.73
−14.7	0.0	29.92

*Amount of pressure exceeding atmospheric pressure.
**Measured with respect to zero (absolute) vacuum; in a vacuum system, absolute pressure (psia) is equal to the negative gauge pressure (psig) subtracted from the atmospheric pressure.

In an example where the volume of the mold and piping is 4 cu ft, the vacuum pump can pull about 29 in. Hg, so the surge tank pressure is 0.5 psi. The desired working pressure is 2.42 psi in the tank, and the initial mold pressure is 14.7 psi. Thus:

$$V_o \times 0.5 + 4 \times 14.7 = (V_o + 4) \times 2.42$$
$$0.5\, V_o + 58.8 = 2.42\, V_o + 9.68$$
$$V_o = 25.58 \text{ cu ft (191 gal or 723 L)}$$

When a lower pressure of 20 in. Hg is used, which is 4.88 psi in the tank:

$$V_o \times 0.5 + 4 \times 14.7 = (V_o + 4) \times 4.88$$
$$0.5\, V_o + 58.8 = 4.88\, V_o + 19.52$$
$$V_o = 8.97 \text{ cu ft (671 gal or 2540 L)}$$

In thermoforming it is sometimes necessary to prestretch (or preblow) the hot sheet before final forming. Usually 3 to 5 psi compressed air is used, which results in a greater amount of air being at atmospheric pressure than in the processing of nonprestretched parts. In the above formula add the volume of the prestretched bubble to the volume of the mold and the pressure differential needed for blowing the bubble to the initial atmospheric pressure in the mold.

The objective is to have the vacuum surge tank as close as possible to the forming station and the vacuum control valve. Use of flexible vacuum hose with connections eliminates elbows, tees, and tubing reducers. All valves must be capable of operation at the full open position. To utilize fully the rapid vacuum capability provided by the surge tank, the mold must be able to take advantage of all vacuum pressure available. Vacuum holes should be drilled as large as possible, and a maximum number should be used.

Back drilling of large holes (to 0.125 in.) on the underside can be used when smaller holes are required on the part side. Male molds can be mounted on a vacuum plate with thin washers or shims, and large vacuum holes can be drilled under the mold. Narrow slots also can be used, and they offer much less resistance than holes when air is evacuated through the mold. Flat areas, segmented sections, or male portions of a mold can be joined with shims, providing long slots.

TEMPERATURE CONTROL

Even though TPs have specific processing heats, forming requires thorough, fast, and uniform radiant heat from the surface to the core to the surface. To achieve these conditions sheets, plastics over 0.040 in. (1.02

mm) should use sandwich-type (bottom and top) heater banks. To ensure that sufficient heat is used, heaters should have capacities of at least 4 to 6 kW/sq ft. Various type of radiant heating elements and their performances are shown in Table 5–2.

The cycle time is controlled by the heating and cooling rates, which in turn depend on the following factors: the temperature of heaters and the cooling medium, the initial temperature of the sheet, the effective heat transfer coefficient (Table 1–6), the sheet thickness, and thermal properties of the sheet material. Different materials absorb radiant heat most efficiently at various wavelengths, which in turn are affected by the temperature of the emitting heater. The most appropriate wavelengths for TPs fall within the infrared spectrum of 6 μm (400°F) to 3.2 μm (1,200°F). For example, ABS, PE, and HIPS absorb radiant heat most efficiently when the heating elements emit 3.5 to 3.3 μm, whereas PC requires 3.4 μm.

Typical material and process heats for a variety of plastics are given in Table 5–3. The normal forming heat should be attained throughout the sheet, and should be measured just before the mold and sheet come together. Shallow draw projects with fast vacuum and/or pressure forming allow somewhat lower sheet heats and thus a faster cycle. Slightly higher

Table 5–2. Types of Radiant Heating Elements.*

	EFFICIENCY, PERCENT			
ELEMENT	WHEN NEW	AFTER 6 MONTHS	AVERAGE LIFE, HOURS	PERFORMANCE
Ceramic panel	65	55	12,000–15,000	Best buy; heats uniformly and is efficient and capable for profiling heat.
Quartz panel	58	50	8,000–10,000	Same as ceramic heaters.
Coiled nichrome wire	18 to 20	8 to 10	1,500	Initially lowest cost; is very inefficient, and heats nonuniformly with use.
Tubular rods**	45	20	3,000	Inexpensive; heats nonuniformly with use and is difficult to screen or mask for profiling heat.
Gas-fired infrared	40 to 45	25	5,000–6,000	Lowest cost to operate; has many disadvantages including wavelength variations and frequent maintenance.

*Steel clamping frames should be nickel–copper–chrome-plated to reflect heat to sheet edges. After 6 months' use, consider replacing side and back reflectors in order to regain 4 to 8 percent efficiency.
**Sanding and polishing oxidized tubular heaters can improve their efficiency 10 to 15 percent.

Table 5–3. Guide to Thermoforming Processing Temperatures (°F).

PLASTIC	MOLD HEAT (1)	LOWER PROCESSING LIMIT (2)	NORMAL FORMING HEAT (3)	UPPER LIMIT (4)	SET HEAT (5)
HDPE	160	260	295	330	180
ABS	180	260	325	380	200
PMMA	190	300	350	380	200
PS	185	260	295	360	200
PC	265	335	375	400	280
PVC	140	210	275	300	160
PSU	320	390	475	575	360

(1) The mold temperature is important in the forming process. High mold heats provide high-quality parts with high impact strength, low internal stress, and good detail, material distribution, and optics (clarity and lack of distortion). However, thin gauge materials frequently can be thermoformed on molds at lower heat, such as 35–90°F, as the additional stresses produced are not pronounced in the thin gauges and do not interfere with product performance.

(2) The lower processing limit represents the lowest heat at which the sheet can be formed without undue stresses. This means that the sheet should touch every corner of the mold prior to reaching this lower limit; otherwise problems develop such as stresses/strains that can cause warpage, brittleness, or other physical changes in the part.

(3) The normal forming temperature is the heat at which the sheet should be formed under normal operation. This temeprature should be reached throughout the sheet. Shallow draws with fast vacuum and/or pressure forming will allow somewhat lower sheet heat and thus a faster cycle. Higher heats are required for deep draws, prestretching, detailed mold decorations, etc.

(4) The upper limit is the heat point where the sheet begins to degrade or becomes too fluid and pliable to form. These temperatures normally can be exceeded only with an impairment of the plastic's physical properties (higher heats obtain for IM and extrusion).

(5) The set temperature is the heat at which the part may be removed from the mold without warpage. Sometimes parts can be removed at higher heats if postcooling fixtures are used.

heats may be required for deep draws, prestretching, and highly detailed molds.

When extrusion and thermoforming are separate operations, the heat energy supplied for extrusion is completely lost by chilling the sheet. Reheating for thermoforming requires additional heat energy. The in-line process offers the advantage of using a high percentage of the energy contained in the sheet to condition it to the forming heat. Actually savings of about 30 to 40 percent can be obtained. The in-line process provides a more even heat distribution, and weight distributions can be reduced without changing physical properties. At equal output rates, an in-line process needs only half the floor space of separate operations.

MOLDS

Molds can range from hardwood for short runs to filled and unfilled high temperature polyester (TS) and epoxy resins, cast solid urethane, sprayed metal, cast aluminum, cast porous aluminum, and machined steels. The most common material is cast aluminum, which provides a good combination of durability, light weight, thermal conductivity, ease of manufacture, and cost.

In tooling design, a male primary mold will allow a deeper draw than a female mold because the plastic can be draped or prestretched over the male mold. However, when a male plug assist is used to prestretch the sheet for a primary female mold, the advantage is nullified. In general, female molds provide easier release, are less likely to get scratched or damaged, produce thicker and stronger rims in containers, can use smaller sheet blanks, and provide the sharpest definition on the outside of the part. Usually, female molds have the disadvantage of producing parts with thin bottoms; however, good plug-assist design and operation can largely eliminate this problem. Male molds are generally lower in cost.

As reviewed, molds used with vacuum or pressure techniques require holes, channels, slits, ducts, and so on, for the evacuation of air or the buildup of pressure. To avoid visible marks on the surface of thermoformed parts, holes should be kept as small as possible, such as 0.010 to 0.025 in. Careful placement of the holes will be helpful in providing fast, efficient air flow during forming. Logic and experience provide guidelines for the placement of openings.

In cast resin molds, vacuum holes can be provided by including greased wires in the casting for later removal. For greater detail such as graining, stitching, and relief work, cast porous aluminum molds should be considered (also used in blow molding molds).

Undercuts can be included by the use of split molds. Some molds use a removable section that pulls out of the mold after forming (or leave it in, if desired, to provide a threaded insert, etc.). In the design of all molds one should consider at least a 2 to 3 degree draft angle per side for the female molds and a 5 to 7 degree angle for male molds (the more draft, the better). A straight-side angle in the direction of the draw makes the parts difficult to release. This is especially true with male molds, where the natural shrink is toward the mold. With advanced forming techniques, such as collapsible molds, parts with zero degrees of draft or even negative drafts can be successfully formed.

Various sheet materials have different mold shrinkage factors, ranging from almost no shrinkage up to as much as $3\frac{1}{2}$ percent. Typical basic shrinkage values are given in Table 5-4. Shrinkage can be changed significantly when additives/fillers are used in the resin blends; it can go from zero to practically any preengineered value. However, the percentage of shrinkage is not as important as the consistency of the factor. Molds can be designed to allow for the shrinkage. For precision parts, careful pretesting is required.

Cooling conditions also affect the rate of shrinkage. Restraining the part, either before or after release, will tend to limit the total shrinkage. The mold heat, cooling speed, and cooling fixtures should remain constant to ensure the uniformity of final part shapes. About 70 to 80 percent of the dimensional change due to shrinkage occurs as the sheet cools from its forming heat to its set heat. Stabilization to the final dimension can take several hours, or even longer. Most of the change may be due to plastic relaxation once forming stresses are removed.

MATERIALS

Practically all TPs can be formed, including TSs. (Although glass fiber reinforced B-staged TS polyester composites have been thermoformed since 1943, TSs are rarely formed.) Plastics (TPs) have certain special characteristics that make good thermoformable materials better than other materials. They should be able to transfer the heating and cooling action quickly and uniformly, to a deep draw, to permit complex contours, to have tear resistance, and to provide a degree of plastic memory. Basically, these materials exhibit hot melt strength and hot melt elongation. Standard tests can be conducted, such as tensile strength and elongation at various heats, to provide comparisons of their formability (Chapter 10).

Some plastic sheets stretch as much as 600 percent and others as little as 15 percent. This behavior directly influences what shapes can be formed and their quality. Those with a puttylike appearance respond to very little pressure; others, which tend to be stiff, require heavier operating equipment. The pressure response is somewhat related to the ability to be stretched while hot, but does not parallel it.

Table 5–4. Shrinkage Guide for Thermoformed Plastics.

PLASTICS	SHRINKAGE, PERCENT
LDPE	1.6–3.0
HDPE	3.0–3.5
ABS	0.3–0.8
PMMA	0.2–0.8
SAN	0.5–0.6
PC	0.5–0.8
PS	0.3–0.5
PP	1.5–2.2
PVC—rigid	0.4–0.5
PVC—flexible	0.8–2.5

The most useful formable TPs do not have sharp melting points (Chapter 1). Their softening with increasing heat is gradual. Each material has its own range of heat, wide or narrow, within which it can be effectively formed. Thus, one plastic may have a forming heat of 275 to 400°F, whereas another may become soft enough for forming at 350°F but melt at 400°F. Also a plastic may stretch well at a given heat but tear easily if heated a few degrees higher or cooled a few degrees. This single property is one of the most important of all the factors involved in forming.

Examples of different behaviors with all formable resins using film (< 0.010 in.) are: PS is unstable with heat and requires extra cooling; PVC and PVDC are excellent, with no restrictions; nylon is difficult; PCTFE is sensitive to heat and pressure fluctuations; HDPE is difficult without a support film; and PP has a very narrow heat range. In fact, PP is extremely unstable within the conventional forming-heat range, so different processing techniques are used for it. Conventional PP has the major deficiency of lacking a rubbery plateau region at the forming heat; it just sags and falls apart. Shell Chemical developed a process to form PP just below its softening point so that no sag problem can occur. PP is forced into the desired shape by the use of mechanical plugs and pressure, in a process known as solid-phase pressure forming (SPPF).

Exxon Chemical researchers changed their PP to overcome its deficiencies. They developed a proprietary catalyst and reactor technology to extrude thermoformable sheet and film. Their material has a rubbery plateau region and a high dynamic modulus so that it is processable in conventional thermoforming machines.

Other plastics that, like PP, are not formable (or easy to form) have undergone similar changes. These plastics inherently have performance and cost characteristics that make them desirable materials for forming. An example that has involved large production quantities is PET from Goodyear Tire and Rubber. To make it formable, researchers produced CPET (crystallized PET). One major user of thermoformed CPET is Campbell Soup Co., where dual-use CPET trays (for conventional ovens and microwaves) replaced aluminum TV dinner trays.

POSTFORMING

A popular forming technique that has provided both performance and cost advantages, principally for long production runs, is applied as the plastic sheet, film, or profile exits an extruder. Upon leaving the die, and retaining heat, the plastic is continuously postformed.

With this type of in-line system, the hot plastic is reduced only to the desired heat of forming. All it may require is a fixed distance from the die

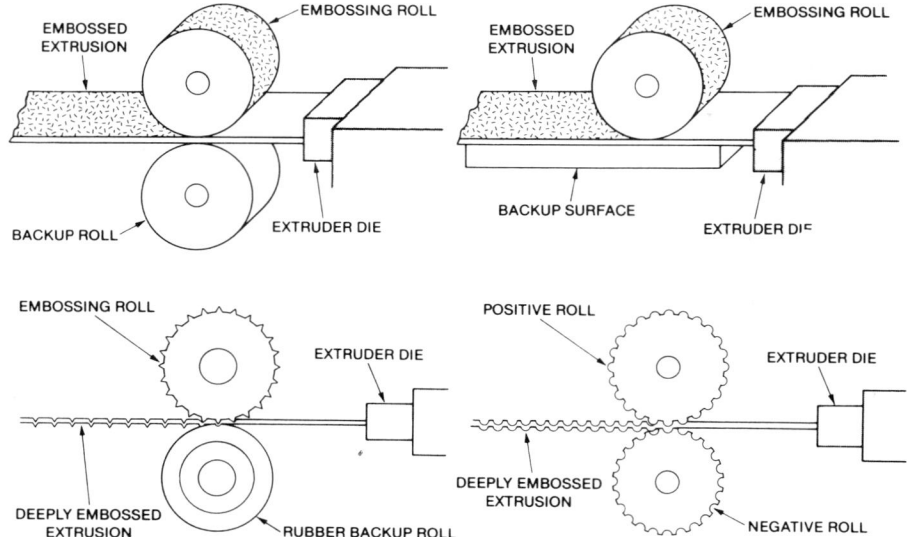

Fig. 5-16. In-line postforming with extruder: embossing one or both sides with shallow or deep patterns.

opening. Cooling can be accelerated with blown air, a water spray, a water bath, or their combinations. Examples of postforming techniques are shown in Figs. 5-16 to 5-20. This equipment, like others, requires precision tooling with perfect registration.

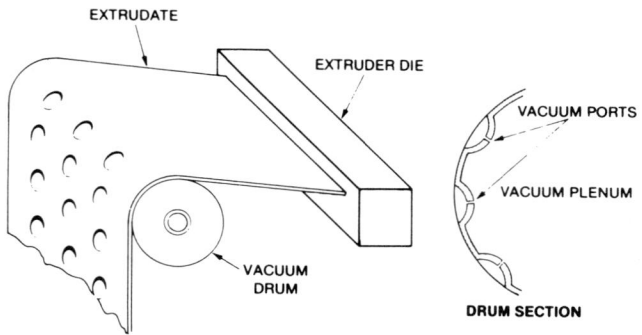

Fig. 5-17. In-line vacuum forming embossing roll with water-cooled temperature control.

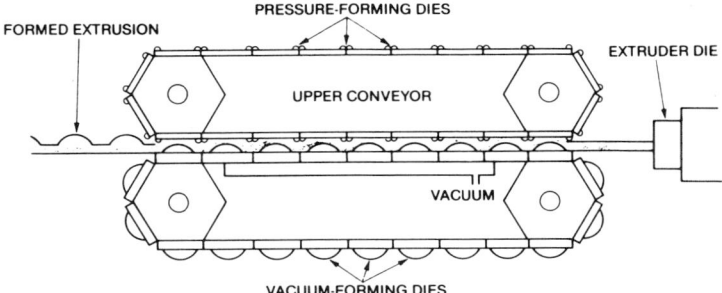

Fig. 5-18. In-line vacuum/pressure former for plastic sheet with matched, water-cooled, forming molds on a continuous conveyor system. This system can be used with different profiles, such as small or large extruded tubes producing corrugated tube/pipe. Moving molds would use corrugated tubular cavities with vacuum/pressure/water cooling lines.

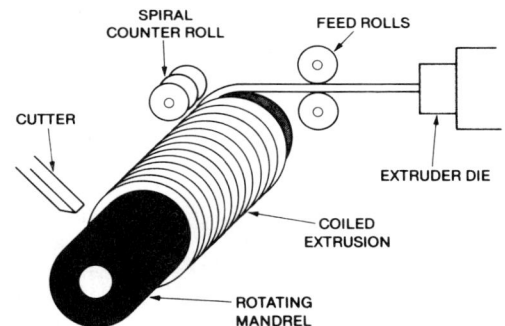

Fig. 5-19. In-line coil former, which can produce telephone cords, springs, and so on, using extruded round, square, hexagonal, and other shapes.

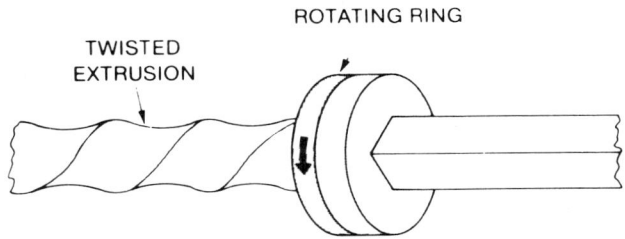

Fig. 5-20. In-line fixed/rotating rings used to twist extrudate.

Table 5–5. Troubleshooting Guide for Thermoforming.

PROBLEM	CAUSE(S)	SOLUTION(S)
Blisters or bubbles	Overheating	Lower the heater temperature
		Increase distance of heater from sheet
		Attach masks or baffles
	Wrong sheet type or formulation	Obtain correct formulation
	Poor storage conditions	Do not remove material from moistureproof wrap until ready to use
Blush or change in color intensity	Insufficient heating	Lengthen heating cycle
	Mold is too cool	Warm the mold
	Assist is too cool	Warm the assist
	Sheet cools before it is completely formed	Speed the drape action
		Add vacuum holes
	Too deep a draw	Use heavier-gauge sheet
	Poor mold design	Use mold of proper design
Sticking to the mold	Rough or improperly designed mold	Make mold smoother
		Increase the taper on male plugs
		Use mechanical release assists
		Use air pressure to blow piece from mold
		Use mold-release agents
Incompletely or improperly formed pieces	Sheet is cold	Lengthen the heating cycle
		Bring heater closer to sheet
	Insufficient vacuum	Check vacuum system
	Vacuum holes are plugged up	Clean, relocate, or add vacuum holes
Warped or distorted pieces	Poor mold design	Redesign mold using proper tapers and ribs
	Sheet removed while too hot	Increase cooling cycle
		Use water-cooled molds
Webbing or bridging	Insufficient vacuum	Check vacuum system
		Add more vacuum holes
	Sheet is overheated	Shorten heating cycle
		Increase heater distance from the sheet
	Long parallel molds with extrusion direction parallel	Move sheet 90° in relation to mold
	Poor mold layout or design	Use mechanical drape or plug assists
	Sharp corners on deep draw	Increase radius
Bad surface markings	Markoff (due to trapped air)	Slow draping action
		Add more vacuum holes
	Markoff (due to accumulation of plasticizer on mold)	Use a temperature-controlled mold

Table 5-5.

PROBLEM	CAUSE(S)	SOLUTION(S)
		Have mold as far away from the sheet as possible during the heating cycle
		Shorten the heating cycle (if too long)
		Wipe the mold
	Mold is cool	Warm the mold
		Bring the heater closer to the sheet
	Mold is too hot	Provide cooling for mold
	Improper mold composition	Avoid phenolic molds with clear transparent sheet
	Mold surface too highly polished	Remove high surface gloss from mold
	Mold surface too rough	Smooth surface
Excessive post shrinkage	Sheet removed from mold while still hot	Increase cooling time
Pinholing or rupturing	Vacuum holes too large	Partially plug up holes with wood or solder or completely plug and redrill
	Uneven heating	Attach baffles to the top clamping frame

TROUBLESHOOTING

Like other processes, forming is subject to many variables that influence appearance, performance, and cost. All the variables are controllable, and logical steps can be taken to manage them, as reviewed in Chapters 2 through 4.

Major influences are sheet thickness, plastic viscosity, and melt index (Chapter 1), regrind (Chapter 9), sheet orientation (Chapter 1), draw ratio, forming temperature and pressure, and surface blemishes, blisters, blushing, scratch marks, and so on. A guide to troubleshooting the thermoforming process is given in Table 5-5.

Chapter 6

COMPRESSION AND TRANSFER MOLDING

INTRODUCTION

Compression and transfer molding (CM and TM) are the two main methods used to produce molded parts from thermoset (TS) resins. Compression molding (CM) was the major method of processing plastics during the first half of this century because of the development of a phenolic resin (TS) in 1909 and its extensive use at that time. By the 1940s this situation began to change with the development and use of thermoplastics (TPs) in extrusion and injection molding (IM) processes. CM originally processed about 70 percent (by weight) of all plastics, but by the 1950s its share of total production was below 25 percent, and now that figure is about 3 percent. This change does not mean that CM is not a viable process; it just does not provide the much lower cost to performance of TPs, particularly at high production rates. In the early 1900s resins were almost entirely TS (95 percent by weight); that proportion had fallen to about 40 percent by the mid-1940s, and now is about 3 percent.

During this century, TSs experienced an extremely low total growth rate, whereas TPs expanded at an unbelievably high rate. Regardless of the present situation, CM and TM are still important, particularly in the production of certain low-cost parts as well as heat-resistant and dimensionally precise parts. CM and TM are classified as high pressure processes, requiring 2,000 to 10,000 psi molding pressures (Fig. 6-1). Some TSs, however, require only lower pressures of down to 50 psi or even just contact (zero pressure). These low pressure systems will be reviewed in Chapter 7.

CM is the most common method of molding TSs. In this process, material is compressed into the desired shape using a press containing a two-part closed mold, and is cured with heat and pressure. This process is not generally used with TPs (1–7, 11–14, 30–37, 230–235).

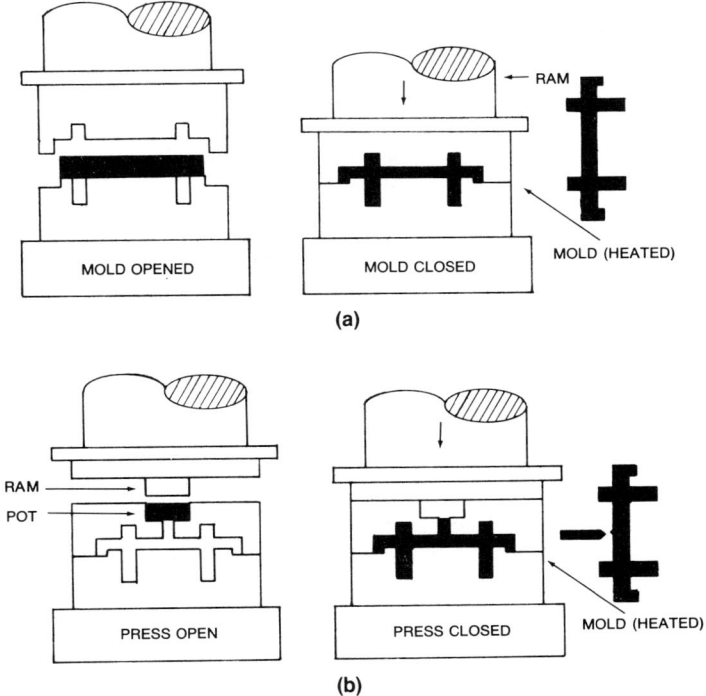

Fig. 6-1. Schematics of (a) compression and (b) transfer molding.

THERMOSET COMPOUNDS

A TS resin is one that is capable of being changed into a substantially infusible part when cured by the application of heat and/or chemical means. Curing, as the term is used by the polymer chemist, is the carrying out of a variety of chemical reactions whose purpose is to bring about certain changes in the physical properties of a polymer. These changes ordinarily include an increase in hardness and strength and a decrease in fluidity (Chapter 1).

Curing may be brought about by heat, irradiation, and/or a curing agent, which may be either a reactive curing agent (including vulcanizing agents and molecular chain extenders) or a catalyst. To control the storage life of TS compounds and control the speed of chemical reaction during the curing cycle, different inhibitors, modifiers, stabilizers, accelerators, activators, and so on, may also be present. The physical changes associated with curing are generally the result of cross-linking between plastic molecules (Chapter 1). The chemical composition of TSs, as produced by resin suppliers, depends on the molder's performance requirements and the types of fillers and/or reinforcements to be included. A wide variety of TSs are available, to meet extensive performance requirements; there are many grades, with resins designed to be: chemically resistant, self-extinguishing, electrical grade, heat-resistant, weather-resistant, light-stabilized, and so on.

Fillers/Reinforcements

Most resins that are processed require some type of fillers and/or reinforcements in order to function. Typical fillers and reinforcements are listed in Table 6–1, which shows the properties they influence. (See Chapter 7 for details on reinforced plastics.) Basically fillers and pigments are added in molding resins to reduce shrinkage, minimize crazing, lower the material cost, impart color or opacity, improve the surface finish, and so on. The most commonly used fillers are calcium carbonates, diatomaceous earths, and clays. The excessive use of fillers should be avoided, as the fluidity of the resin can be reduced to the point where it is difficult to process.

The use of physical and chemical thickening agents in a resin system may be desirable for specific requirements, producing a condition known as thixotropy—the property of certain colloidal gels to coagulate (becoming stiff and jellylike) when at rest, but to become fluid when agitated or otherwise subjected to stress. The practical application of thixotropic agents is to permit the use of resin systems on vertical surfaces without excessive drain or runoff. The addition of colloidal silica (2 to 3 percent by weight) to a TS polyester resin will give it a greaselike consistency.

Most parts are easily released from the mold after completion of the curing cycle if the mold is pretreated with waxes or silicones to lubricate the surface, and/or a mold release agent (zinc stearate, paraffin wax, etc.) is added to the resin to facilitate the removal of parts. With certain release agents, either on the mold surface or in the mix, problems such as difficulties in decorating or bonding parts after molding can occur. This is particularly true with certain silicones, which can also cause corrosion or erosion of metallic and nonmetallic surfaces when used as inserts (electrical connector, etc.).

Types of Resins

Different TS resins undergo CM and TM, such as phenolics, polyesters (TSs), epoxies, ureas, melamines, and silicones, all with their own molding parameters and performance properties. There are two basic types of polyesters, namely TS and TP, the largest volume of the IM and extruded polyesters being TPs. In this chapter, however, the emphasis is on TS polyesters, which have completely different processing characteristics and properties.

With most TS resins there exists a wide range of flow characteristics, cure times, and ultimate mechanical, chemical, and electrical properties. These molding compounds are mixtures of constituents, of usually different size and shape; so the compounds themselves present the greatest number of variables that must be understood and properly applied. Knowledge of these resins is principally gained through molding experience.

From a processing point of view, the viscosity–time curve is often the most critical characteristic. Most compounds are granular at room temperature; and when exposed to higher curing heats, the granules melt and become fluid. Under continued heat, cross-linking occurs, and the plastic compound solidifies. As the material goes from solid to fluid to solid (Fig. 1–12), its viscosity changes accordingly. Generally, at some optimum heat for a given part in a specific mold, the viscosity of the fluid plastic will be at a minimum for the period of time needed to ensure filling of the cavities at an acceptable pressure. If the heat is too low or too high, the filling of the cavities could be problematic even if higher pressures were used. Ideally, proper pressure is applied when the lowest viscosity occurs, resulting in good surface finish and relatively stress-free parts.

The majority of TS compounds are heated to about 300°F for optimum cure. Higher heats could degrade their performance or could cause them to solidify rapidly, particularly in TM, where material could solidfy before the cavity completely filled. Lower heats extend the cycle time. The molds are heated by electricity, steam, or hot circulating heat-transfer fluids.

Preheating

As compounds generally have good heat insulation, preheating is often used to reduce the molding cycle; it can aid in providing even heat through the material and can cause a more rapid rise in heat than occurs in the mold. Preheating may be accomplished by a warm surface plate, infrared lamps, a hot air oven, screw/barrel preheater, or—by what usually is the best and quickest means of spreading heat through the material—high frequency (dielectric) heating. Usually preheating is done at 150 to 300°F, followed by quick transfer to a mold cavity. The actual heat depends on the material, the heater capability, and the speed of transfer. Circular preforms are used with dielectric heaters so they can be rotated to obtain uniform heating. Compressed compound "pills" are used to produce preforms to reduce the bulk factor, facilitate handling, and control the uniformity of charges for mold loading. Preforms can be of the shape desired in the mold cavity, and

Table 6–1a. Guide on the Use of Fillers and Reinforcements for Composites

PROPERTIES IMPROVED

FILLER OR REINFORCEMENT	CHEMICAL RESISTANCE	HEAT RESISTANCE	ELECTRICAL INSULATION	IMPACT STRENGTH	TENSILE STRENGTH	DIMENSIONAL STABILITY	STIFFNESS	HARDNESS	LUBRICITY	ELECTRICAL CONDUCTIVITY	THERMAL CONDUCTIVITY	MOISTURE RESISTANCE	PROCESSABILITY	RECOMMENDED FOR USE IN*
Alumina, tabular	•	•	•	•		•	•					•	•	S/P
Aluminum powder										•	•			S
Aramid	•	•	•	•	•	•	•	•	•				•	S/P
Bronze							•	•		•	•			S
Calcium carbonate	•	•	•	•		•	•	•					•	S/P
Carbon black		•				•	•			•	•		•	S/P
Carbon fiber				•		•	•			•	•			S
Cellulose				•		•	•	•						S/P
Alpha cellulose			•			•								S
Coal, powdered	•											•		S

236

Material	*Type
Cotton	S
Fibrous glass	S/P
Graphite	S/P
Jute	S
Kaolin	S/P
Mica	S/P
Molybdenum disulfide	P
Nylon	S/P
Orlon	S/P
Rayon	S
Silica, amorphous	S/P
Sisal fibers	S/P
Fluorocarbon	S/P
Talc	S/P
Wood flour	S

*P = thermoplastic, S = thermoset

237

Table 6-1b. Examples of Various Reinforcing Fibers and Fillers Used with Thermoset Resins.

THERMOSETS	ALUMINA	CALCIUM CARBONATE	CARBON BLACK	CLAY	COTTON FLOCK	GLASS BUBBLES	GLASS FIBERS	GRAPHITE	MICA	QUARTZ	TALC	WOOD FLOUR
Alkyds	•	•		•			•		•		•	•
Diallyl phthalate	•	•	•	•			•				•	
Epoxy	•	•	•	•		•	•	•				
Phenolic	•	•	•	•	•	•	•	•		•	•	
Polyester		•	•	•	•	•	•	•	•			•
Melamine					•	•	•		•			•
Urea				•		•	•				•	
Silicone	•			•			•	•		•		•
Urethane	•	•	•	•			•	•		•		

they can include different reinforcements, such as woven cloth, short fiber mats, and so on.

Outgassing

Many TSs, particularly phenolics, ureas, and melamines, emit a gas during curing that must be allowed to escape. (A major exception is polyesters.) Some of the gases can escape through clearances at the two mold halves and around ejection pins. To ensure that these gases do not become entrapped and weaken the part or cause surface blemishes, during the start of the cure cycle the mold is slightly opened (about 0.001–0.005 in.) to allow the gases to escape. This action is called breathing or bumping. If the cure cycle permits, it is best to repeat this bumping cycle.

The length of time for single or multiple bumping depends on the material used and the size and configuration of the part. This dwell time with release of gases can actually reduce the total cure cycle time. Outgassing also can occur after parts are molded. With certain materials this condition can last for months or years. If parts are to be metallized or plated, outgassing tendencies can actually rupture the plated film.

Postcuring

TS molded parts frequently are postcured by postbaking at a material supplier's recommended times and heats, to enhance mechanical and thermal properties and dimensional stability and eliminate outgassing. Improved creep resistance and reduction in stresses also result. This postcuring is also used with certain TPs after IM or extrusion to improve their performance.

The postcuring heat is usually below the actual molding heat. Postcuring should be done with a multistage heat cycle such as is used for glass fiber and/or mineral reinforced phenolic compounds: (1) for parts $\frac{1}{8}$ in. or less in thickness, times are 2 h at 280°F, 4 h at 330°F, and 4 h at 375°F; and (2) for parts exceeding $\frac{1}{8}$ in. thickness, the time periods are doubled for each $\frac{1}{16}$ in. of thickness. The use of longer times is more effective than simply increasing the heat.

The reinforcement system of the compound will, to a large extent, dictate the heat/time cycles. Parts molded from compounds using organic reinforcements are postcured at lower heats than those using glass and mineral reinforcements. Parts of uneven thickness will exhibit uneven shrinkage. This shrinkage effect is included in the mold design.

COMPRESSION MOLDING

CM machines can be discussed in three major categories: platen frame, driving means, and controls. Two or more platens can be used so that molds are placed between platens. Most of these CM (and TM) machines have the platens moving vertically to simplify locating and feeding the molds. As in other press operations (IM, blow molding, etc.), the platens and their supporting frames must have sufficient strength to minimize deflections. Beware of "lightweight" machines; there may not be enough steel to contain molds so that they do not bend, twist, and so forth.

The driving means to move platens and apply clamping pressures is usually a double acting hydraulic ram; that is, it uses high pressure oil, both to bring the platens together and also to separate them. Electric-motor-driven pumps provide 2,000 to 3,000 psi oil pressure. The driving means consists of an oil reservoir, valving, and so on, resembling that of IM machines. Most CM is performed in machines that only provide relatively constant, high clamping pressure. For better operation, the processor should consider those that provide variable pressure loading similar to what is used in IM. The result is significant improvement in melt flow during curing. The reader should see Chapter 2 for information on machine construction/operation and maintenance, particularly for oil filtration systems, that can be used to improve CM (and TM).

Machine controls, as with IM, include manual, semiautomatic, and fully automatic types. (See Chapter 2 regarding the operation of controls.) Unfortunately, most CM machines operating in the United States are manual. There is plenty of opportunity to apply control systems to improve productivity output as well as product quality.

TRANSFER MOLDING

In TM, the mold halves are brought together under pressure, as in CM. The charge of molding compound is then put into a pot, and is driven from the pot through runners and gates into the mold cavities by means of a plunger. Its basic construction, driving system, and controls are similar to those of CM except for the additional action required by the transfer pot (Fig. 6–1b).

The process differs from CM in that the plastic is heated to a point of plasticity in the pot before it reaches and is forced into the closed mold. This procedure facilitates the molding of intricate parts with small holes, numerous metal inserts, and so on. Less force is used, and less melt action occurs in the cavity (Fig. 6–2 and Table 6–2).

Automation of the transfer/plunger concept is accomplished by the addition of a hopper-fed screw plasticator, in a system called screw transfer, which can replace the preform and preheat operations and automatically load the pot in CM (and is used to load a CM mold cavity). Conventional screw injection molding (IM) is another so-called transfer system (Chapter

Fig. 6-2. Front view of a 64-capacity (integrated circuits) mold, showing layout, top-transfer pot, and plunger used in a transfer molding system. (Courtesy of Hull Corp.)

2). As explained in the IM chapter, for TSs the processor uses a screw compression ratio of one, the barrel heat is kept relatively low (below the curing/hardening heat), and the mold is at a higher heat to permit final cure.

MOLDING

In the process of loading material into a compression mold, the charge of material to be molded is somewhat smaller in length and width but is thicker than the final part and is loaded near the geometric center. The closing of the mold spreads the preform to fill out the cavity. Instead of preforms, loose granules can be used. With SMC (sheet molding compound—see Chapter 7), the material is cut to a predetermined size and weight and then placed on the lower half of an open mold in a specific arrangement called the charge pattern.

Placing material in an open mold rather than injecting it into a closed

Table 6–2. Comparison of Compression and Transfer Molding.

CHARACTERISTIC	COMPRESSION	TRANSFER
Loading the mold	1. Powder or preforms. 2. Mold open at time of loading. 3. Material positioned for optimum flow.	1. Mold closed at time of loading. 2. RF heated preforms placed in plunger well.
Material temperature before molding	1. Cold powder or preforms. 2. RF heated preforms to 220–280°F.	RF heated preforms to 220–280°F.
Molding temperature	1. One step closures—350–450°F. 2. Others—290–390°F.	290–360°F.
Pressure via clamp	1. 2,000–10,000 psi (3,000 optimum on part). 2. Add 700 psi for each inch of part depth.	1. Plunger ram—6,000–10,000 psi. 2. Clamping ram—minimum tonnage should be 75% of load applied by plunger ram on mold.
Pressure in cavity	Equal to clamp pressure	Very low to maximum of 1,000 psi
Breathing the mold	Frequently used to eliminate gas and reduce cure time	1. Neither practical nor necessary. 2. Accomplished by proper venting.
Cure time (time pressure is being applied on mold)	30–300 s—will vary with mass of material, thickness of part, and preheating.	45–90 s—will vary with part geometry.
Size of pieces moldable	Limited only by press capacity.	About 1 lb maximum.
Use of inserts	Limited—inserts apt to be lifted out of position or deformed by closing.	Unlimited—complicated. Inserts readily accommodated.
Tolerances on finished products	1. Fair to good—depends on mold construction and direction of molding. 2. Flash—poorest, positive—best, semi-positive—intermediate.	Good—close tolerances easier to hold.
Shrinkage	Least.	1. Greater than compression. 2. Shrinkage across line of flow is less than with line of flow.

mold, as in transfer or injection molding, helps to preserve the integrity of SMC reinforcing fibers. For this reason, SMC parts usually have better strength when made in an open mold. The pattern used to mold a part can have a critical effect on both quality and performance.

A key feature of compression molds for SMC is a telescoping shear edge, which is located around the periphery of the mold cavity (Fig. 6–3). Its function is to seal off the mold when closed and to vent air and gases from the mold cavity. The shear edge also allows the cavity half of the mold to slide over the core half. Because of this feature, full molding pressure can be applied to each part and the plastic retained within the cavity. The closed position of the mold can vary, depending on the charge volume and the pressure on the part. This variance in turn causes an equal variation in the thickness of the molded SMC part; so it is very important to control the SMC charge size (thickness, length, and width), weight, and volume. One should use the same approach with other materials, such as granule and BMC (bulk molding compound; see Chapter 7).

Venting

All closed molds, as in TM, IM, blow molding, and so on, require a means by which air and volatiles (for certain TSs and TPs) are evacuated from the cavities. As reviewed above, bumping is used, but the usual technique is to

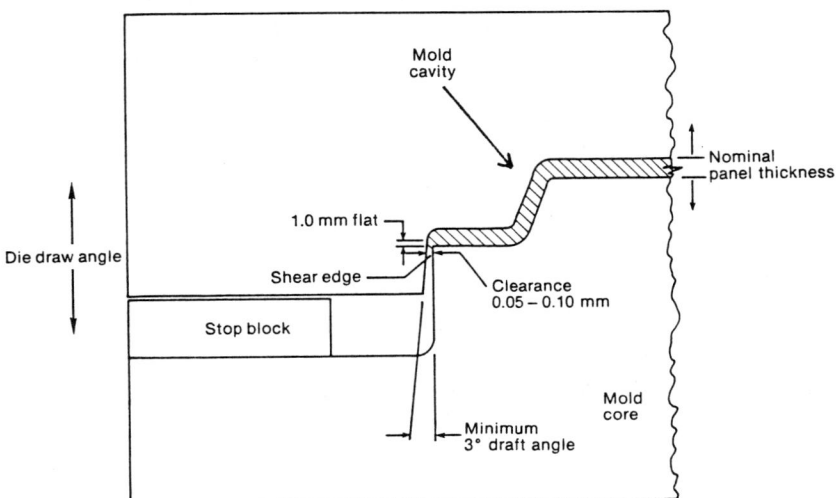

Fig. 6–3. Schematic of closed mold using a shear edge around the cavity periphery.

incorporate "openings," usually located at the mold parting line. Their size depends on the plastic compound's viscosity, but they are usually 0.25 in. wide × 0.001 to 0.003 in. deep, located where the cavity will be filled last. The vent-opening location also depends on the heating pattern of the mold, particularly if the heat flow pattern was not logically planned. CAD programs are available to provide the proper heating pattern. With excessive heat in one section of the mold, the viscosity of the compound could be low enough to require a vent opening in that area. Knockout pins often provide a means for venting, and they may require recessed sections, such as "flats" ground on the OD, that will allow venting.

Flash

During the heat curing cycle, the low viscosity of TSs tends to cause flash, particularly in CM. Basically there is no flash with TPs, as viscosity is not so low, and closed molds are used. The inherently more costly removal of flash from the parting line and holes with CM, as well as slower molding cycles, difficulty in molding side holes or sections, and problems in molding flash-free metal inserts, have all helped to reduce the use of CM in favor of closed-mold TM and IM.

Flash removal is most often accomplished with tumbling machines, where, as the name implies, parts are tumbled against each other to break off the flash. The simplest tumblers are merely wire baskets driven by a small electric motor and pulley belt. Tumblers also can be very elaborate, involving not only the tumbling operation, but often "blasting" with granulated peach or apricot pits, walnut shells, plastic pellets, and so on, to provide additional flash removal. This type of action also can provide a means of improving surface polish and/or toughness for certain molding compounds. A steam jet sometimes is used to minimize the accumulation of static charges. Tumblers can be of the batch type or provide continuous movement of parts to more accurately control the time of flash removal on each part.

Pressure

The curing pressures required for TSs depends on the formulation of the compound. With CM, the cavity depth can have a significant effect on pressure, as well as on the benefits of using preheated rather than cold material delivered into the cavity (Table 6–3).

Cavity Plating

TS molds are usually hard-chrome-plated on areas that are exposed to the molding compound during the molding process, such as the plunger, pot,

Table 6-3. Examples of Pressure Required to Compression Mold Based on Cavity Depth using Phenolic Molding Compounds (psi Pressure).

DEPTH OF MOLDING, IN.	CONVENTIONAL PHENOLIC		LOW PRESSURE PHENOLIC	
	DIELECTRIC PREHEAT	NOT PREHEATED	DIELECTRIC PREHEAT	NOT PREHEATED
0–¾	1,000–2,000	3,000	350	1,000
¾–1½	1,250,–2,500	3,700	450	1,250
2	1,500–3,000	4,400	550	1,500
3	1,750–3,500	5,100	650	1,750
4	2,000–4,000	5,800	750	2,000
5	2,250–4,500	*	850	**
6	2,500–5,000	*	950	**
7	2,750–5,500	*	1,050	**
8	3,000–6,000	*	1,150	**
9	3,250–6,500	*	1,250	**
10	3,500–7,000	*	1,350	**
12	4,000–8,000	*	1,450	**
14	4,500–9,000	*	1,550	**
16	5,000–10,000	*	1,650	**

*Add 700 psi for each additional inch of depth. Beyond 4 in. depth it is desirable, and beyond 12 in. essential, to preheat.
**Add 250 psi for each additional inch of depth. Beyond 4 in. depth it is desirable, and beyond 12 in. essential, to preheat.

sprue bushing, runners, gates, actual cavity, and any mold area subject to the clamp force. Plating may be done after the mold has been through initial trial runs and approved for production. With some compounds, the mold surface is coated with material that will not permit plating (poor adhesion would occur). One should consult the material supplier about this potential problem, particularly when using additives in the compound such as silicones, fluorocarbons, and so on. Once cavity surfaces have been so treated, it is difficult, probably impossible, to remove an unwanted coating. This problem also affects secondary operations such as decorating, painting, and so on.

An allowance is provided in the cavity for tight part tolerance requirements and plating. As plating highlights the mold surface, it is imperative that cavity polishing prior to plating be done precisely. Plating will not cover up a poorly polished surface.

Some of the reasons for plating are: (1) to provide an excellent part finish; (2) to enhance compound flow, producing more uniform and denser parts; (3) to lengthen mold life by improving wear resistance, particularly when coarse materials are included in the compound, such as glass, metals, and so on; and (4) to provide resistance to mold staining, permitting longer production runs with less mold cleaning time. The cost of hard-chrome plating is minor when compared to the savings in the molded part.

Shrinkage

As reviewed in the other chapters, shrinkage with TPs can be rather substantial, but with TSs it is less so, principally because of mixing with inorganic mineral additives that are not affected by heat changes. The degree and direction of shrinkage will depend on the molded part density, part configuration, ratio of resin to additives/reinforcements, and/or orientation of fibers (Chapter 2).

Thermoplastics

Even though CM and TM are used principally with TSs, these processes are sometimes used to mold conventional TPs for short runs with low-cost molds. They also are used with TPs that are difficult to melt, have too short a heat melt range for IM or other processes, and require high pressures. A typical example is ultrahigh molecular weight polyethylene (UHMWPE).

MOLD CONSTRUCTION

The information presented on molds in Chapter 2 applies to compression and transfer molds (1, 6, 11, 230). In TM, the transfer pot and plunger should have a diametrical clearance of about 0.001 in./in. of diameter. With one or two small half-round O-ring type grooves cut circumferentially around the plunger, considerable plunger wear can be tolerated without leakage of material. The grooves fill with material and remain filled during any production run. The pot should be hardened, polished, and plated, and the plunger should be well polished and hardened.

Runners—Cold and Hot

The runner size depends on the material being processed and whether it is a TS cold or hot runner. As reviewed in Chapter 2 and Fig. 2-7, with TPs a hot runner solidifies with the injection molded part. If a cold runner is used (with TP), only the molded parts solidifies; there is no runner scrap.

With TS closed molds (TM, IM, etc.), the complete mold can be at the maximum heat, which cures and solidifies both the runner and the molded parts. With TSs, this runner is called a hot runner. If the mold is designed so that the maximum heat surrounds only the cavity and the lower heat (similar to the lower pot heat) surrounds the runner, that is called a cold runner. In this case, material in the runner remains a fluid, so that the next shot into the mold cavity receives the runner material before material flows from the pot in TM.

Unfortunately, solidified TS runners, flash, and defective parts cannot be recycled as they are with TPs. Some operations can use this material as a filler after it has been granulated; but in practically all molding plants using TS, the material is lost and has to be considered in the part cost. One should consider using "cold runners" with TSs. Materials must be of a type that will permit their use; for example, the time cycle should permit material to remain in the fluid state. As different hardeners, accelerators, and so on, can be used, it is practical to use cold runners. They have limited use in TM but are used more widely in IM.

Design Analysis

With experience, logic, and trial-and-error runs, molds can be designed to maximize the efficiency of material flow and heat control. The result is that parts are molded to meet performance requirements at the lowest cost. Generally heat flow/control in molds is rather inefficiently evaluated, so that there is excess or unnecessary heat, and part performance may not be maximized. The cycle time could be affected.

The flow and heating analyses that have been developed into CAD (computer aided design) systems for IM (Chapter 2) are now also being applied in CM and TM. They facilitate the construction of molds, by locating cavities, sizing runners, and locating heater lines (electric, steam, etc.) and other devices to meet cost/performance objectives (1, 82, 110–112).

Chapter 7

REINFORCED PLASTICS/COMPOSITES

INTRODUCTION

When methods of processing plastics are discussed, the category of reinforced plastics (RP)/composites represents many different fabrication processes. In this chapter the usual processes (IM, CM, etc.) are included, as well as specialized techniques. Some of these processes are listed in Fig. 7-1.

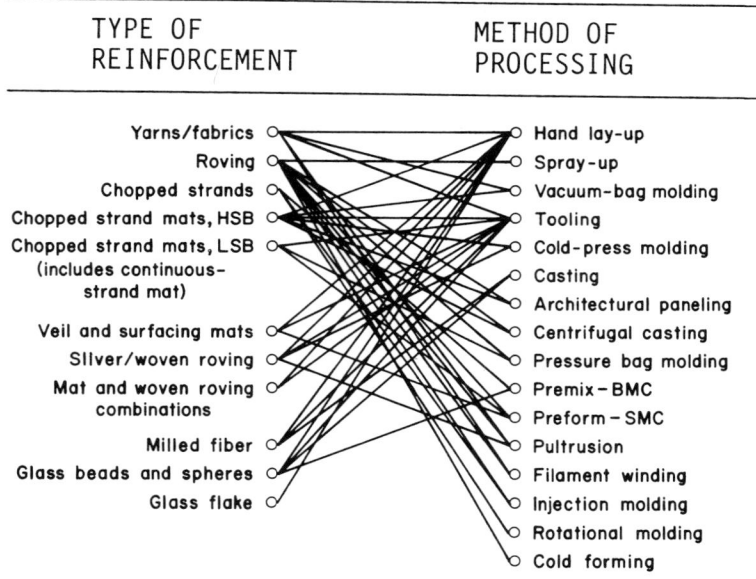

Fig. 7-1. Examples of reinforcements and processes used for reinforced plastics.

The terms reinforced plastics (RP) and composites refer to combinations of plastic materials and reinforcing materials, usually in fiber form (chopped fibers, porous mats, woven fabrics, continuous fibers, etc.; see Fig. 7–1). Both thermoset (TS) and thermoplastic (TP) resins are used. When modern RP industry started in 1940, glass-fiber-reinforced unsaturated polyester (TS), low pressure or contact pressure, curing resins were used. Today about 60 percent of the plastics industry uses many different forms of glass fiber-polyester composites. In this chapter the abbreviation RP will be used, and in references to polyester resin it will refer only to TS, as relatively little TP polyester is used in RPs.

Based on SPI Composites Institute data on U.S. production of composites, in 1987 about 560,000 lb of the total of 2,536,000 lb of composites produced were reinforced TPs. The total figure includes all types of reinforcements and resins. At least 80 percent of composites were glass fiber reinforced, and 60 percent were polyester (TS) types.

A processor can produce products whose mechanical properties in any direction are both predictable and controllable. This is done by carefully selecting the resin and the reinforcement in terms of both composition and orientation, followed by the appropriate process. All types of shapes can be produced: flat and complex shapes, solid and tubular rods/pipes, molded shapes/housings/complex configurations, structural shapes (angles, channels, box and I-beams, etc.), and so on. RPs can basically produce the strongest materials in the world, as summarized in Fig. 7–2 (1–3, 24, 30–33, 52, 54, 102, 188, 230–266).

REINFORCEMENTS

The molder has a variety of alternatives to choose from, regarding the kind, form, and amount of reinforcement to use (Figs. 7–1 and 7–3 and Tables 6–1 and 7–1). With the many different types and forms (organics, inorganics, fibers, flakes, etc.), practically any performance requirement can be met, molded into any shape. Shapes range from extremely small to extremely large, from simple to extremely complex.

The reinforcement type and form (woven, braided, chopped, etc.) depend on performance requirements and the method of processing the RP. Fibers can be oriented in many different patterns to provide the directional properties desired (Fig. 7–4). Depending on their packing arrangement, different reinforcement-to-resin ratios are obtained. In its simplest presentation, using glass fiber with epoxy resin, if the fibers were packed as close as possible (like stacked pipe) the glass would occupy 90.6 percent of the volume. With a "square" packing (fibers directly on top of and alongside each other) the

Strength

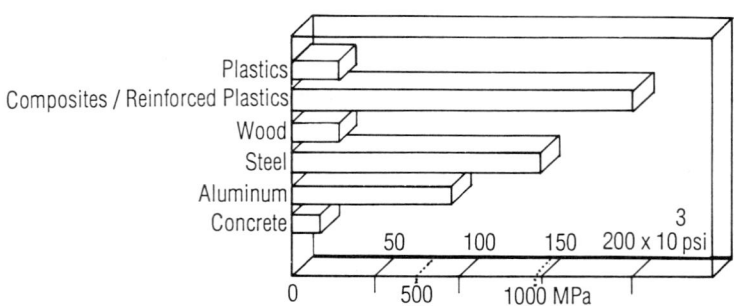

Modulus of Elasticity

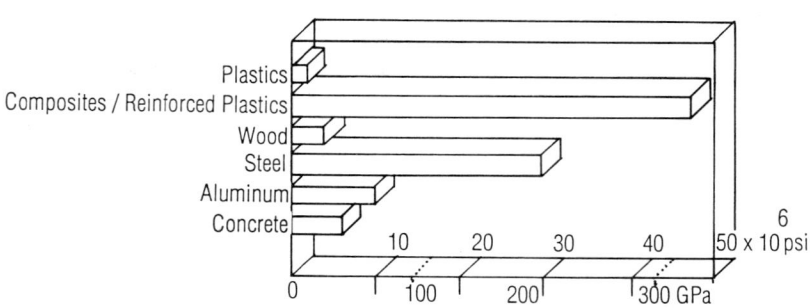

Specific Gravity

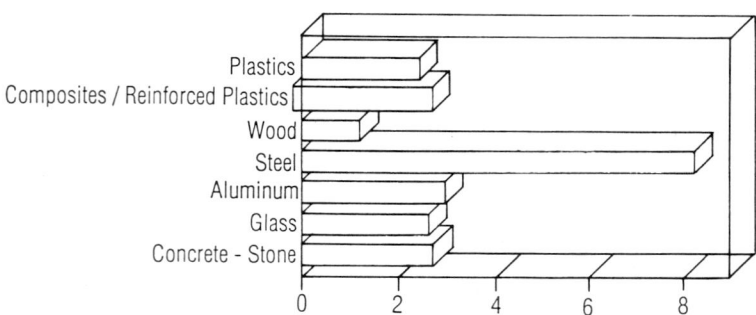

Fig. 7–2. Comparison of properties of reinforced plastics/composites with those of other materials.

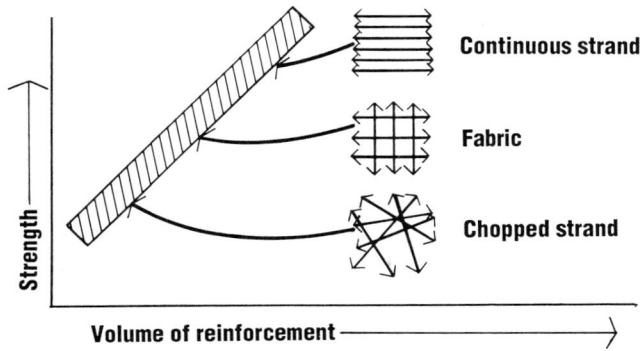

Fig. 7-3. Strength-to-volume relationship for reinforcements used in composites.

glass volume would be 78.5 percent. Glass fibers and most other reinforcements require special treatment to ensure maximum performance—such as selecting materials compatible with the resins used, protecting individual filaments during handling and/or processing, and so on.

The acceptance and use of nonwoven fabrics as reinforcements has led to the development of major products. These reinforcements include felts and paper structures, which usually contain a binder that retains these structures and is compatible with the resin matrix. Combinations of different chopped fibers (glass and aramid, etc.) are also used, including long filaments, woven fabrics, and so on. The combinations provide unique properties, and, in most cases, permit the molding of different shapes that otherwise would not be possible. The longer fibers are best for optimizing mechanical properties. With short, chopped fiber structures, the fiber length can range from extremely short (0.001 in.) to at least 0.5 in., and on up to 2 in. The length used usually depends on processing and performance requirements. Basically, to obtain the best mechanical performance with fibers in a properly molded part, it is only necessary for them to have an aspect ratio (length over diameter) of about ten (242).

The use of conventional industrial cutting methods on these reinforcements usually is not the best procedure or will not work (what cuts glass will not cut aramid, etc.). Woven and other constructions can present a variety of problems. Methods applicable to each type are available, and it is best to contact the material supplier for recommendations.

**Polar
Directional Properties**

- Orthotropic or Unidirectional
- Bidirectional
- Isotropic or Planar
- Unreinforced Plastics

Different Fiber Orientations

Tensile Fracture Characteristics

Orthotropic or Unidirectional
Variations in Properties with Angles of Stress

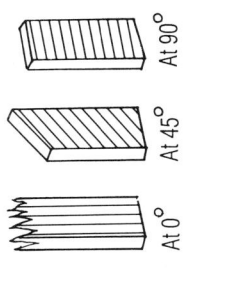

At 0°   At 45°   At 90°

Bidirectional
Variations in Properties with Angle of Stress

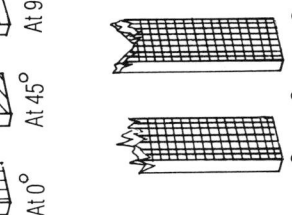

At 0° and 90°   At 45°

Isotropic or Planar
Properties Independent of Angle of Stress

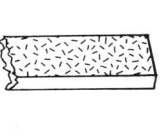

At any angle

Stress vs. Strain Diagrams at Various Angles

0° Orthotropic

0°•90° Bidirectional

all directions Isotropic

45° Bidirectional

45° Orthotropic

90° Orthotropic

Unreinforced Plastics

High — Stress

Strain — High

Fig. 7-4. Examples of performance of composites with different orientations of fiber reinforcements.

253

Table 7-1. Examples of Fiber Reinforcements Used in Reinforced Plastics.

TYPE OF FIBER REINFORCEMENT	SPECIFIC GRAVITY	DENSITY LB/IN^3	TENSILE STRENGTH 10^3 PSI	SPECIFIC STRENGTH 10^6 IN.	TENSILE ELASTIC MODULUS 10^6 PSI	SPECIFIC ELASTIC MODULUS 10^8 IN.
Glass						
E Monofilament	2.54	0.092	500	5.43	10.5	1.14
12-end roving	2.54	0.092	372	4.04	10.5	1.14
S Monofilament	2.48	0.090	665	7.39	12.4	1.38
12-end roving	2.48	0.090	550	6.17	12.4	1.38
Boron (tungsten substrate)						
4 mil or 5.6 mil	2.63	0.095	450	4.74	58	6.11
Graphite						
High strength	1.80	0.065	400	6.15	38	5.85
High modulus	1.94	0.070	300	4.29	55*	7.86
Intermediate	1.74	0.063	360	5.71	27	4.29
Organic						
Aramid	1.44	0.052	400	7.69	18	3.46

*Also commercially available up to 100×10^6 psi.

Note: The principal reinforcement, with respect to quantity, is glass fibers. Many other types are used (cotton, rayon, polyester/TP, nylon, aluminum, etc.). Of very limited use because of cost and processing difficulty are "whishers" (single crystals of alumina, silicon carbide, copper, or others), which have superior mechanical properties.

PLASTICS

Practically all resins (TS and TP) are used in RPs, but a few predominate (Tables 7-2 to 7-5), with TS polyesters being the major type. Typical polyesters are summarized in Table 7-4. The polyester RPs are used in all processes, but their principal use is in the low pressure methods (spray-up, hand lay-up, pressure bag molding, casting, pultrusion, rotational molding, filament winding, and compression molding.

RP MATERIALS

RPs can be processed in different ways. The individual components (reinforcement and resin) can be put together by the processor. Although TP resins generally require no additional material, the TSs usually require the addition of different additives and fillers, such as those reviewed in Chapter 6.

Most of the TPs used in RP are injection molded with compounds prepared by material suppliers. It is estimated that more than half of TSs are prepared by the processor. There are compounds available from material suppliers, as well as in some processing plants, that are ready to be processed, the most popular being SMCs and BMCs.

Sheet Molding Compound

An SMC is a reinforced plastic compound in sheet form. Most SMCs combine glass fiber with a polyester (TS) resin. Any combination of reinforcement and resin can be produced. The reinforcements can have continuous long fibers or any size of chopped fibers laid out in a different orientation from that of the resin. The different orientation makes it feasible to use SMCs on flat to complex-shaped molds. These SMCs will contain various additives and fillers to provide a variety of processing and performance properties (Table 6-1).

SMCs are made to meet the shelf life required. These B-staged compounds are usually used in a few weeks or months. Some have a shelf life of six months, for example. Suppliers' recommendations should be followed in keeping these compounds at a low temperature, or a curing action will occur (Chapter 6).

TPs are also used in sheet form with different reinforcements and resins (Table 7-2). They are called stampable sheets rather than SMCs. These compounds provide unique properties with a quick and easy processing capability.

Table 7–2. Mechanical, Thermal, and Processing Properties of Glass Fiber Reinforced Plastics. (Courtesy of Owens-Corning Fiberglas Corp.)

MATERIAL		GLASS FIBER, WT%	SPECIFIC GRAVITY
			D792
Glass fiber-reinforced thermosets (RTS)	polyester SMC, compression	30.0	1.85
	polyester SMC, compression	20.0	1.78
	polyester SMC, compression	50.0	2.00
	polyester BMC, compression	22.0	1.82
	polyester BMC, injection	22.0	1.82
	epoxy filament wound	80.0	2.08
	polyester, pultruded	55.0	1.69
	polyurethane, milled fibers (RRIM)	13.0	1.07
	polyurethane, flaked glass (RRIM)	23.0	1.17
	polyester spraying/lay-up	30.0	1.37
	polyester, woven roving, lay-up	50.0	1.64
Glass fiber-reinforced thermoplastics (RTP)	acetal resin	25.0	1.61
	nylon-6,6	30.0	1.48
	polycarbonate	10.0	1.26
	polypropylene	20.0	1.04
	poly(phenylene sulfide)	40.0	1.64
	acrylonitrile–butadiene–styrene terpolymer (ABS)	20.0	1.22
	poly(phenylene oxide) (PPO)	20.0	1.21
	styrene–acrylonitrile copolymer (SAN)	20.0	1.22
	poly(butylene terephthalate)	30.0	1.52
	poly(ethylene terephthalate)	30.0	1.56
Unreinforced thermoplastics (TP)	acetal resin		1.41
	nylon-6,6		1.13
	polycarbonate		1.20
	polypropylene		0.89
	poly(phenylene sulfide)		1.30
	acrylonitrile–butadiene–styrene terpolymer (ABS)		1.03
	poly(phenylene oxide) (PPO)		1.10
	styrene–acrylonitrile (SAN)		1.05
	poly(butylene terephthalate)		1.31
	poly(ethylene terephthalate)		1.34
Metals	ASTM A-606 HSLA steel, cold rolled		7.75
	SAE 1008 low carbon steel, cold rolled		7.86
	AISI 304 stainless steel		8.03
	TA 2036 aluminum, wrought		2.74
	ASTM B85 aluminum, die cast		2.82
	ASTM AZ91B magnesium, die		1.83
	ASTM AG40A zinc, die cast		6.59

See the Appendix for English to metric conversions.
ASTM test methods used.

THERMAL COEFFICIENT OF EXPANSION	HEAT DEFLECTION AT 1.8 MPa, C°	THERMAL CONDUCTIVITY, W/(m·K)	SPECIFIC HEAT, J/(kg·K)^b	TENSILE STRENGTH, MPa	TENSILE MODULUS, GPa
D696	D648	C177		D638	D638
	200+		1.26	83	11.7
	200+		1.26	36.5	11.7
9.4	200+		1.26	158	15.7
6.6	260	8.37	1.26	41.3	12.1
6.6	260	8.37	1.26	33.5	10.5
2.0	200+	1.77	0.96	552	27.6
5.0		6.92	1.17	207	17.2
78.0	29			19.3	
53.1				30.4	
12.0	200+	2.60	1.30	86.2	6.9
4.0	200+			255	15.5
4.7	161			128	8.6
1.8	254	2.60	1.26	159	8.3
1.8	141	7.97	1.21	83	5.2
2.4	132	14.5		45	3.7
1.1	266	3.47	1.05	152	14.1
				76	6.2
2.1	99	2.42			
2.0	143	6.57	0.84–1.67	100	6.3
2.1	102	4.84		100	8.6
1.4	213	12.1	0.46	131	8.3
1.7	216	11.2		145	9.0
4.7	110	2.80	1.46	81	2.6
4.5	75	2.94	1.26	79	2.8
3.7	132	2.34	1.26	66	2.3
3.8	46–60	2.10	1.88	34	0.7
	135	2.89		66	3.3
				41	2.1
3.2	93–104	1.61			
68.0	100	1.59	0.84–1.67	54	2.6
36.0	104	1.21	1.38	66	2.8
4.5	50–85	1.76–2.89		57	1.9
	38–41	1.51	1.42	59	2.8
				448	207
6.8		43.3	0.46		
				331	207
6.7		60.6	0.42		
9.6		16.3	0.50	552	193
13.9		159	0.88	338	70
11.6		91.8		331	71
14.0		72.5	1.05	228	448
15.2		113	0.42	283	75

(*continued*)

Table 7-2. Continued.

MATERIAL		ELONGATION, %	FLEXURAL MODULUS, GPa
		D638	D790
Glass fiber-reinforced thermosets (RTS)	polyester SMC, compression	<1.0	11.0
	polyester SMC, compression	0.4	9.7
	polyester SMC, compression	1.7	13.8
	polyester BMC, compression	0.5	10.9
	polyester BMC, injection	0.5	9.9
	epoxy filament wound	1.6	34.5
	polyester, pultruded		11.0
	polyurethane, milled fibers (RRIM)	140.0	0.26–0.37
	polyurethane, flaked glass (RRIM)	38.9	1.0
	polyester spraying/lay-up	1.3	5.2
	polyester, woven roving, lay-up	1.6	15.5
Glass fiber-reinforced thermoplastics (RTP)	acetal resin	3.0	7.6
	nylon-6,6	1.9	5.5
	polycarbonate	9.0	4.1
	polypropylene	3.0	3.6
	poly(phenylene sulfide)	3.0	13.1
	acrylonitrile–butadiene–styrene terpolymer (ABS)	2.0	6.0
	poly(phenylene oxide) (PPO)	5.0	5.2
	styrene–acrylonitrile copolymer (SAN)	1.8	7.6
	poly(butylene terephthalate)	4.0	8.1
	poly(ethylene terephthalate)	6.6	8.6
Unreinforced thermoplastics (TP)	acetal resin	30.0	2.7
	nylon-6,6	60.0	2.9
	polycarbonate	110.0	2.3
	polypropylene	200.0	0.9–1.4
	poly(phenylene sulfide)	1.0	3.8
	acrylonitrile–butadiene–styrene terpolymer (ABS)	5.0	2.4–2.8
	poly(phenylene oxide) (PPO)	50.0	2.3–2.8
	styrene–acrylonitrile (SAN)	0.5	3.8
	poly(butylene terephthalate)	50.0	2.3–2.8
	poly(ethylene terephthalate)	50.0	2.4–3.1
Metals	ASTM A-606 HSLA steel, cold rolled	22.0	
	SAE 1008 low carbon steel, cold rolled	37.0	
	AISI 304 stainless steel	40.0	
	TA 2036 aluminum, wrought	23.0	
	ASTM B85 aluminum, die cast	2.5	
	ASTM AZ91B magnesium, die	3.0	
	ASTM AG40A zinc, die cast	10.0	

See the Appendix for English to metric conversions.
ASTM test methods used.

COMPRESSIVE STRENGTH, MPa	IMPACT STRENGTH IZOD AT 22°C, J/m	HARDNESS	WATER ABSORPTION IN 24 H, %	MOLD SHRINKAGE, %
D695	D256	D785	D570	D955
166	854	Barcol 68	0.25	
159	438	Barcol 68	0.10	0.002
221	1036	Barcol 68	0.50	
138	227	Barcol 68	0.20	0.001
	154	Barcol 68	0.20	0.004
310	2400	M98	0.50	0.008
207	1335	Barcol 50	0.75	
		Shore D65-75		
	112			
152	690–800	Barcol 50	1.30	
186	1760	Barcol 50	0.50	
117	96	M79	0.29	0.004
183	117	M95	0.50	0.002
97	107	M80	0.14	0.005
172	59	R103	0.05	0.003
145	80	R123	0.01	0.002
97	64	R107	0.30	0.002
121	96	R107	0.24	0.003
121	59	R122	0.06	0.002
124	96	R118	0.06	0.003
172	96	R120	0.05	0.003
90	32	R119	1.3–1.9	0.005
103	43	R120, M83	1.0–1.3	0.008
86	854	M70	0.15	0.005–0.007
24	50–1000	R50-96	0.03	0.020
110	<27	R123	<.02	0.007
69	160–320	R107-115	0.20–0.45	0.004–0.009
83	270	R115	0.07	0.005–0.007
97	16–24	M80-85	0.20–0.35	
59	43	M68-78	0.08–0.09	0.015–0.020
76	13–35	M94-101	0.1–0.2	0.02–0.025
448		B80		
331		B34-52		
552		B88		
338		R80		
331		Brinell 85		
227		Brinell 85		
283		Brinell 82		

Table 7–3. Example of Properties and Processes for the Major Thermoset Resins Used in Composites.

THERMOSETS	PROPERTIES	PROCESSES
Polyesters	Simplest, most versatile, economical, and most widely used family of resins; good electrical properties, good chemical resistance, especially to acids.	Compression molding, filament winding, hand lay-up, mat molding, pressure bag molding, continuous pultrusion, injection molding, spray-up, centrifugal casting, cold molding, encapsulation.
Epoxies	Excellent mechanical properties, dimensional stability, chemical resistance (especially to alkalies), low water absorption, self-extinguishing (when halogenated), low shrinkage, good abrasion resistance, excellent adhesion properties.	Compression molding, filament winding, hand lay-up, continuous pultrusion, encapsulation, centrifugal casting.
Phenolic resins	Good acid resistance, good electrical properties (except arc resistance), high heat resistance.	Compression molding, continuous lamination.
Silicones	Highest heat resistance, low water absorption, excellent dielectric properties, high arc resistance.	Compression molding, injection molding, encapsulation.
Melamines	Good heat resistance, high impact strength.	Compression molding.
Diallyl o-phthalate	Good electrical insulation, low water absorption.	Compression molding.

Bulk Molding Compound

A BMC is a molding compound that is not produced in sheet form. It basically consists of the mixture used in an SMC, except that it contains only short fibers.

Recycle Scrap

As reviewed throughout this book, TPs can be granulated and recycled, whereas TSs cannot be remelted but could be used as fillers. However, recycling of TPs can degrade performance (Chapters 1–4 and 9). When reinforced TPs are granulated, the length of the fibers is reduced. When they are reprocessed with virgin materials or alone, their processability and performance definitely change. So it is important to determine if the change

Table 7-4. Characteristics of Glass Fiber Reinforced Polyesters (TS).

POLYESTER TYPE	CHARACTERISTIC	TYPICAL USES
General purpose	Rigid moldings.	Trays, boats, tanks, boxes, luggage, seating.
Flexible resins and semirigid resins	Tough, good impact resistance, high flexural strength, low flexural modulus.	Vibration damping; machine covers and guards, safety helmets, electronic part encapsulation, gel coats, patching compounds, auto bodies, boats.
Light-stable and weather-resistant	Resistant to weather and ultraviolet degradation.	Structural panels, sky-lighting, glazing.
Chemical-resistant	Highest chemical resistance of polyester group; excellent acid resistance, fair in alkalies.	Corrosion-resistant applications such as pipe, tanks, ducts, fume stacks.
Flame-resistant	Self-extinguishing, rigid.	Building panels (interior), electrical components, fuel tanks.
High heat distortion	Service up to 500°F, rigid.	Aircraft parts.
Hot strength	Fast rate of cure (hot), moldings easily removed from die.	Containers, trays, housings.
Low exotherm	Void-free thick laminates, low heat generated during cure.	Encapsulating electronic components, electrical premix parts—switchgear.
Extended pot life	Void-free uniform, long flow time in mold before gel.	Large complex moldings.
Air dry	Cures tack-free at room temperature.	Pools, boats, tanks.
Thixotropic	Resists flow or drainage when applied to vertical surfaces.	Boats, pools, tank linings.

will affect the final part performance; if it will, a limit to the amount of regrind mix should be determined.

Void Content

Voids are generally the result of the entrapment of air during the construction of a lay-up, particularly with the use of hand lay-up and very low pressure processing methods. It is possible to have void contents of 1 to 3

Table 7-5. Trade-offs in Thermoplastic Composites. (Courtesy of LNP Corp.)

DESIRED MODIFICATION	HOW ACHIEVED	SACRIFICE (FROM BASE RESIN)		COMMENTS
		AMORPHOUS	CRYSTALLINE	
Increased tensile strength	Glass fibers	Ductility, cost	Ductility, cost	Glass fibers are the most cost effective way of gaining tensile strength. Carbon fibers are more expensive; fibrous minerals are least expensive but only slightly reinforcing. Reinforcement makes brittle resins tougher and embrittles tough resins. Fibrous minerals are not commonly used in amorphous resins.
	Carbon fibers	Ductility, cost	Ductility, cost	
	Fibrous minerals	NA	Ductility	
Increased flexural modulus	Glass fibers	Ductility, cost	Ductility, cost	Any additive more rigid than the base resin produces a more rigid composite. Particulate fillers severely degrade impact strength.
	Carbon fibers	Ductility, cost	Ductility, cost	
	Rigid minerals	Ductility	Ductility	
Flame resistance	FR additive	Ductility, tensile strength, cost	Ductility, tensile strength, cost	FR additives interfere with the mechanical integrity of the polymer and often require reinforcement to salvage strength. They also narrow the molding latitude of the base resin. Some can cause mold corrosion.
Increased heat-deflection temperature (HDT)	Glass fibers	Ductility, cost	Ductility, cost	When reinforced, crystalling polymers yield much greater increases in HDT than do amorphous resins. As with tensile strength, fibrous minerals increase HDT only slightly. Fillers do not increase HDT.
	Carbon fibers	Ductility, cost	Ductility, cost	
	Fibrous minerals	NA	Ductility	
Warpage resistance	5 to 10% glass fibers	NA	Cost	Amorphous polymers are inherently nonwarping molding resins. Only occasionally are fillers such as milled glass or
		NA	Cost	

Function	Additive	Properties	Properties	Comments
	5 to 10% carbon fibers Particulate fillers	Ductility, cost, tensile strength	Ductility, cost, tensile strength	glass beads added to amorphous materials because they reduce shrinkage anisotropically. Addition of fibers tends to balance the difference between in-flow and cross-flow shrinkage usually found in crystalline polymers. When a particulate is used to reduce and balance shrinkage, some fiber is needed to offset degradation.
Reduced mold shrinkage (increased mold-to-size capability)	Glass fibers Carbon fibers Fillers	Ductility, cost Ductility, cost Tensile strength, ductility, cost	Ductility, cost Ductility, cost Tensile strength, ductility, cost	Reinforcement reduces shrinkage far more than fillers do. Fillers help balance shrinkage, however, because they replace shrinking polymer. The sharp shrinkage reduction in reinforced cyrstalling resins can often lead to warpage. The best "mold-to-size" composites are reinforced amorphous composites.
Reduced coefficient of friction	PTFE Silicone MoS_2 Graphite	Cost	Cost	These fillers are soft and do not dramatically affect mechanical properties. PTFE loadings commonly range from 5 to 20%; the others are usually 5% or less. Higher loadings can cause mechanical degradation.
Reduced wear	Glass fibers Carbon fibers Lubricating additives	— — —	— — —	The subject of plastic wear is extremely complex and should be discussed with a composite supplier.
Electrical conductivity	Carbon fibers Carbon powders	Ductility, cost Tensile strength, ductility, cost	Ductility, cost Tensile strength, ductility, cost	Resistivities of 1 to 100,000 ohm-cm can be achieved and are proportional to cost. Various carbon fibers and powders are available with wide variations in conductivity yields in composites.

263

percent. Depending on the application, voids can cause a reduction in part performance, particularly in certain environments and after lengthy outdoor exposures. If voids are undesirable, procedures can be used to reduce or eliminate them, such as applying a vacuum during the process. Another preventive method is to squeeze out air during lay-up by a roller or a spatula.

The following method can be used to estimate void content:

$$\text{Percent voids} = 100 - 100a \left(\frac{d}{c} + \frac{e}{b} + \frac{f}{g} \right)$$

where:

a = Specific gravity of product
b = Specific gravity of fiberglass = 2.55
c = Specific gravity of cured resin = range 1.18 to 1.24
d = Resin content, by weight
e = Glass content, by weight
f = Filler content, by weight
g = Specific gravity of filler

The above method is not exact because the assumption is made that the resin system has the same density with reinforcement as it does in an unreinforced casting. The net result is a possible overstatement of the void content. In addition to air entrapment, entrapment of volatiles can occur with certain resins that release them during processing.

PROCESSES

Choosing the optimum process encompasses a broad spectrum of possibilities. In some situations only one process can be used, but generally there are options. Influencing the process selection are quantity, size, thickness, tolerances, type of material, and performance requirements (Table 7–2). Regarding tolerances, as mentioned in other chapters, resins with fillers and/or reinforcements generally are far more stable in meeting tight tolerances. (In fact, the TSs, whether unreinforced or reinforced, are more dimensionally stable than other resins.)

RP parts are fabricated by processes using pressures that range from contact (or no pressure), through moderate (50–100 psi), on up to thousands of psi. Temperatures can range from room temperature to the usual 250 to 600°F and on up, particularly for certain high performance TPs. The time cycles can range from seconds to minutes, hours, or even days. Parts cover a wide size range, from small parts to those as large as boat hulls, 80 ft long (25 m) or larger. The actual process conditions of pressure, heat, and time depend on the material to be processed and on whether a long cure time is required to form the part into a shape (hand lay-up, etc.).

Information on processing requirements for materials is reviewed throughout this book. See Chapter 6, regarding TS materials, and Fig. 7–5, which shows that the time for the processing flow of TS plastics occurs when the viscosity of the melt is at its lowest. Different plastics can be used, with shorter or longer times at this low viscosity level. When working with new materials, the processor should obtain such details from the material supplier.

Processing may involve equipment that is simple to operate, or it may require extensive specialized equipment. Among the most common processes are contact molding methods (hand lay-up, spray-up, vacuum bag, pressure bag, autoclave, etc.), matched mold methods (compression molding, transfer molding, resin transfer molding, injection molding, compression-injection molding, stamping, etc.), and other methods (filament winding, cold press molding, pultrusion, continuous laminating, centrifugal casting, encapsulation, rotational molding, reaction injection molding, etc.). Tables 7–6 through 7–8 summarize some of these processes.

Hand Lay-up

This is the oldest, and in many ways the simplest and most versatile, process; but it is slow and very labor-intensive. It consists essentially of the hand tailoring and placing of layers of (usually glass fiber) mat, fabric, or both on a one-piece mold and simultaneously saturating the layers with a liquid TS resin (usually polyester). The assemblage is then cured with or without

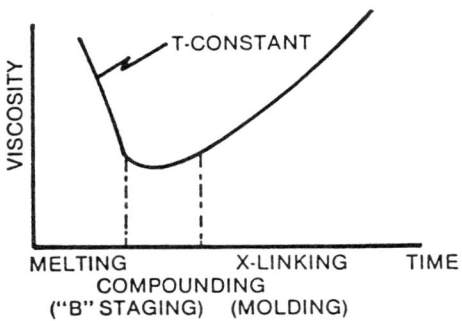

Fig. 7–5. Viscosity change during processing of thermosets; the B-stage represents the start of the heating cycle, which reduces viscosity, followed by a chemical reaction (cross-linking) and solidification of the plastic.

Table 7-6. Process Comparison of Various Reinforced Plastics Manufacturing Techniques. (Courtesy of Owens-Corning Corp.)

| | RESIN TRANSFER MOLDING | OPEN MOLDING | | COLD PRESS MOLDING | COMPRESSION MOLDING | |
		SPRAY-UP	HAND LAY-UP		MAT/PREFORM	SHEET MOLDING COMPOUND
Mold construction	FRP*, spray metal, cast aluminum; gasket seal, air vents, self-sealing injection port	FRP		FRP, spray metal, cast aluminum, pinch (land)	Metal, shear edge	High grade steel; shear edge
Pressure	Pressure feed pumping equipment req'd; mold halves clamped (methods range from clamp frame to pressure pod)	None		Lows pressure press, capable of 50 psi (hydraulic or pneumatic mechanical); resin dispensing equipment not req'd but recommended.	Hydraulic press, normal range of 100–500 psi	Hydraulic; as high as 2,000 psi

Cure system	Room temperature				Heated; normal range of 225-325°F	Heated; normal range of 275-350°F
Resin compounding equipment	High shear type	Not needed			High shear type	
Reinforcement	Continuous strand mat, preform, woven roving	Continuous roving	Chopped strand mat, woven roving, cloth	Continuous strand mat, preform, woven roving	Continuous strand mat, preform, woven roving	Continuous roving (specific orientations for higher strength)
Part trim equipment	Yes				With optimum shear edges, minor trimming only	
Generally expected mold life (parts)	3,000	1,000		3,000	150,000+	150,000+

*FRP = Fiber glass reinforced plastics.

Table 7–7. Comparison of Resin Transfer Molding, SMC Compression, and Injection Molding.

	PROCESS		
	RTM	SMC COMPRESSION	INJECTION
Process operation:			
Production requirement, annual units per press	5,000–10,000	50,000	50,000
Capital investment	Moderate	High	High
Labor cost	High	Moderate	Moderate
Skill requirements	Considerable	Very low	Lowest
Finishing	Trim flash, etc.	Very little	Very little
Product:			
Complexity	Very complex	Moderate	Greatest
Size	Very large parts	Big flat parts	Moderate
Tolerance	Good	Very good	Very good
Surface appearance	Gel-coated	Very good	Very good
Voids/wrinkles	Occasional	Rarely	Least
Reproducibility	Skill-dependent	Very good	Excellent
Cores/inserts	Possible	Very difficult	Possible
Material usage:			
Raw material, cost	Lowest	Highest	High
Handling/applying	Skill-dependent	Easy	Automatic
Waste	Up to 3%	Very low	Sprues, runners
Scrap	Skill-dependent	Cuts reusable	Low
Reinforcement flexibility	Yes	No	No
Mold:			
Initial cost	Moderate	Very high	Very high
Cycle life	3,000–4,000 parts	Years	Years
Preparation	In factory	Special mold-making shops	
Maintenance	In factory	Special machine shops	

heat, commonly without pressure. Alternatively, preimpregnated B-staged, partially cured dry material (such as SMC) may be used, but in this case heat is applied with the probability of applying low pressure.

Fabrication begins with a pattern from which a mold is made. The mold may be of any low-cost material, including wood, hard plaster or hydrostone, concrete, a metal such as aluminum or steel, and glass fiber reinforced polyester or epoxy. If only a few parts are to be made, a single mold will suffice; otherwise multiple molds may be required. If the volume is large enough and speed is important, heating elements such as lines for steam or other fluids, or electrical heat units, may be incorporated. Automated equipment also may be installed (Fig. 7–6). The mold may be male

Table 7-8. Guide to Compatibility of Materials and Processes.

	THERMOSETS					THERMOPLASTICS									
	POLYESTER	POLYESTER SMC	POLYESTER BMC	EPOXY	POLYURETHANE	ACETAL	NYLON 6	NYLON 6/6	POLYCARBONATE	POLYPROPYLENE	POLYPHENYLENE SULFIDE	ABS	POLYPHENYLENE OXIDE	POLYSTYRENE	POLYESTER
Injection molding	•		•	•	•	•	•	•	•	•	•	•	•	•	•
Hand lay-up	•			•											
Spray-up	•			•											
Compression molding	•	•	•	•	•						•				
Preform molding	•			•											
Filament winding	•			•											
Pultrusion	•			•											
Resin transfer molding	•			•											
Reinforced reaction injection molding	•			•	•		•								

269

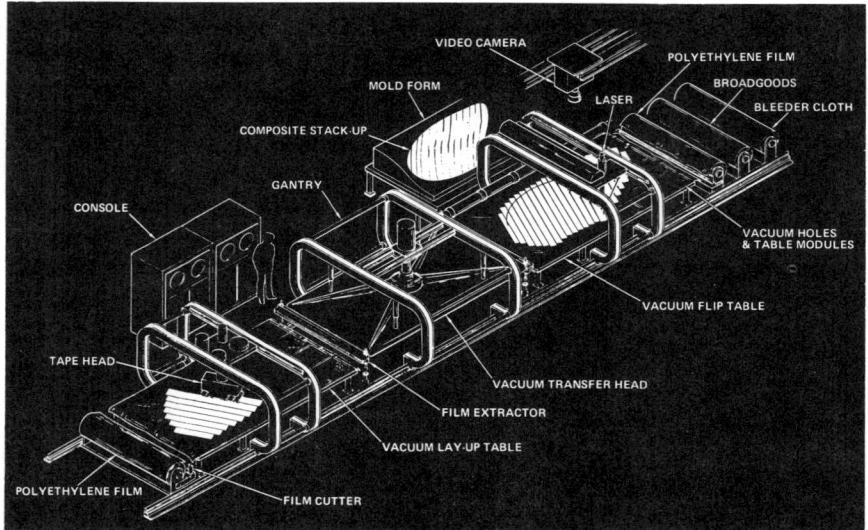

Fig. 7-6. Automated–integrated reinforced plastics lay-up process that uses TS preimpregnated reinforced sheet.

(plug) or female (cavity), depending upon which side of the formed part is to have the accurate configuration (the other side will be rough).

Prior to the actual lay-up the mold must be sealed if it is porous, such as wood, and coated with a mold release agent to prevent sticking of the molded part. Waxes, silicones, thin films, and other agents are used. Lay-up consists of tailoring the sheet materials to fit, and placing them in layers on the mold, saturating the layers by brush, spray, or any other suitable means, and working them with a serrated roller to consolidate the layers, reducing or eliminating voids and porosity.

Frequently, to provide resistance to weathering, erosion, or chemical attack, a resin-rich surface coat (gel coat) is added. This surface layer is applied first and then allowed to stiffen into a tough layer (not cured) before additional layers are applied. It usually is reinforced with a surface veil using C-glass (rather than the usual E-glass) if chemical resistance is required, or a synthetic fiber veil may be used for resistance to weather, particularly sunlight. The resin may be a special formulation that includes TP to improve the surface appearance. Subsequent layers of mat, fabric, or combinations are then applied.

Inserts, strengthening ribs (of wood, metal, or glass fiber shapes), and other devices can be incorporated. They are placed in the lay-up.

Spray-up

An air spray gun includes a roller cutter that chops glass fiber rovings to a controlled length before they are blown in a random pattern onto a surface of the mold simultaneously with a spray of catalyzed resin (Fig. 7-7). The chopped fibers are coated with resin as they exit the gun's nozzle. The resulting, rather fluffy, mass is consolidated with serrated rollers to squeeze out air and reduce or eliminate voids. As in hand lay-up, the first layer of a gel coat may be applied over the mold, followed by successive passes of the sprayed-on composite before a final gel coat is applied. If required, inserts, and so on, can be included during the spraying operation. Thixotropic agents (Chapter 6) may be employed in the resin.

Spray-up as hand lay-up, usually results in little material waste. One can tailor the formation; the charge follows the contours of three-dimensional shapes very easily. There is no practical limit to size, local reinforcement is readily provided by building up the thickness or incorporating reinforce-

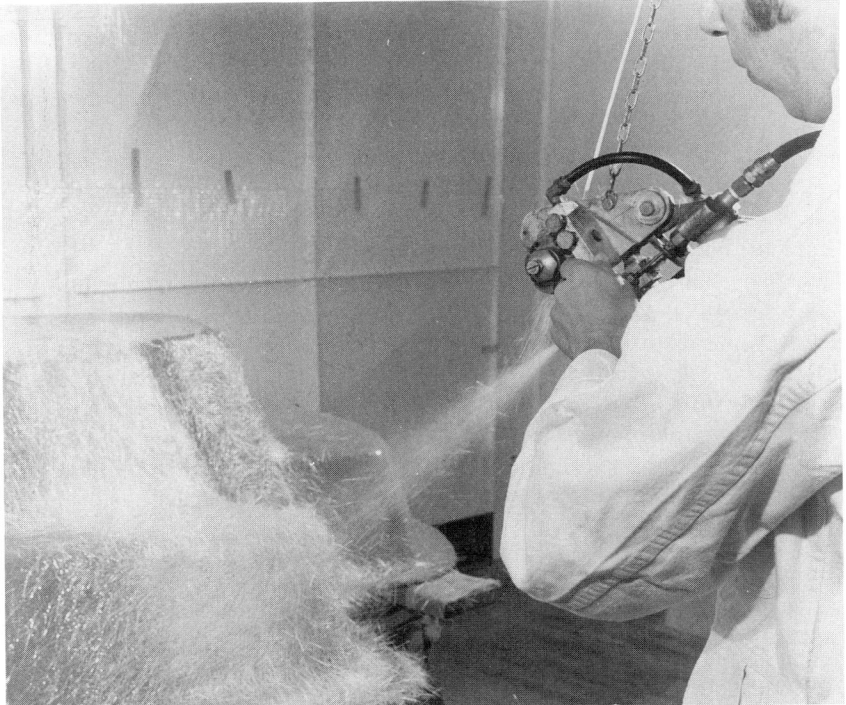

Fig. 7-7. Spray-up process for composites.

ments, and the cost of the methods is about the same. The production speed is usually less with spray-up, but its thickness control is less efficient, and its strength is likely to be lower and more variable than with hand lay-up. Spray-up depends more on the skill and care of the operator than does hand lay-up. To help overcome or improve on these negative factors, spray-up can be automated, which is a practical approach in long production runs.

Vacuum Bag

A molded part made by hand lay-up or spray-up is allowed to cure without the application of external pressure. For many applications, this approach is sufficient, but maximum consolidation usually is not achieved with its use. There is some porosity; fibers may not fit closely into internal corners with sharp radii but tend to spring back; and resin-rich or resin-starved areas may occur because of drainage, even with thixotropic agents. With moderate pressure these defects can be overcome, with an improvement in mechanical properties and better quality control of parts.

One way to apply such moderate pressure is to enclose the "wet-liquid resin" composite and mold in a flexible membrane or bag, and draw a vacuum inside the enclosure. Atmospheric pressure on the outside presses the bag or membrane uniformly against the wet composite. Pressures commonly range from 10 to 14 psi (69–97 kPa). (See Chapter 5 regarding vacuum pressure.) Withdrawal of the air inside the bag not only causes external pressure, but it tends to draw air bubbles out of the wet material, thus reducing porosity. Hand working over the bag with rollers, when vacuum is applied, helps to consolidate the structure.

For a bag or a membrane, flexible materials that are used include silicones, neoprene, natural rubber, PVC, PVOH, and others that are not affected by the resin. Care must be taken to arrange the bag or membrane with the vacuum hoses in such a way as to avoid local dams that entrap air instead of allowing it to escape. Bleeder mats or porous sheets should be placed around the edges of the molded part and/or over the mat assemblage under the bag. One should consider using a bleeder with vacuum line at any place on the part that permits a rough surface at the point of contact but improves the removal of trapped air. As in lay-up and spray-up, curing can be accelerated by heating the mold.

Pressure Bag

If more pressure is required than what is available with the bag system, a second envelope can be placed around the whole assemblage and air pressure admitted between the inner bag and outer envelope. This method is also called the vacuum-pressure bag process.

Autoclave

Still higher pressures can be obtained by placing the vacuum assemblage in an autoclave (Fig. 7–8). Air or steam pressures of 100 to 200 psi (690–1,380 kPa) are commonly achieved. If still higher pressures are required and the danger of extremely high air pressures is to be avoided, a hydroclave may be used, employing water pressures as high as 1,000 psi (6,900 kPa). The bag must be well sealed to prevent infiltration of high pressure air, steam, or water into the molded part. In these processes an initial vacuum may or may not be employed.

Pressure Bag Molding

In this process, seamless containers, tanks, pipes, and other products can be made. A preform is used that is made by a spray-up, a mat, or a com-

Fig. 7–8. Autoclave process for composites.

bination of materials (using a perforated screen to produce the preform). Only enough resin is included in the preform to hold the fibers in place. Usually it amounts to about 0.5 percent by weight, and is compatible with the matrix resin.

An inflatable elastic pressure bag is positioned within the preform, and the assembly is put into a closed mold. (The mold could be a drum, etc.) Resin is injected into the preform, and the pressure bag is inflated to about 50 psi (345 kPa). Heat is applied, and the part is cured within the mold. When cure is complete, the bag is deflated and pulled through an opening at the end of the mold, and the part removed.

Foam Reservoir Molding

This process, also known as elastic reservoir molding, consists of making basically a sandwich of resin-impregnated open-celled flexible polyurethane foam between the face layers of fibrous reinforcements. When this composite is placed in a mold and squeezed, the foam is compressed, forcing the resin outward and into the reinforcement. The elastic foam exerts sufficient pressure to force the resin impregnated reinforcement into contact with the mold surface.

Resin Transfer Molding

RTM is a closed-mold, low pressure process in which a preplaced dry reinforcement preform is impregnated with a liquid resin (usually polyesters, although epoxies and phenolics may be used) in an injection or transfer process, through an opening in the center of a mold (similar to the setup of Fig. 6–1b). The preform is placed in the mold, and the mold is closed. A two-component resin system (including catalyst, hardener, etc.) is then mixed in a static mixer (Chapter 3) and metered into the mold through a runner system. The air inside the closed mold cavity is displaced by the advancing resin front, and escapes through vents located at the high points or the last areas of the mold to fill (as in injection molding, Chapter 2). When the mold has filled, the vents and the resin inlets are closed. The resin within the mold cures, and the part can be removed.

During the 1940s and 1950s a similar system used vacuum. There was a dam around the outside opening of the two-part mold that contained the preform. This dam was filled with the mixed resin, and in the center of the mold, there was an opening that drew a vacuum. Thus resin could be drawn through the reinforcement, producing a cured part subjected to a maximum pressure of up to 14 psi. This type of vacuum at the vents or parting line is sometimes used with RTM to aid resin flow.

Advantages of RTM are that the molded part has two finished surfaces, and the overall process may emit a lower level of styrene vapor if the polyester resin used contains styrene. The mold, unlike a compression or TP stamping mold, is completely closed to defined stops prior to final part formation/curing. This procedure provides a more reproducible part thickness and tends to minimize trimming and deflashing of the final part.

Use of a reinforcement preform allows the preplacement of a variety of reinforcements in precise locations. The preforms remain in position during mold closing and resin injection. If large amounts of random reinforcements are used, consideration must be given to minimizing the washing or movement of the fibers due to resin flow near the resin inlet gates. A low injection pressure is another characteristic. Simple parts with a low proportion (10–20 percent, by volume) of reinforcement will fill rapidly at pressures of 10 to 20 psi (70–140 kPa). More complex parts, with 30 to 50 percent reinforcement, may require resin injection pressures in the 100 to 200 psi (700–1,400 kPa) range for rapid mold filling.

If low pressure is a requirement, as with large panels, a low injection pressure generally can be maintained and the fill time extended. In cases where low pressure and a fast fill time are required, the preform construction must be carefully tailored to promote a rapid low-pressure fill without fiber movement. For a high volume, the cycle time will vary, depending on the complexity of the part and the degree of part integration achieved. For a simple component, a 1 min. cycle is commercially achievable at a production rate of 1.2 million parts per year. For complex parts, a cycle time of 6 min. or longer could be needed.

Reaction Injection Molding

RIM is very similar to RTM. In the reinforced RIM (RRIM) process a dry reinforcement preform is placed in a closed mold. Next a reactive resin system is mixed under high pressure in a specialty designed mix head. Upon mixing, the reacting liquid flows at low pressure through a runner system to fill the mold cavity, impregnating the reinforcement in the process. Once the mold cavity is filled, the resin quickly completes its reaction. The complete cycle time required to produce a molded part can be as little as 1 min. (see Chapter 8 and Fig. 8–1).

Advantages of RRIM are similar to those listed for RTM. However, RRIM uses preforms that are less complex in construction and lower in reinforcement content than those used in RTM. The RRIM resin systems currently available will build up viscosity rapidly, which results in a higher average viscosity during mold filling. This action follows the initial filling with a low viscosity resin.

Compression and Transfer Molding

Compression and transfer molding processes are high-volume, high-pressure methods suitable for molding simple to complex parts. (See Chapter 6 for details.)

Coating Molded Parts. As there is a combination of rather rigid glass fibers with "soft" plastics in compression molding, smooth surfaces can be difficult to obtain. Ripples, sink marks, and other imperfections can occur, particularly with a high fiber content. To produce a smooth surface, various methods are used, such as blending TP resins with the base TS resin (polyester), using special coating veils, and so on. There is also a system where a surfacing resin (usually polyurethane) is injected in a compression mold just prior to the final cure. At the appropriate time in the cycle, the mold is "cracked" open a few thousandths of an inch, and within a second the resin is injected (Fig. 7–9). The mold closes, and the cure cycle is completed. The coating provides a smooth surface, as well as excellent adhesion to the composite.

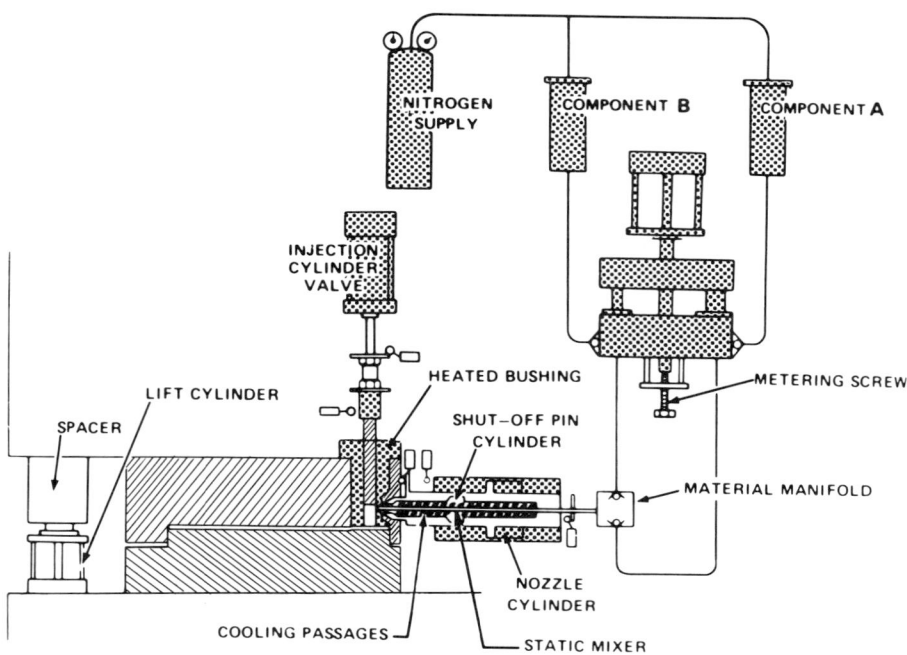

Fig. 7–9. Schematic of a complete system to coat compression molded parts.

Cold Press

This process is an economical press molding method for manufacturing an intermediate number of parts, such as 200 to 2,000. It uses low-pressure, room-temperature curing resins, and inexpensive molds. Cold press is similar to compression molding except that the resin curing action occurs via its own exothermic heat of reaction (after resins are mixed with catalyst, etc.). Pressures are moderate, usually 20 to 50 psi. Thus molds can be made of relatively inexpensive metals, plaster, or reinforced plastics. The edges need not be trimmed. Ribs, bosses, and other fairly complex shapes are not easily produced. Two good mold surfaces are obtained.

Stamping

In the stamping process, a reinforced thermoplastic sheet material is precut to required sizes. The precut sheet is preheated in an oven, the heat depending on the TP used (such as PP or nylon, where the heat can range upward from 520°F or 600°F). Dielectric heat is used to ensure that the heat is quick, and, most important, to provide uniform heating through the thickness and across the sheet. After hearing, the sheet is quickly formed into the desired shape in cooler matched-metal dies using conventional stamping presses or SMC-type compression presses. Stamping is a highly productive process, capable of forming complex shapes with the retention of the fiber orientation in particular locations as required. The process can be adapted to a wide variety of configurations, ranging from small components to large box-shaped housings and from flat panels to thick, heavily ribbed parts.

Reinforcements of different types and layouts are used. Typically, to make a stampable sheet two layers of reinforcing sheet material are laminated to extruded sheets of thermoplastic (Chapter 3). The resultant sheet is homogeneous and has uniform mechanical properties.

Pultrusion

In contrast to extrusion, in pultrusion a combination of liquid resin and continuous fibers is pulled continuously through a heated die of the shape required for continuous profiles. Shapes include structural I-beams, L-channels, tubes, angles, rods, sheets, and so on, and the resins most commonly used are polyesters with fillers. Other resins such as epoxies and urethanes are used where their properties are needed. Longitudinal fibers are generally continuous rovings. Glass fiber material (mat or woven) is added for cross-ply properties.

There are six key elements to a pultrusion process, three of which precede the use of the pultrusion machine. The line starts with a reinforcement handling system (referred to as a creel, as used in a textile weaving operation), a resin impregnation station, and the material forming area. The machine consists of components designed to heat, continuously pull, and cut the profiles to a desired length. With machines producing profiles, line speeds range at least from 1 to 15 ft/min (with large to small cross sections).

The process starts when reinforcements are pulled from the creels and through a resin bath where they are impregnated with formulated/mixed resin. In some operations, previously resin-impregnated reinforced tapes replace the creel and bath stations. The resin impregnated fibers are usually preformed to the shape of the profile to be produced. This composite then enters a heated steel die that has been precision-machined to the final shape of the part. Wheel pullers, clamps, or other devices continuously pull, and when the profile exits the die/mold, it is cured. The profile finishes its cooling in ambient air, water, and/or forced air as it is continuously pulled. The product emerges from the puller mechanism to be cut to the desired length by an automatic device such as a flying cutoff saw.

Control devices must be used to ensure that the proper resin impregnation occurs and is held within the required limits. Simple devices, such as "doctor" rolls or squeeze rolls, are usually sufficient. It is important to control the resin viscosity. The most difficult part to set up is the shape of the opening in the die/mold. Experience and/or trial and error are required.

Continuous Laminating

Flat and corrugated, translucent to opaque, panels can be made in presses, but they are more commonly made by continuous laminating. A layer of liquid polyester is deposited on a moving belt (covered with a film) and passed under a chopper that cuts glass fiber rovings into lengths, commonly 1.5 to 2 in. (37.5–50 mm) and deposits them into a random mat on the resin. The fibers and resin are compacted and covered with an upper film, such as cellophane or PE, and passed through squeeze rolls. The composite envelope may be passed through corrugators or remain flat. Next it passes through an oven to cure. Subsequently, the cover films are stripped, and the continuous sheet is cut into the desired lengths.

Centrifugal Casting

Cylindrical shapes such as pipe, tubing, and tanks can be made by placing chopped-strand mat and/or directional chopped rovings against the inner wall of a hollow mold (such as a cylinder, etc.). The mold is heated and rotated, and liquid resin is applied against the fiber. The resin can be delivered through a pipe that moves in the center of the mold, horizontally forward and backward at a controlled rate. Centrifugal force distributes and compacts the resin and fibers against the wall. Hot air may be blown through the mold to accelerate the cure. When the part is cured, the mold is stopped and the part removed.

Encapsulation

This extremely simple process lends itself to high-volume production and automation at a very low cost. Inserts of any size, shape, and quantity can be encapsulated. There is little—or generally no—material waste. Milled fibers or short chopped glass strands are combined with slightly exothermic polyesters, epoxies, silicones, and other materials, and mixed prior to being poured into open molds. Molds can be made in simple forms to complicated shapes, using many different materials (wood, plaster of paris, concrete, Kirksite, aluminum, etc.). Depending on the mix, cures can be achieved at room temperature or heat. The cure cycle can be controlled so that an unblemished part is produced. If required, a vacuum is used with the composite during mixing and/or during curing to eliminate air bubbles or voids.

Filament Winding

In filament winding (FW), continuous filaments are wound onto a mandrel after passing through a resin bath, unless preimpregnated (prepreg) filaments or tapes are used, which eliminate the resin bath. The shape of the mandrel is the internal shape of the finished part. The configuration of the winding depends upon the relative speed of rotation of the mandrel and the rate of travel of the reinforcement-dispensing mechanism. The three most common types are: helical winding, in which the filaments are at a significant angle with the axis of the mandrel (Fig. 7–10); circumferential winding, in which the filaments are wound like thread on a spool; and polar winding, in which the filaments are nearly parallel to the axis of the mandrel, passing over its ends on each pass.

Different configurations can be employed on successive passes, and the orientation of the filaments tailored to the stresses set up in the part. For example, with pipe, continuous helical winding can be employed on a segmental mandrel, on an extruded mandrel, or on release film placed on a stationary mandrel. Other filament winders include braiding machines, loop wrappers, small to very large storage tank machines, rectangular box-frame machines, and many different special fiber-placement machines with several degrees of freedom for intricate shapes.

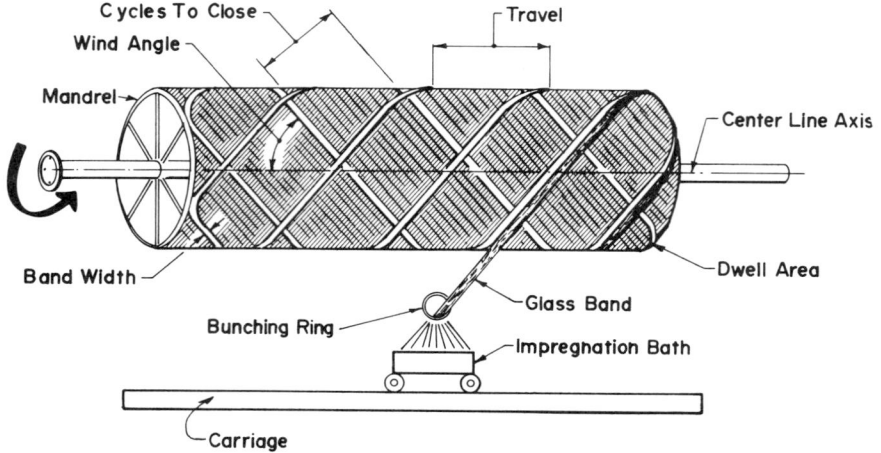

Fig. 7-10. Schematic of helical filament winding.

Curing can be performed by ovens, autoclaves, vacuum bags, heating lamps, and so on. Room-temperature cure systems also are used. In some applications, after being wrapped around the mandrel and prior to curing, the very high-glass-content, preimpregnated–oriented composite is removed from the mandrel. The removal is usually done by cutting in the axial direction. This sheet can be cut to required sizes and placed in a heated two-part compression mold that provides the shape and the curing stage, in a procedure similar to the stamping of TP sheets. The compression mold can be open or (usually) closed. In a matched two-part closed mold, the cutting of trim edges can be used to provide a finished part (Fig. 6-3). To take advantage of the FW orientation capability, TP resins may be used instead of the usual TS resin. In turn, the TP sheet is used in the conventional stamping procedure (see above).

The FW mandrel must be strong enough to withstand rather high accumulated tension loads due to the filament winding, and must be stiff enough not to sag between end supports. At the same time, it must be possible to remove the mandrel from the finished part after curing, which may require the use of an intricate collapsible mandrel.

As filaments are continuous and tightly packed, they permit a high filament-to-resin ratio. This capability often results in products having the highest strength-to-weight ratio obtainable in any structures.

Even though most FW uses glass filaments, all types of filaments can be used (Table 7-1). Precautions must be observed if superior properties are to be achieved. Glass fibers are strong, but as glass they are subject to a severe loss in strength with surface abrasion. They must be carefully handled and processed to avoid such deterioration. In a lay-up for FW, as well as others, plastic abrasion-resistant fibers or (usually) film can be included. This construction permits parts to operate in severe load environments, such as vibrations, twisting, and so on, and eliminates or significantly reduces glass-to-glass abrasion where a high fiber-to-resin ratio exists. Other types of fibers should be studied to determine whether fiber damage can occur when the part is in service. Certain fibers, with or without resins, might be brittle, and other problems could develop. The designer of the part should have knowledge of potential problems. If problems do develop, steps can be taken during processing to overcome them. If unwanted porosity occurs, liners (gel coatings, elastomeric materials, etc.) can be included during FW.

Injection Molding

As explained in Chapter 2, most IM parts are made from TP, and some of the TP uses milled glass fibers to improve part performance. Other fibers have seen limited use to date. TS compounds usually include reinforcements. Details on IM and factors that influence machine and mold performance, such as wear and abrasion, are reviewed in Chapter 2.

TS reinforced molding compounds processed in IM require water- or oil-cooled barrels, rather than the electrically heated barrels used for TP. The heat of the TS during screw plastication has to be kept lower than its curing heat. The mold heat is higher than that of the resin, causing the TS finally to react and solidify. If there is excess heat in the barrel, the resin cross-links and the machine stops operating. A cleanup is required, which can cause machine downtime to be extensive. With a liquid heat control, better control of melt heat is maintained. Also the compression ratio of the screw used to process the TS is "one," which helps to keep the heat low and, more important, under control.

IM machines are available that can process BMC using stuffer mechanisms in the hopper (as in certain extrusion operations; see Chapter 3). Depending on the "stiffness" of the BMC, these machines use a plunger rather than a screw.

Coining

See Chapter 8 for details on the coining process. With reinforced plastic this process (also called injection-compression molding) provides a means of controlling fiber orientation, and so on.

Table 7-9. Troubleshooting RP Processes.

PROBLEM	POSSIBLE CAUSE	SOLUTION
Nonfills	Air entrapment	Additional air vents and/or vacuum required.
	Gel and/or resin time too short	Adjust resin mix to lengthen time cycle.
Excessive thickness variation	Improper clamping and/or lay-up	Check weight and lay-up and/or check clamping mechanisms such as alignment of platens, etc.
Blistering	Demolded too soon	Extend molding cycle.
	Improper catalytic action	Check resin mix for accurate catalyst content and dispersion.
Extended curing cycle	Improper catalytic action	Check equipment, if used, for proper catalyst metering. Remix resin and contents; agitate mix to provide even dispersion.

TROUBLESHOOTING

Each of these processes has common as well as specific approaches to setting up a troubleshooting guide. Throughout this book (in Chapter 2 and elsewhere) this subject is reviewed, and those discussions can provide approaches and ideas for setting up guides. Here are some general suggestions (see also Table 7-9):

1. Check mixing and/or pumping equipment. Adjust the resin mix or pumping equipment to achieve the proper time period.
2. Determine whether the mold requires proper preparation to achieve part release, using wax, PVA, and so on.
3. For applicable processes, check to determine whether the reinforcement is properly located so as not to interfere with mold stops, seals, and/or bleed ports.
4. For applicable processes, see that the clamping frame does not interfere with mold closing.
5. Where heating and/or cooling systems are used, determine that they are operating properly. Check instruments to ensure that they are recording properly.

Chapter 8

OTHER PROCESSES

REACTION INJECTION MOLDING

The RIM process involves the high-pressure impingement mixing of two or more reactive liquid components, and injection of the mixture into a closed mold at low pressures (Fig. 8-1). Large and thick parts can be molded using fast cycles with relatively low-cost materials. Its low energy requirements with relatively low investment costs make RIM attractive (1, 24, 267–269).

Different materials can be used, such as nylon, polyester (TS), and epoxy, but TS polyurethane (PUR) is predominantly used. Almost no other plastic has the range of properties of PUR—a modulus of elasticity in bending of 200 to 1,400 MPa and heat resistance from 90°C to over 200°C (the higher values are for chopped glass fiber reinforced RIM, or RRIM; see Chapter

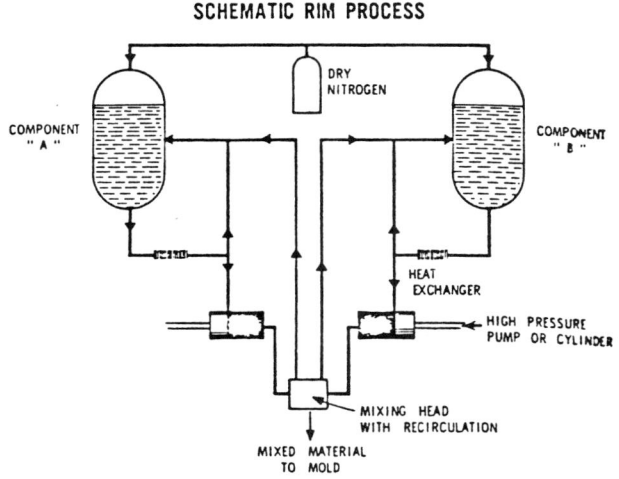

SCHEMATIC RIM PROCESS

Fig. 8-1. Schematic of PUR-RIM process.

7). With PUR, a chemical reaction between the mixed liquid components, polyol chain extender, and isocyanate takes place inside the mold after each shot, enabling the properties of the moldings to be predetermined and controlled. Depending on the mix, the cycle times can be reduced. In the past ten years, cycle times for typical PUR-RIM or PUR-RRIM have been reduced from 3 min. to about 30 s because of the development of a faster reaction speed (glycol chain extenders were replaced by amine chain extenders).

A major development has been to provide easy release of parts from the mold. Special mold treatments and careful physical removal of parts were required, increasing the cost. PURs now incorporate an internal mold release (IMR) agent that improves part performance and process economics. However, thin-walled, complex parts may require the use of release agents on the mold. Material developments are under way to rectify this situation.

Table 8-1 compares PUR-RIM and injection molding of unreinforced and glass fiber reinforced thermoplastics, in the production of parts with large surface areas. RIM also is comparable to resin transfer molding (RTM;

Table 8-1. Comparison of RIM and Injection Molding of Unreinforced and Reinforced Plastics, in the Production of Parts with Large Surface Areas.

	PUR-RIM	INJECTION MOLDING
Plastic temperature, °C	40–60	200–300
Plastic viscosity, Pa·s	0.5–1.5	100–1,000
Injection pressure, bar	100–200	700–800
Injection time, s	0.5–1.5	5–8
Mold cavity pressure, bar	10–30	300–700
Gates	1	2–10
Clamping force, t	80–400	2,500–10,000
Mold temperature, °C	50–70	50–80
Time in mold, s	20–30	30–80
Annealing	30 min. @ 120°C	Rarely
Wall/thickness ratio	1/0.8	1/0.3
Part thickness, typical maximum, cm	10	1
Shrinkage, %		
Unreinforced	1.30–1.60	0.75–2.00
Reinforced—glass		
parallel to fiber	0.25	0.20
vertical to fiber	1.20	0.40
Inserts	Easy	Costly
Sink marks around metal inserts	Practically none	Distinct
Mold prototype, months	3–5 (epoxy)	9–12 (steel)
Mold alterations	Cost-effective	Costly

see Chapter 7) in regard to the processing of TS resins and the molding of large surface areas; both processes offer the ability to tailor the reinforcement to the application. RRIM generally delivers faster cycles than other processes, but needs much more expensive high-pressure dispensing equipment to handle the fast-reacting resin systems.

Even though epoxy molds can be used with RIM for small parts and small production quantities, to obtain high surface quality a steel mold must be used. To make RIM more profitable than IM, molds must be designed to utilize their technical advantages, such as proper gating, controlled venting, rheological mold lay-out, mold rigidity, separating edge definitions, proper mold heat profiling, and automatic separation of gates. For example, gates should be designed to conserve material (the PUR used is not recyclable, as it is a TS. (However, the choke-bar gate is still the most generally accepted version.) Designs are no longer found by empirical means during the start-up of molding. Requirements must be learned through experience or trial and error according to the basic theory. Venting requires an approach similar to that of gating.

LIQUID INJECTION MOLDING

LIM has been in use longer than RIM, but the processes are practically similar. The advantages it offers in the automated low pressure processing of (usually) thermoset resins—fast cycles, low labor cost, low capital investment, energy saving, and space saving—may make LIM competitive to potting, encapsulating, compression transfer, and injection molding, particularly when insert molding is required (1).

Different resins can be used, such as polyester, silicones, polyurethanes, nylon, and acrylic. A major application for LIM with silicones is encapsulation of electrical and electronic devices.

LIM employs two or more pumps to move the components of the liquid system (such as catalyst and resin) to a mixing head before they are forced into a heated mold cavity. In some systems, screws or static mixers are used. Only a single pump is required for a one-part resin, but usually two (or more)-part systems are used. Equipment is available to process all types of resin systems, with unsophisticated or sophisticated control systems. A very critical control involves precision mixing. If voids or gaseous by-products develop, vacuum is used in the mold.

FOAMING

Many different foamed plastic products are produced, and practically all processes reviewed in this book can be used to make them, particularly extrusion, RIM, and injection molding. Almost all plastics can be used to make these cellular core structures, which range from flexible to very rigid objects. Basically, the resin is mixed with a blowing agent, which can be a solid, liquid, or gaseous substance that imparts the cellular character to the product. Blowing agents are classified as either physical or chemical.

The physical blowing agents include compressed gases and volatile liquids. The most widely used physical type is compressed nitrogen, which is usually injected in the hot plastic melt before the melt enters a cavity. The advantages of N_2 are that it is inert, leaves no decomposition residue, and is not limited to a specific decomposition heat range (1, 270). The volatile liquids are generally hydrocarbons such as hexane or pentane, as well as other aliphatic hydrocarbons and fluorochlorocarbons. The liquids act as a source of gas by changing their physical state from liquid to gas during processing. However, they are not extensively used because they do not provide the processing and performance flexibility of the gas types.

Chemical blowing agents (CBAs) are generally solid materials that decompose when heated to a specific temperature, yielding one or more gases and a solid residue. The CBAs also can be divided, into organic and inorganic types. The most common inorganic CBA is sodium bicarbonate. Its major advantage is low cost, whereas its major disadvantage is that it decomposes over a very broad heat range as compared to organic CBAs, so that its decomposition cannot be controlled as readily as that of the organics. The primary criterion used to select a CBA is the processing heat of the plastic to be foamed.

Sodium borohydrides are inorganic chemical blowing agents that decompose in the presence of a proton donor and thus are applicable over wide processing temperatures. These agents give off hydrogen gas, and can be used to produce parts that are paintable sooner than parts foamed with other CBAs because of rapid hydrogen diffusion out of the part.

Direct injection of nitrogen requires special equipment that will allow the N_2 to be injected into a hot melt, such as an extruder or injection barrel, whereas a CBA does not require this type of equipment. The most common means of using CBAs is by drum tumbling with the resin. A wetting agent such as white mineral oil is commonly used to ensure good adhesion to the resin pellets; this is particularly important if the mix is to be air-conveyed to the hopper. Another mixing method is to use one of the many hopper metering and blending units available, which eliminate the labor required in drum tumbling and a potentially "poorer" mix. A third popular method uses a liquid dispersion of a CBA that is compatible with the resin being foamed. An especially popular approach involves putting the CBA in a resin concentrate. A variety of resins are used and provide excellent dispersion of the CBA.

Any blowing agent selected must fill the usual requirements for plastic additives; that is, the blowing agent as well as its decomposition products should not negatively influence processing conditions, part performance, and toxicological properties of the plastic. It cannot be corrosive or have other side effects. Most blowing processes affect the surface of the foam molded part; the surface may not be smooth or could have a swirl effect. Techniques can be used to provide smooth surfaces, depending on the process being used (1). Use of a hotter mold (and a longer time cycle) tends to improve the surface finish.

With conventional injection molding where a solid part is being molded, CBAs also can be useful. With about 0.5 percent CBA, by weight, surface imperfections such as voids, sink marks, and so on, can be eliminated. The CAB has no effect on part density.

Expandable Polystyrene

EPS molding illustrates the use of blowing agents. Resin beads containing a blowing agent are supplied to the molder in solid form. Each about 0.1 to 0.2 mm in diameter, these beads or spheres contain a small amount of a hydrocarbon liquid, usually pentane, that is used as the blowing agent. Because of its unusual processing and because it represents a large market for plastics (packaging, insulation, etc.), EPS is reviewed in the following paragraphs.

The process involves two major steps. The first step consists of a pre-expansion of the virgin beads by heat (steam, hot air, radiant heat, or hot water). Steam is the most used medium, as it is the most practical, most economical, and so on. In the pre-expansion step, the beads are brought to almost the density required in the molded part, and then they are stored for 6 to 12 h to allow them to reach equilibrium.

The next step conveys these beads, usually through a transport tube by air, to the two-cavity mold. Final expansion occurs in the mold, usually with steam heat, either by having live steam go through perforations in the mold itself or by means of steam probes that are withdrawn as the beads are expanding. During expansion, the beads melt together, adhering to each other and forming a relatively smooth skin, filling the cavity or cavities. With small parts, multiple cavities can be used. After the heat cycle, the cooling cycle starts. Because the EPS is an excellent thermal insulator, it takes a relatively long time to remove the heat prior to demolding, or the part will distort. Cooling is usually done by a water spray on the mold. To facilitate removal, particularly for complex shapes, mold-release agents are used.

An outstanding property of EPS is its extremely low density (when compared to other processes), which—by alteration of the preforming treatment—can be varied according to the end use. Other types of plastics are employed to produce expandable plastic foam (EPF), including PE, PP, PMMA, and ethylene–styrene copolymers. They can use the same equipment with only slight modifications. These plastics have different properties from those of PS, and open up new markets; they provide improved sound insulation, resistance to additional heat deformation, better recovery of shapes in moldings, and so on.

It is important to choose the correct type of pre-expansion to ensure reliable production. There are continuous, single-stage, multi-stage, or discontinuous preforming systems. The advantages of the continuous type over the others are lower equipment costs, higher throughput, simpler construction, easier maintenance, and more reliable production. Its disadvantage is the time needed to change over to a different bulk density, which, depending on the size and throughput, can be from 5 to 15 min. During this time the bulk density is pre-expanded between the old and new settings.

This time disadvantage is avoided in the discontinuous unit, where within certain limits the change is made directly. Equipment manufacturers provide units with different capabilities based on production needs.

Sophisticated control systems are available to manage existing variables. They are needed to handle rapid tooling changes, withdrawal and stacking devices, interactive control systems, and vacuum cooling incorporating steam condensers and/or a central vacuum system.

The most reliable method for filling cavities has been to use a filling gun operating on an injector principle. A method gaining popularity uses compressed air in the hopper. Of the two approaches, the injector method provides a lower capital cost, smaller air requirement, and lower material unit cost, whereas compressed air provides a significant shorter fill time.

Cooling by means of a controlled quantity of water and consolidation under vacuum using condensers now represents the state of the art. With rapid tool-changing systems, the changeover time can be reduced from the usual 1 to 2 h to 15 to 20 min. Another important means of reducing the processing cost is the use of devices that remove parts from the mold and stack them.

ROTATIONAL MOLDING

RM is a simple, basic, four-step process that uses a thin-walled mold with good heat transfer characteristics. This closed mold requires an entrance for insertion of plastic and, most important, the capability to be "opened" so that cured parts can be removed. These requirements are no problem. Liquid or dry-powder plastic, equal to the weight of the final part, is put into the mold, which rotates simultaneously about two axes located perpendicular to each other. With slow rotation about each axis, the material inside the mold tumbles to the bottom, creating a continuous path that covers all mold surfaces equally.

The next step involves heating the mold while it is rotating. Molds can be heated by: a heated oven; a direct flame; a heat transfer liquid, which is either in a jacket around the mold or sprayed over the mold; or electric-resistance heaters, placed around the mold. With uniform heat transfer through the mold, the resin melts to build up a layer of molten plastic on the mold's inside surface.

After the required heat–time cycle is completed, the mold is ready for cooling, which is accomplished with the mold rotating continually. Cooling is usually done by air from a high-velocity fan or by a fine water spray over the mold.

After cooling, the final step is to remove the solid part and reload the mold with plastic.

RM machines can be built very inexpensively, but they are labor-intensive, and their heating is hard to control. Most that are being built have horizontal rotating arms with closed, recirculating, high-velocity, hot-air ovens, with total automation of the complete cycle. Many of these machines also are computer-programmed to obtain consistent part quality. The most common units combine recirculating hot-air, gas-fired ovens with either cast aluminum or fabricated sheet metal molds. For fast operation, machines have four positions—load, heat, cool, and unload—and use four arms, each holding a mold. Thus each position is constantly in use.

This process is capable of molding small to large hollow items with very uniform wall thicknesses, using certain plastics. Production rates, compared to those of other processes, can be low. However, the total cost of equipment and the production time for moderate-sized and, especially, large parts are also low. Large parts range up to 22,000 gal (85,200 L) in size, with a wall thickness of 1.5 in. (0.6 cm). One tank used 5,300 lb of XHDPE; the first charge was about 3,300 lb, followed by 1,000 lb and finally another 1,000 lb.

For a plastic to be suitable for RM, it must flow well enough to cover the mold surface evenly. As a powder, it must have a small-enough particle size and a low-enough melt viscosity to flow without bubbles or voids. Because it will be in contact with air on the inside, it must have good thermal stability. PE usually has a melt index of at least 2 to 3, in order to flow adequately. Solid plastics generally must be reduced to a uniform powder (via grinders, etc.) that will go through a 35 mesh screen, which is about 0.020 in. (500 μm) in diameter.

Molds can be of any shape and can include corrugated or rib constructions to increase their stability and stiffness. The thickness of their walls is limited to allow heat penetration. Figure 8-2 is an example of a mold shape with its rotating mechanism.

COINING

This process, which has been used at least since the early 1940s (1), combines the best of injection molding and compression molding. The process also is called injection-compression or injection-stamping. Basically it involves using an IM machine (Chapter 2) to melt a plastic (unreinforced or reinforced TP or TS) and direct a fixed amount of the melt into a compression mold (Chapter 6). As shown in Fig. 8-3, a compression mold has a male plug that fits into a female mold.

When melt enters the mold, it is not completely closed. Thus the melt literally flows unrestricted in the cavity and is stress-free. After injection is completed, the mold is closed, with the pressure on the melt very uniform. The result is practically a stress-free solid part held to very tight tolerances. This process can result in a faster molding cycle than that of conventional compression or even injection molding. Its disadvantage is the extra cost for the closing mechanism and its control, which becomes insignificant when all the performance requirements that are gained are evaluated.

POWDER COATING

Powder coating is a solventless coating—a coating that is not dependent upon a sacrificial medium such as a solvent, but is based on the performance constituents of solid TP or TS resins. It can be a homogeneous blend

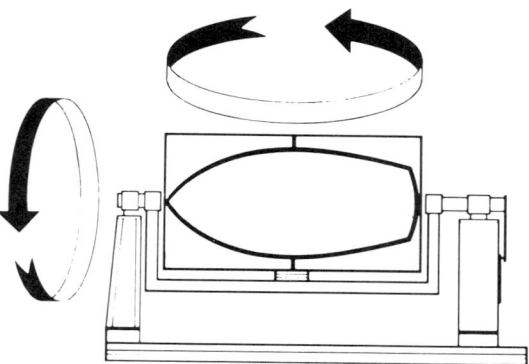

Fig. 8-2. Illustration of a mold and its rotating mechanism, as used in rotational molding.

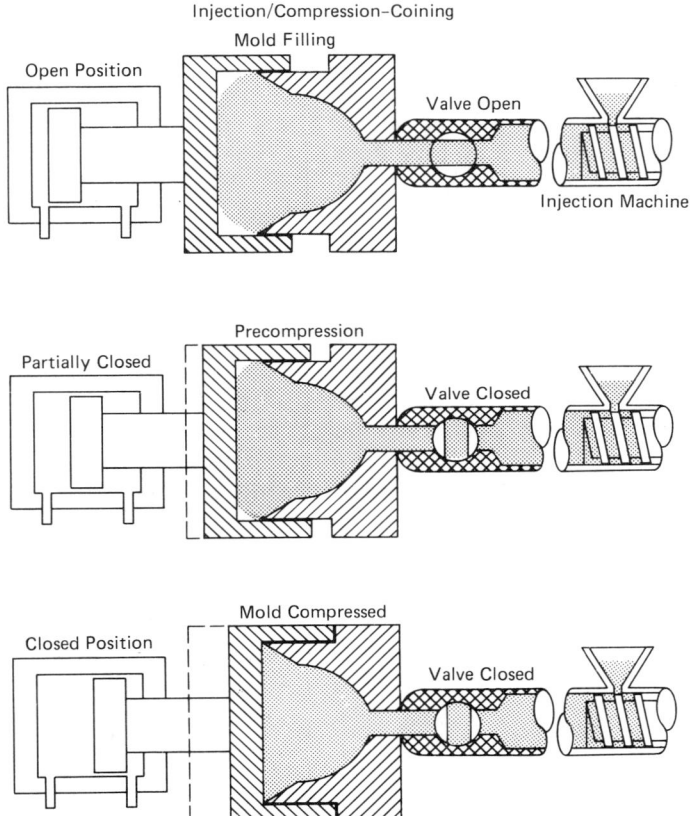

Fig. 8-3. Schematic of the coining process, which combines injection molding with compression molding.

of the plastic with fillers and additives in the form of a dry, fine-particle-size compound similar to flour.

Advantages of the process include its minimizing air pollution as well as water contamination, and increased part performance with coating, resulting in cost savings. This is basically a chemical coating, so it has many of the same problems as solution painting. If not properly formulated, the coating may sag at high thicknesses, show poor performance when not completely cured, show imperfections such as craters and pinholes, and have poor hiding with low film thickness.

The three basic methods are the fluidized bed, electrostatic spray, and electrostatic fluidized bed processes. The earliest powders were applied by blowing or sprinkling the powder onto a hot object in a container of powder. With a fluidized bed, the powder is placed in a bed (container) with a false bottom, which is a flat porous membrane. The bottom does not allow the powder to fall through but does allow an upflow of dry air to distribute it evenly. The result is a fluid action, of air and powder, that allows an object to be coated when entering the "bed" and to be subjected to an even distribution of powder.

The part to be coated is preheated to a temperature above the powder's melting point, and the preheated part then is immersed in the fluidized powder. The powder particles melt and fuse together, forming a film on the part. A post-heat is sometimes required to give the film more flow, a better appearance, and/or a complete cure. The final film thickness is determined by the preheat temperature, the part's mass and its ability to retain heat, the immersion time, the part's movement in the bed, and the velocity of the fluidizing air. The thickness usually ranges from 0.004 to 0.015 in. (0.015–0.060 mm). The particle size of the powder is also important. A particle size that is too fine (about 30 microns) will cause fluidizing problems in the bed and unwanted dusting over the bed.

Advantages of this system are simplicity of the equipment, low cost, and ease of applying the rather heavy film thicknesses. There are also disadvantages: it is not possible to apply film less than 0.004 in. thick, the part size may have to be limited to retain preheat, the size of the bed limits the part size, and color changes involve considerable cleaning unless multiple beds are used.

Electrostatic spraying is successful because most powders are insulators with relatively high volume resistivity values. They will accept a charge (of positive or negative polarity) and are attracted to a grounded or an oppositely charged object.

The system consists of a powder reservoir, a powder feed mechanism, a gun design, a powder generator, an application booth, and powder recycling equipment. The powder reservoir usually is a fluidized bed or a mechanically agitated hopper. From the reservoir, the powder is fed to the gun—a step accomplished by a venturi air pump. Arriving at the gun, the powder/air mixture is charged. The corona charge occurs internally at the tip of the gun or at an electrode close to the powder exit. The charge from the power pack varies from 60 to 120 kV and 200 to 400 microamps. Also located at the tip of the gun is a diffuser, to direct the powder and to shape the powder cloud.

The thickness of applied powder is dependent on the part's speed through the cloud, the cloud pattern, the powder feed rate, the air currents in the booth, the gun arrangements, and the resistivity of the powder. The primary advantage of this system is the low film thickness of coatings applied to a cold part. It is also readily automated, even if there are various-size parts. The initial transfer efficiency varies from 50 to 80 percent, but may reach 98 percent powder usage with an efficient powder recycling system.

Changing the powder can present a problem with this system because the procedure is long and difficult compared to procedures used with other coating systems. Another disadvantage is spray formation, which prevents powder deposition in small openings and tight angles. Also, because of the resistivity of the powder, thick films, in excess of 0.008 in. (0.03 mm), can be difficult to deposit.

The electrostatic fluidized bed is a combination of the electrostatic spray and fluidized bed methods. A current of air passing upward through the bed fluidizes and suspends the powder, which is charged by electrodes in the permeable membrane. The voltage applied determines the density of the cloud when a ground is passed over the bed. A cold and grounded object is passed over the bed, not immersed in it. The electrostatic principles that apply to powder spray are pertinent here also. Coating thicknesses can range from the low values of electrostatic spraying to the high values of the fluidized spray method.

VINYL DISPERSION

Vinyl dispersions are fluid suspensions of special fine-particle-size polyvinyl chloride resins in plasticizing liquids. When the PVC is heated to about 148 to 180°C (300–355°F), fusion or mutual solubilization of the resin and the plasticizer takes place. The dispersion turns into a homogeneous hot melt. When the melt is cooled below 50 to 60°C (122–140°F), it becomes a tough vinyl product.

With vinyl dispersions, the processor can use convenient liquid handling techniques such as spraying, pouring, spread coating, and dipping. This system permits products to be made that otherwise would require costly and heavy melt processing equipment.

The term "plastisol" is used to describe a vinyl dispersion that contains no volatile thinners or diluents. Plastisols often contain stabilizers, fillers, and pigments along with the essential dispersion resin and the liquid plasticizer. All ingredients exhibit very low volatility under the processing and use conditions. Plastisols can be made into thick fused sections with no concern for solvent or water blistering, as with solution or latex systems; so, they are described as being "100 percent solids" materials.

It is convenient in some instances to extend the liquid phase of a dispersion with organic volatiles, which are removed during fusion. The term "organosol" applies to these dispersions.

CASTING

Some TPs and TSs begin as liquids that can be cast and polymerized into solids. In the process, various ornamental or utilitarian objects can be embedded in the plastic. By definition, casting applies to the formation of an object by pouring a fluid monomer–polymer solution (Chapter 1) into an open mold where it completes its polymerization. Casting can also lead to the formation of film or sheet, made by pouring the liquid resin onto a moving belt or by precipitation in a chemical bath. Casting differs from many of the other techniques described in this book in that it generally does not involve pressure or vacuum, although certain materials and complex parts may require pressure or vacuum casting.

The starting material is usually in liquid form, rather than the granular or powdered forms that go into most molding systems. The material is generally a monomer rather than a polymer (Chapter 1). Chemically a monomer is a relatively simple compound that reacts to form the plastic. The plastics that are used for most of the other molding processes have already been polymerized, except in RIM. Some takeoffs on the casting process, such as rotational molding (which is also known as rotational casting), have developed into highly individual processing methods with their own unique technology. For example, acrylic sheet can be made by casting a liquid monomer between two polished sheets (usually glass) to produce an extremely smooth, highly polished, and very accurately controlled thickness. The sheets also have two surfaces that are extremely parallel, resulting in very accurate light transmission.

CALENDERING

The calendering process, used in the production of plastic films and sheets, converts plastic into a melt and then passes the pastelike mass through nips of a series of heated and rotating speed-controlled rolls into webs of specific thickness and width. The web may be polished or embossed, and may be either rigid or flexible.

A wide variety of plastics can be used although about 80 percent is PVC and 15 percent ABS. (Calendering consumes about 6 percent of total plastics consumption for all processes.) Other plastics used are HDPE, PP, and styrenes. The basic plastic limitation of the calendering process is the need to have a sufficiently broad melt index to allow a heat range for the process. This permits the material to have a relatively high viscosity in the banks of the calender (banks indicating where two rolls meet, or the nip of the rolls). As a result of the viscosity, a shear effect can be developed throughout the process, and especially between the calender rolls. Thus, the calender forms the web as a continuous "extrusion" between the rolls. Unlike the process in an extruder (Chapter 3) or in injection molding (Chapter 2), the plastic mass cannot be confined when being calendered. Because of the lack of confinement, the shear effect and a broad melt band are essential aspects of calendering.

The blending or compounding of the plastic with different additives and fillers is a critical part of the process. The blending must produce a uniformly colored and stabilized product, in powder form. After blending, the rate of consumption dictates the temperature of the melt. Because the plastic is processed between the process heat and its critical heat of degradation, the time at heat becomes extremely critical and an important part of the process. For example, the processor will minimize the amount of melt in the bank (nip) of the rolls. The residence time of the plastic flux at high heat must be limited. PVC is especially heat-sensitive.

Handling of scrap and cold trim from the product line poses a potentially very difficult problem. The scrap and trim could represent from 10 to 40 percent of the mix with virgin material, depending on the width of the calender in relation to the sheet width. The flux rate and energy required to remelt the scrap are considerably less than that required to flux the virgin plastic, so there is some danger of decomposition of the material whenever scrap is processed. However, for optimum uniformity, it is necessary to prepare the compound, blending both new and old material to obtain product standardization. Reprocessed material thus is best added to the blender when a standard can be established, although in some plants it is fed directly to the fluxing unit. Careful control of the scrap percentage in the total mix is essential to obtain a good-quality product.

These lines are very expensive to set up and operate, so they are most economically operated on long runs. They compete with film and sheet extruders. Extreme care is required to ensure that the material or the processing area is not contaminated. Any foreign matter can severely damage the expensive and highly polished rolls, and cause other serious damage.

GAS INJECTION

Parts can be molded by gas injection using injection molding machines (Chapter 2). This process is most effective and economical when used for large parts. It offers a way to mold parts with only 10 to 50 percent of the clamp tonnage that would be necessary in conventional IM (1).

The technique—practiced in several variations, with some patented—involves the injection of an inert gas, usually nitrogen, into the melt as it enters the mold. This is not structural foam, as no foam core is produced; instead, the gas forms a series of interconnecting hollow channels in the thicker sections of the part. The gas pressure is maintained through the cooling cycle. In effect, the gas packs the plastic into the mold without a second-stage high-pressure packing in the cycle as used in IM, which requires high tonnage to mold large parts.

Molded-in stresses are minimal. The thick but hollow sections provide rigidity and do not create sink or warpage problems. The cycle time is reduced because the thick sections are hollow. As the gas is not mixed with the melt, there is no surface splay, which is typical of low-pressure structural foam molding (1). Gas injection is being used with commodity and engineering resins.

Chapter 9

AUXILIARY EQUIPMENT AND SECONDARY OPERATIONS

INTRODUCTION

As emphasized throughout this book, many different types of auxiliary equipment and secondary operations can be used to maximize overall processing plant productivity and efficiency. Their proper selection, use, and maintenance are as important as the selection of the processing machines (injection molder, extruder, etc.). The processor must determine what is needed, from upstream to downstream, based on what the equipment has to accomplish, what controls are required, ease of operation and maintenance, safety devices, energy requirements, compatibility with existing equipment, and so on. This chapter provides examples of this selection procedure and its importance in evaluating all the equipment required in a processing line. Details on all the equipment that is available can be obtained from plastics industry trade publications, usually compiled in an annual issue. These and other pertinent publications are included in the reference section (1–4, 33, 271–289).

MATERIAL HANDLING

In most processes, for either small or large production runs, the cost of the plastics used compared to the total cost of production in the plant may be at least 60 percent. The proportion might be only 30 percent, but it is more likely to exceed 60 percent; so it is important to handle material with "care" and to eliminate unnecessary production problems and waste. Where small-quantity users or expensive engineering resins are concerned, containers such as bags and gaylords are acceptable; but for large commercial and custom processors, these delivery methods are bulky and costly. Resin storage in this form is also expensive.

Any large-scale resin handling system has three basic subsystems: unloading, storage, and transfer. For a complete system to work at peak efficiency, processors need to write specifications that fully account for the unique requirements of each subsystem. The least efficient component, no matter how inconsequential it may seem, limits the overall efficiency of the entire system. The guidelines presented here will help one to specify material handling systems from the time of delivery to the plant on to processing machines that will maximize both efficiency and capacity.

Railcars can be unloaded in many different ways, but some suppliers simply recommend inexpensive back-to-back flexible hose assemblies to unload them. To save time at the loading dock, one should use a stationary pipe manifold arrangement along the length of the rail siding with connections about every 15 ft. Thus transport is achieved with one short flex hose from car to car instead of the handling of as much as 100 ft of heavy flex hose. Labor and unloading time will be saved, and the pressure drops inherent in long multiple-connection flex hose runs will be avoided.

The easiest method of unloading uses a vacuum pump/dust collector, which can be located in the silo skirt. The pump induces a vacuum in the line, drawing resin from the railcar into a vacuum loader. When the vacuum loader fills, the pump stops, and the resin dumps into the silo. This on–off batching effect keeps transfer rates relatively low, typically 6,000 to 7,000 lb/h. If the manifold pickup is far from the silo, the transfer rates will drop.

If unloading requires high speeds or transfer over long distances, one should consider a push/pull system; resin is pulled from the railcar by negative pressure, and then pushed by positive pressure to its final location. Some equipment suppliers recommend a one-blower system, but that limits transfer capacities, requires more work-hours, and can actually degrade resin through excessive heat transfer within the air stream. One-blower systems can also lead to line blockages, as they lack purging capabilities.

The ideal way to unload railcars and transfer resin to storage silos is to use a two-blower system. One blower handles railcar unloading through negative pressure; the other transfers resins to silos using positive pressure. Splitting up the unloading and transfer of resin between two blowers permits peak efficiency in each operation. An improper pressure balance across a single-blower system could greatly reduce transfer rates at a critical point in the system, making the single blower work harder, increasing the heat of the conveying air, and causing resin degradation, particularly for thermally sensitive resins.

A two-blower system times purging to virtually eliminate any line blockages. With this arrangement, the pressure blower continues operating to ensure complete transfer to silos, while the vacuum blower remains out of the transfer loop.

Silos and other containers provide more than just a place to stow away resins until it is ready for use; silos protect resins from environmental damage caused by excess moisture, atmospheric pollutants, and solar radiation. Moreover, silos should require a minimum amount of maintenance; they should not leak, nor should they introduce rust or other contaminants in the resin. They should resist structural damage from environmental corrosion.

There are three types of silo construction: (1) spiral, with silos fabricated from spun aluminum; (2) bolted, with silos made of carbon steel, stainless steel, or aluminum; and (3) welded, also with silos constructed of carbon steel, stainless steel, or aluminum. Silos intended for pellet storage usually have 45-degree discharge cones at the bottom. Powders and hard-to-flow resins usually have a 60-degree cone to guarantee that the cone angle is greater than the resin's angle of repose.

Each type of silo has both advantages and disadvantages. Spirals require the least maintenance. Bolted construction can cost less than spirals, but with erection costs that can be three times higher. Because rubber gaskets are used, bolted silos have a greater tendency to leak. Welded silos have the least manufacturing cost of all, but their installation could be problematic and costly. So it is important to study available storage systems in order to choose the supplier best able to meet individual specification requirements.

If computer-integrated resin handling systems are considered, one must compare their operating procedures with one's process requirements. These process requirements describe the flow of resin and product through the system, which determines the system's electronic architecture. Pertinent considerations include batch vs. continuous operations, the type and number of conveying lines, resin storage and distribution, quality control means and procedures, inventory control, the type and quantity of process parameter sensors, the type and quantity of controlled devices, modes (automatic, semiautomatic, manual, and/or shutdown modes), process information, process management controls, and centralized vs. local operation (Fig. 9–1).

Energy Conservation

Energy conservation is only one of many factors that should be considered in the selection of an automated materials conveying system (as well as all equipment used in the processing line). Fortunately, any steps taken to save energy will also save money in most cases. The traditional arguments favoring the silo are savings on resin costs, labor savings through the elimination of handling bags and cartons, the saving of costly warehouse inside floor space, and energy savings. For example, if a plant used a large quan-

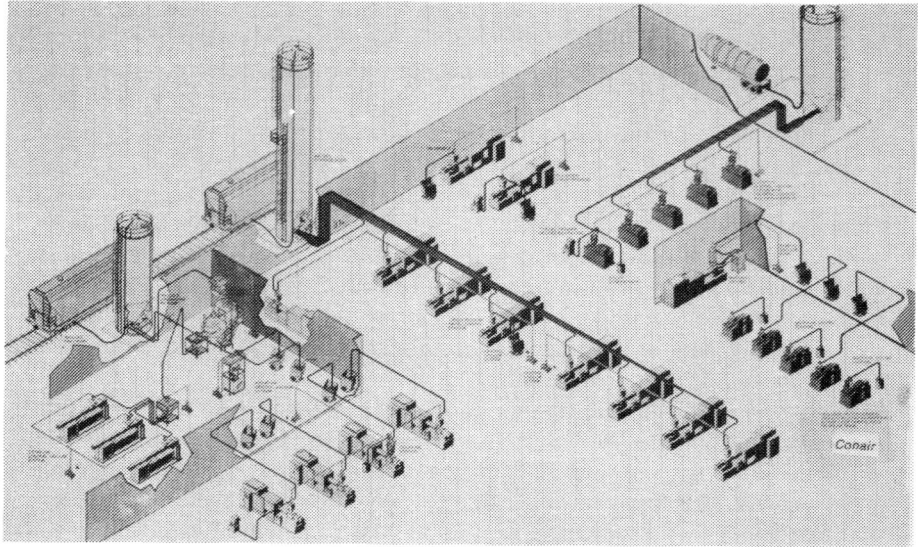

Fig. 9–1. Schematic of bulk resin handling; a route to better production economics.

tity of resins and did not use silos, during the winter months bags or gaylords would be delivered repeatedly through "open" delivery doors, and warm air would be lost.

With automatic delivery from silos, all resin handling lines are kept as short as possible. There is no reason for lines to conform to the right angles of the walls; they should follow a straight line from the resin's source to where it has to be delivered. There are graphs from systems suppliers that show the relationship between the length of conveyor lines and power requirements (1). The graphs also show the horsepower required, based on different factors—such as the length and diameter of the delivery pipe, the position of the pipe, the type of resin being conveyed, the size of the hopper at the machine, and lb/h that can be delivered.

A graph will show, for example, that with an average pellet size of ⅛ in., 35 lb/cu ft bulk density, and a conveying vacuum of 12 in. of mercury (Hg), PE can be moved in different ways. With a 25 HP vacuum hopper unit, a line will convey 18,000 lb/h of PE if it is only 100 ft long. Using the same power and a 450-ft line, less PE will be moved. Suppliers' data will show the horsepower required to move a material in the fastest way with the least energy consumption. To convert energy consumption to electric-bill charges, one uses the following formula: 1 HP = 0.745 kW, and 1 kW × 3,413 = Btu (1).

With a long pipeline, a 25 HP vacuum pump could be used. If the line could be shortened, a 10 HP unit would convey the same amount of plastic, resulting in power savings. One foot of vertical height in the line equals 2 ft of horizontal distance in its effect on conveying rates. Bends in the pipe add a considerable amount of equivalent footage.

If the lines cannot be kept short or relatively free of bends, then pressure drops will exceed 12 in. Hg, requiring more horsepower. Where the pressure drop is greater than 12 in. Hg, which is the normal operating and limiting pressure for most vacuum systems, it is necessary to use a positive pressure system. All these factors must be considered initially to obtain the best delivery system most economically.

PARTS HANDLING

The logic and approach used in materials handling also applies to the use of handling equipment to move processed parts. Parts handling equipment (PHE) does not resemble the humanoids of science fiction. Robots are blind, deaf, dumb, and limited to a few preprogrammed motions; but in many production jobs that is all that is needed. They are solutions looking for a problem. Most plants can use some degree of PHE, and it can substantially increase productivity.

Use of PHE can range from simple operations, as reviewed, to rather complex operations with very sophisticated computer controls. Although the concept of automatic operations is very appealing, the ultimate justification for PHE (like material handling, process controls, etc.) must be made on the basis of economics. At times it may provide the solution to "handling" a part that otherwise would be damaged. Tables 9–1 and 9–2 provide information on PHE for injection molding.

CLEAN PELLETS/PLANT

When plastics are extruded and pelletized, varying amounts of oversized pellets and strands are produced, along with some fines. When the plastics are dewatered/dried or pneumatically conveyed, more fines, fluff, and streamers may be generated. In many cases, these "problem products" must be eliminated, significantly reduced, or removed during processing in order to produce good-quality products. The process or product requirement may not be influenced, but with high throughputs, these interferences can present all kinds of problems. Even if acceptable products are produced, costs will be increased by the greater processing time, energy requirements, process controls, and reject rate that result, and/or other problems that may develop.

Table 9-1. Example of Parts Handling Equipment Functions.

	TYPE	COLLECT	REMOVE OR PICK	PLACE	ORIENT	COUNT/WEIGHT	ACCUMULATE
Not integrated with IMM function	Manual	X	X	X	X	X	X
	Box	X	X				X
	Conveyor	X	X				X
	Unscramble/orient	X	X	X	X	X	X
Integrated with IMM	Sweep		X				
	Extractor	X	X	X			
	Cavity separator	X	X	X	X	X	X
	Robot/bang-bang	X	X	X	X	X	X
	Robot/sophisticated	X	X	X	X	X	X

Fines or dust are created when small particles are torn away from the pellet and hit a rough surface at high speed (rough pipe wall, etc.). With smooth pipe walls, the pellet will slide, generating frictional heat, and could become sticky before it deflects (depending on the melting point). The pellet's surface can actually smear off, leaving a film streak on the surface of the pipe. In time, this film can cover the whole inside of the pipe, and it can tear off in the form of fluff (short threads or hairlike particles) and subsequently streamers (long ribbonlike particles, also called angel hair or snakeskins). In addition to causing dirty products, the streamers can form into balls and create blockages through the rest of the system.

The entire action that produces these contaminants is based on a combination of conditions of heat, weight, and velocity. The first appearance of fluff indicates that an operational threshold has been crossed. As conditions worsen, longer streamers form large quantities of this material.

These fines and other undesirable materials can cause a variety of handling and processing problems; so if they exist, they must be removed. Suppliers have different removal systems such as filters, cyclones, and so on.

Some in-plant systems, such as in-line filters, are high-maintenance items. As the filter element fills, the pressure increases while the filtering system's efficiency decreases. The filter must be checked and cleaned on a regular schedule to maintain its efficiency.

To remove dust or fines, consider using a compressed-air, backwash-type filter/receiver. Electronically controlled pulsing jets effectively shake fines

Table 9-2. Example of Parts Handling Equipment Growth Rate.

TYPE	PERCENTAGE USED WITH IMM		NO. OF MOLD CAVITIES	PART SIZE	COST/UNIT DOLLARS
	CURRENT	FUTURE			
Manual	20	12	any	any	Does not include PHE
Box/collector	30	15	any	Sm, Med	50 to 500
Conveyor	30	30	any	any	500 to 3,000
Unscramble/orient	10	18	2 to 24	Med	2,000 to 40,000
Sweep	3	5	1 to 16	Sm, Med	200 to 1,500
Extractor	4	7	1 to 24	Sm, Med	500 to 5,000
Cavity separator	½	2	12 to 96	Sm	5,000 to 50,000
Robot/bang-bang	2	8	4 to 10	Med	5,000 to 25,000
Robot/sophisticated	½	3	1 to 12	Lge	25,000 to 150,000

from filter media automatically, eliminating human error in maintaining the filter system.

The need for cleanliness of the pellets, as well as the plant, is easy to understand. No casual dirt or dust, contaminants, different compounds, metal chips, or other foreign matter can be tolerated. This type of care also protects the processing equipment.

DRYING PLASTICS

There is much more to drying resins for processes such as IM, extrusion, and blow molding than blowing "hot air" into a hopper. Effects of excessive moisture cause degradation of the melt viscosity and in turn can lead to surface defects, production rejects, and even failure of parts in service (Table 9–3). First one determines from a supplier, or experience, the resin's moisture content limit. Next one must determine which procedure will be used in determining water content, such as "weighing, drying, and reweighing." This procedure has definite limitations. Fast automatic analyzers, suitable for use with a wide variety of resin systems, are available that provide quick and accurate data for achieving in-plant control of this important parameter.

Drying or keeping moisture content at designated low levels is important, particularly for hygroscopic resins (nylon, PC, PUR, PMMA, ABS, etc.). Most engineering resins are hygroscopic and must be "dry" prior to processing. Usually the moisture content is < 0.02 percent, by weight. In practice, a drying heat 30°C below the softening heat has proved successful in preventing caking of the resin in the dryer. Drying time varies between 2 to 4 h, depending on moisture content. As a rule of thumb, the drying air should have a dewpoint of −30°F and be capable of being heated up to 250°F. It takes about 1 cfm of air for every lb/h of resin processed when using a desiccant dryer. The pressure drop through the bed should be less than 0.02 in. H_2O per inch of bed height.

Simple tray dryers or mechanical convection hot air dryers, while adequate for some resins, simply are not capable of removing water to the degree necessary for proper processing of hygroscopic resins, particularly during periods of high ambient humidity. The most effective and efficient drying system for these resins is one that incorporates an air-dehumidifying system in the material storage/handling network. It has to consistently and adequately provide moisture-free air in order to dry the "wet" resin.

Although this type of equipment is expensive initially, it results in improved production rates and aids targets of zero defective parts. There are a variety of manufacturers and systems from which to choose. All the sys-

Table 9-3. Moisture Troubleshooting Guide. (Courtesy of *Plastics World*.)

SYMPTOM	POSSIBLE CAUSES	SOLUTIONS
Silver streaks, splay	Wet material due to improper drying, high percent regrind, over wet virgin resin	Follow resin manufacturer's drying instructions and dryer manufacturer's operating and maintenance instructions Use desiccant dryer
Brown streaks/burning	Contamination	Purge barrel/screw and clean dryer/auxiliary equipment Check resin Check molding equipment settings and controls
	Overheating of material	Check resin manufacturer's instructions about processing temperatures
Bubbles	High moisture content	Check each step of drying process Check dried resins exposure to air
	Trapped air	Force air out of feed vent Increase screw speed and/or back pressure
Brittle parts	Wet resin or overdried resin	Check drying instructions and conditions Increase melt temperature/reduce injection pressure
	Molded-in stresses	Review part design
	Poor part design	Review design for notches and other stress concentrators
Flash	Wet material	Check drying procedures
	Insufficient clamp tonnage	Use larger machine
	Excessive vent depth	Change mold design
	High injection pressure	Decrease injection pressure
	Damaged mold	Repair damage
	Misaligned platen	Realign platen
	Material temperature too high	Decrease material temperature by lowering cylinder temperature, decrease screw speed/lower back pressure (screw machine)
	Cycle time too long	Decrease overall cycle time
	Plunger pushing forward too long	Decrease plunger forward time

(continued)

Table 9-3. Continued.

SYMPTOM	POSSIBLE CAUSES	SOLUTIONS
Material in drying hopper caking or meltdown occurring	Process temperature set too high	Check resin data sheet for meltdown temperature Make sure operators know correct process temperature set point
Dew point reading too high	Dirty process/auxiliary filter(s)	Clean or replace filters (see Note 1)
	Desiccant saturated	Dry cycle machine for several complete cycles (This is common with equipment which is not operated on a continual basis)
	Material residence time in hopper too short	Replace with larger hopper (see Note 2)
	Return air temperature too high	Add "after-cooler" to return air line
	Heaters burned out	Replace
	Bad heater thermostat or thermocouple	Replace
	Cycle timer malfunctioning	Adjust or replace
	Air control valves not seating properly	Adjust
	Contaminated or wornout desiccant	Replace (see Note 3)
	Incorrect blower rotation	Check and correct rotation
	Regeneration heating elements inoperative	Check electrical connections; replace elements if needed
	Desiccant assembly not transferring	If valve system, check and repair valve/drive assembly If rotational system, adjust drive-assembly. Check electrical connections on motor and replace motor if needed
	Moist room air leaking into dry-process air	Check hopper lid, all hose connections, hoses and filters. Tighten, replace, repair as needed
	Dew point meter incorrect	Check meter and recalibrate
Dew point cycling from high to low	Electrical malfunctions	Check electrical connections on heaters/controller. Repair/replace
	Desiccant bed(s) contaminated	(see Note 3)
Process air temperature too high	Incorrect temperature setting	Reset for correct temperature
	Thermocouple not properly located	Secure thermocouple probe into coupling at inlet of hopper
	Electrical malfunctions	Check electrical connections and replace if necessary Insulate hopper and hopper inlet air line

Problem	Possible Cause	Remedy
Excessive changeover temperature	Insufficient reactivation airflow	See "insufficient air flow"
	Malfunctioning cycle time	Adjust or replace
	Blades of blower wheel dirty	Clean
Process air temperature too low	Incorrect temperature setting on-controller	Reset for correct temperature
	Controller malfunctioning	Check electrical connections. Replace/repair if needed
	Process heating elements	Check electrical connections
		Replace/repair if needed
	Thermostat malfunction	Replace or repair
	Voltage differentials	Check supply voltage
	Inadequate air flow	Check/clean filters, check blower rotation and correct, check and repair air flow meter
	Hose connections incorrect	Check connections. Delivery hose should enter hopper at bottom
	Inadequate insulation	Insulate hopper and hopper inlet air line
	Dryer inadequate for required temperatures	Replace with high temperature dryer
Insufficient air flow (Dew point reading could be good but resin is still wet.)	Process or auxiliary filter(s) blocked	Clean or replace (see Note 1)
	Blower rotation incorrect	Check manufacturer's electrical instructions, and change blower rotation
	Air ducts blocked	Remove obstruction
	Air flow meter incorrect	Disengage line exiting dryer repair if needed
	Desiccant bed contaminated	Replace desiccant (see Note 3)
	Tightly packed material in hopper	Increase hopper size or drain hopper and refill
Heater burn out	Excessive vibration	Relocate dryer, reduce vibration
	High voltage condition	Reduce voltage, relocate dryer, or use heaters rated for actual voltage
	Malfunction in heater thermostat	Adjust or replace
	Blades on blower wheel dirty	Clean

Note 1: An inexpensive pressure-differential switch, common option for almost every brand of dehumidifying dryer, will signal when a filter is restricting air flow.
Note 2: Since drying systems tend to be designed for a specific material, different materials may need longer residence times or higher drying temperatures.
Note 3: Plastic dust contaminants, because of their flash-point temperature, can ignite during regeneration of the desiccant bed causing a fire inside the dehumidifier.

tems are designed to accomplish the same end results, but the approaches to regeneration of the desiccant beds vary.

Hygroscopic resins are commonly passed through dehumidifying hopper dryers before they enter a screw plasticator. However, except where extremely expensive protective measures are taken, the drying may be inadequate, or the moisture regain may be too rapid to avoid product defects unless barrel venting is provided (Chapters 2 and 3 review venting). In some plants, to ensure proper drying for "delicate" parts such as lenses, both drying prior to entry into the barrel and venting are used. Although it is much less hygroscopic than the usual resin (ABS, PC, etc.), PS also is usually vented during processing to protect against surface defects.

FEEDER/BLENDER

As with other process equipment, suppliers of dryers, blenders, and metering equipment are increasing the feeding accuracy of the equipment by the incorporation of microprocessor-based controllers that can be easily interlinked with main plant computer facilities. Most processors deal with a wide variety of resins and additives, including regrind, color concentrates, and so on. As the quality and cost of the product depend directly upon how carefully processors measure and mix components, blending represents a critical stage in material handling (such as meeting tight requirements according to a target; see Fig. 9–2).

To mix components easily and economically, processors may use blenders mounted on machine hoppers. In this operation, which offers great flexibility, precisely predetermined proportions of each ingredient flow into a mixing chamber, which should discharge a very accurate mix. The key to process success lies in carrying it out *precisely*. To do this, feeders meter by volume or by weight.

In volumetric blending, variable-speed metering augers feed multiple components into the mixing chamber. A microprocessor provides accurate, repeatable programming and closed-loop control over the variable speed of each metering auger. A single master speed control can also increase or decrease throughput while maintaining the desired blend ratios. Operators can calibrate the actual volume easily by occasionally diverting the component to a sampling chute and weighing a sample from each auger. The microprocessor monitors auger speeds at regular intervals to maintain constant feed rates regardless of variations in the metering torque.

Gravimetric blending improves accuracy and requires less operator involvement in calibration, particularly in running processes where great accuracy is needed. As a result of metering by weight, the overfeeding of expensive additives is eliminated. The principle of gravimetric feeding with

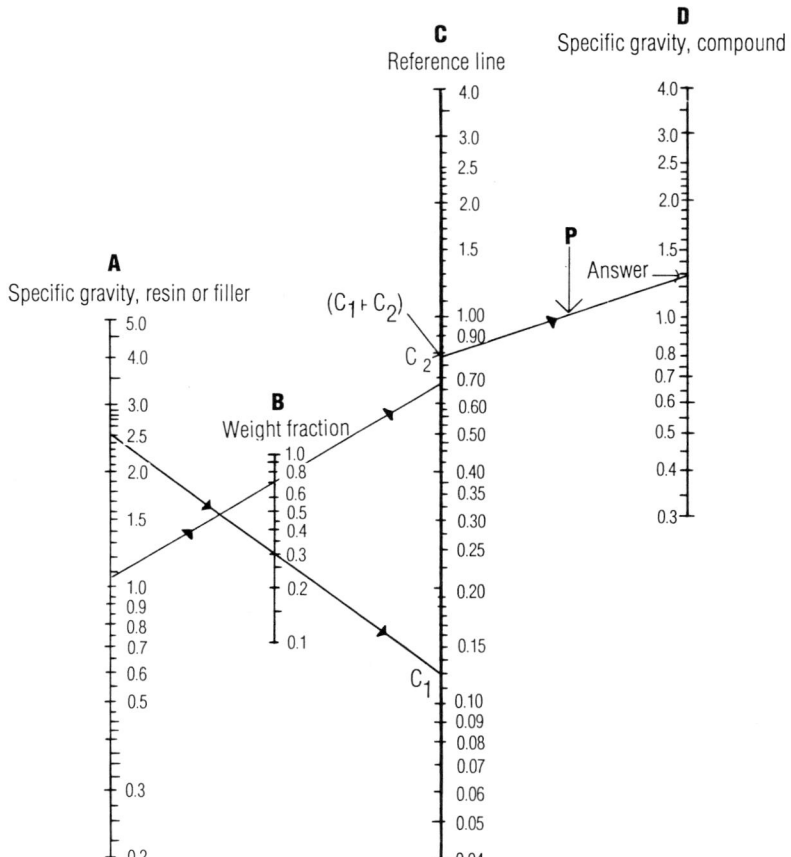

Fig. 9–2. Nomograph for finding the specific gravity of filled compounds using various filler-to-resin ratios.

throughput or metered weight control is well established. Equipment suppliers can assure an accuracy of ± 0.25 to 0.50 percent for ingredient and blend-ratio setpoints at 2 sigma. By comparison, volumetric and quasi-gravimetric blenders that use batch operation usually have accuracy variations of 2 to 10 percent or more.

Basically, each hopper is mounted on a separate weigh-load cell specially designed and isolated to minimize the effects of on-machine vibration and electrical interference. This unique load cell approach updates the controller with weight information so that it can adjust the motor speed for any necessary blend corrections. If the bulk density of any component changes, the blend is not influenced, as the unit only works on weights. Controllers are also capable of refilling individual component hoppers automatically with no mixing delays.

Gravimetric metering is particularly advantageous in coextrusion. In many cases it is not possible to continually determine the individual film thickness or thickness profile of the product. This system offers a very simple means of constantly maintaining the average thickness of the individual films and the overall thickness at better than ± 0.5 percent.

This mixing and metering activity is part of the overall plant conveying process. Conveying includes not only mechanical and pneumatic transport throughout the plant or within each individual processing line, but also filtration of the exhaust air and safety engineering to prevent dust explosion. With so much mixing required, color changes are becoming increasingly frequent. Table 9–4 explains the advantages of coloring at the hopper throat. As explained throughout this book, the processor should determine what is required and obtain equipment to meet his/her specification. An individual evaluation is the best basis for taking advantage of processing line developments.

GRANULATING

In practically all processing plants, it is necessary to reclaim reprocessable thermoplastic scrap, flash, rejected parts, and so on. If possible, the goal is to eliminate "scrap" because it has already cost money and time to go through the process; granulating just adds more money and time. Different types of granulators are available from many different suppliers, and selection of a granulator depends on such factors as the type of plastic used, the type of reinforcement, product thickness and shape (Fig. 9–3), and so on. There are some serious units in which subsystem granulators incrementally reduce the size from a large one to the required small size.

An easy "cutting" unit is required, to granulate with minimum friction, as too much heat will destroy the plastic. General-purpose types have definite limitations. As stressed throughout this book, blending with virgin material definitely influences, and can significantly change, melt processing conditions and the performance of the end product. Recycled material is denser and usually has a variable-size regrind that could affect the product's properties, as shown in Figs. 9–4 and 9–5.

FINISHING/DECORATING

The finishing of plastics includes different methods of adding either decorative or functional surface effects to a plastic product (Tables 9–5 and

Table 9–4. Comparison of Central Blending vs. Coloring at the Hopper Throat. (Courtesy of Maguire Products, Inc., Media, PA 19063.)

CENTRAL BLENDING	AT-THE-THROAT COLORING
1. If the color blend is not correct (parts too light or more color than necessary in the mix), the mix must be emptied from the hopper and re-blended, a costly and time-consuming process. Even then, parts may not have the exact depth of color you desire.	1. Adjustments to color are made with immediate results. If molded parts are light in color, you can make an adjustment and see the results in minutes.
2. While some parts, as a cost-saving measure, could be molded with less color than others, the mixing of several batches with different let-down ratios is generally considered impractical.	2. Every molded part can be custom blended while in production for optimum color usage and cost savings. Settings are recorded and used for exact repeatability on future runs.
3. Conveying of blended material often results in separation of the color pellets from the natural material due to the much heavier bulk density of color pellets. The result is inconsistent coloring of parts.	3. Material is conveyed without color; separation is not a factor.
4. Normal machine vibration and flow patterns in the hopper bring about some separation of the color pellets during residence time in the hopper. Again, the result is inconsistent coloring of parts.	4. Color is metered only as parts are molded. There is no residence time for separation to occur.
5. When a production run is over, any inventory of blended material left in the hopper must be held for use at another time. Storage space, contamination, spillage, mis-labeling and tracking of this valuable inventory are all problems.	5. All inventory of blended materials is ELIMINATED. Unused color is returned to the container it came from—uncontaminated.
6. When a production run for one color is over, your hopper must be emptied in preparation for the next color. Time is lost. Sometimes material is lost. If hopper and conveying system are not adequately cleaned, color contamination and rejected parts will result.	6. Color changes require no emptying of the main material hopper. Production does not stop. Transition time to the new color is several minutes.
7. If your material must be dried before use, the drying of color blends presents many logistics problems; more driers, careful planning for transition to the next color, cleanout of dryers, etc.	7. You dry only natural material without having to concern yourself about color changes.

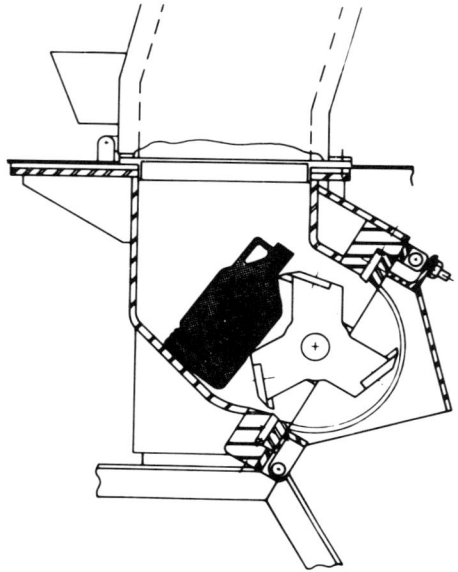

Fig. 9–3. Different types of granulators are available, including those used to properly granulate blow-molded bottles.

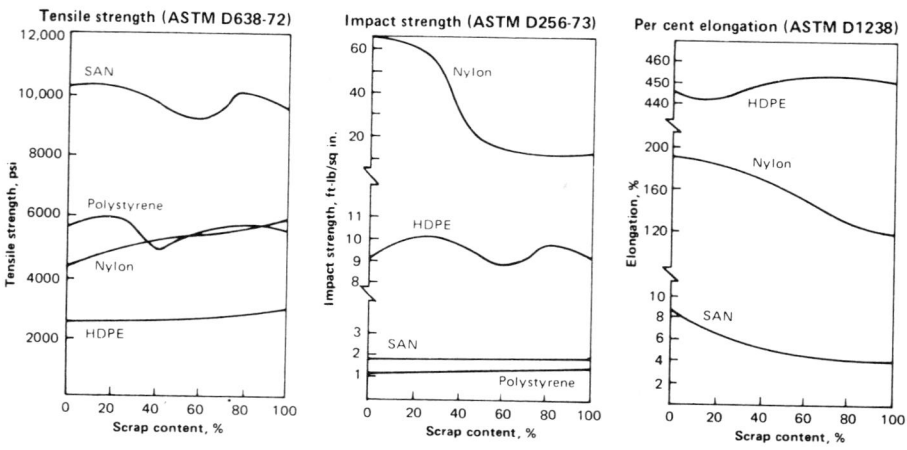

Fig. 9–4. How regrind levels affect mechanical properties of certain formulations of plastics, "once through" and blended with virgin plastics.

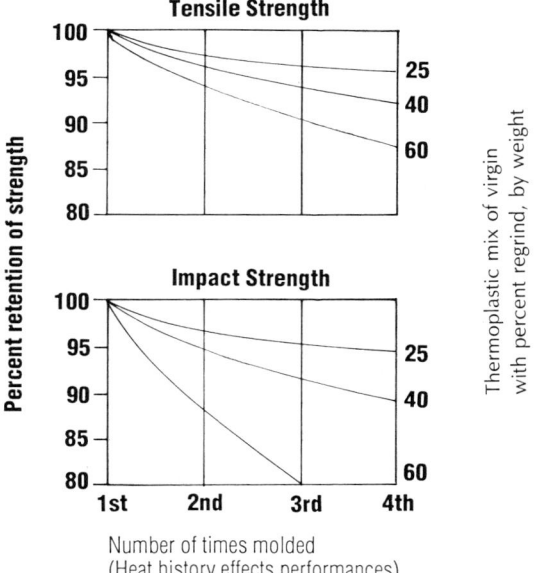

Fig. 9-5. Examples of potential effects of regrind on performance of injection molded TP, mixed with virgin material.

9-6). Plastics are unique, in that color and decorative effects can be added prior to and during processing (Chapter 3, etc.). Two (or more)-color plastic parts are easily processed. Decorative surface textures can be incorporated on the surface of practically any product during or after processing. Decorative foils (plastic, aluminum, etc.), placed on the surface during injection molding, extrusion, blow molding, and so on, can also improve the performance of the plastic.

JOINING/ASSEMBLING

These techniques vary considerably in terms of the plastic and the application. For example, adhesives are widely used. There are solvent systems for most TPs, but not thermosets. Monomeric or polymerizable cements can be used for most TPs and TSs. There are certain plastics with outstanding chemical resistance, such as the polyolefins, that preclude the use of many cements; they generally require some form of surface treatment prior to adhesion, such as flame treatment.

Solvent bonds work because they react chemically with the plastic. However, they literally destroy it so it is important to limit such factors as the

Table 9-5. Printing and Decorating Systems. (Courtesy of *Plastics Technology*.)

THE PROCESS	WHAT IT'S ABOUT	EQUIPMENT	APPLICATIONS	EFFECT
Painting				
1. Conventional Spray	Paint's sprayed by air or airless gun(s) for functional or decorative coatings. Especially good for large areas, uneven surfaces or relief designs. Masking used to achieve special effects.	Spray guns, spray booths, mask washers often required; conveying and drying apparatus needed for high production.	Can be used on all materials (some require surface treatment).	Solids, multi-color, overall or partial decoration, special effects such as wood-graining possible.
2. Electrostatic Spray	Charged particles are sprayed on electronically conductive parts; process gives high paint utilization; more expensive than conventional spray.	Spray gun, high-voltage power supply; pumps; dryers. Pretreating station for parts (coated or preheated to make conductive).	All plastics can be decorated. Some work, not much, being done on powder coating of plastics.	Generally for one-color, overall coating.
3. Wiping	Paint is applied conventionally, then paint is wiped off. Paint is either totally removed, remaining only in recessed areas, or is partially removed for special effects such as wood-graining.	Standard spray-paint setup with a wipe station following. For low production, wipe can be manual. Very high-speed, automated equipment available.	Can be used for most materials. Products range from medical containers to furniture.	One color per pass; multi-color achieved in multistation units.

Process				
4. Roller Coating	Raised surfaces can be painted without masking. Special effects like stripes.	Rolle applicator, either manual or automatic. Special paint feed system required for automatic work. Dryers.	Can be used for most materials.	Generally one-color painting, though multicolor possible with side-by-side rollers.
Screen Printing	Ink is applied to part through a finely woven screen. Screen is masked in those areas which won't be painted. Economical means for decorating flat or curved surfaces, especially in relatively short runs.	Screens, fixture, squeegee, conveyorized press setup (for any kind of volume). Manual screen printing possible, for very low-volume items.	Most materials. Widely used for bottles; also finds big applications in areas like tv and computer dials.	Single or multiple colors (one station per color).
Hot Stamping	Involves transferring coating from a flexible foil to the part by pressure and heat. Impression is made by metal or silicone die. Process is dry.	Rotary or reciprocating hot stamp press. Dies. High-speed equipment handles up to 6000 parts/hr.	Most thermoplastics can be printed; some thermosets. Handles flat, concave or convex surfaces, including round or tubular shapes.	Metallics, wood grains or multicolor, depending on foil. Foil can be specially formulated (e.g., chemical resistance).
Heat Transfers	Similar to hot stamp but preprinted coating (with a release paper backing) is applied to part by heat and pressure.	Ranges from relatively simple to highly automated with multiple stations for, say, front and back decoration.	Can handle most thermoplastics. A big application area is bottles. Flat, concave or cylindrical surfaces.	Multi-color or single color; metallics (not as good as hot stamp).
Electroplating	Gives a functional metallic finish (matte or shiny) via electrodeposition process.	Preplate etch and rinse tanks; Koroseal-lined tanks for plating steps; preplating and plating chemicals; automated systems available.	Can handle special plating grades of ABS, PP, polysulfone, filled Noryl, filled polyesters, some nylons.	Very durable metallic finishes.

(continued)

Table 9-5. Continued.

THE PROCESS	WHAT IT'S ABOUT	EQUIPMENT	APPLICATIONS	EFFECT
Metallizing 1. Vacuum	Depositing, in a vacuum, a thin layer of vaporized metal (generally aluminum) on a surface prepared by a base coat.	Metallizer, base- and top-coating equipment (spray, dip or flow), metallizing racks.	Most plastics, especially PS, acrylic, phenolics, PC, unplasticized PVC. Decorative finishes (e.g., on toys), or functional (e.g., as a conductive coating).	Metallic finish, generally silver but can be others (e.g., gold, copper).
2. Cathode Sputtering	Uniform metallic coatings by using electrodes.	Discharge systems—to provide close control of metal buildup.	High-temperature materials. Uniform and precise coatings for applications like microminiature circuits.	Metallic finish. Silver and copper generally used. Also gold, platinum, palladium.
3. Spray	Deposition of a metallic finish by chemical reaction of water-based solutions.	Activator, water-clean and applicator guns; spray booths, top- and base-coating equipment if required.	Most plastics. For decorative items.	Metallic (silver and bronze).
Tamp Printing	Special process using a soft transfer pad to pick up image from etched plate and tamping it onto a part.	Metal plate, squeegee to remove excess ink, conical-shaped transfer pad, indexing device to move parts into printing area, dryers, depending on type of operation.	All plastics. Specially recommended for odd-shaped or delicate parts (e.g., drinking cups, dolls' eyes).	Single- or multi-color—one printing station per color.

Method	Description	Equipment	Plastics Used	Color
In-the-Mold Decorating	Film or foil inserted in mold is transferred to molten plastics as it enters the mold. Decoration becomes integral part of product.	Automatic or manual feed system for the transfers. Static charge may be required to hold foil in mold.	Most plastics, especially polyolefins and melamines. For parts where decoration must withstand extremely high wear.	Single- or multi-color decoration.
Flexography	Printing of a surface directly from a rubber or other synthetic plate.	Manual, semi- or automatic press, dryers.	Most plastics. Used on such areas as coding pipe and extruded profiles.	Single- or multi-color.
Offset Printing	Roll-transfer method of decorating. In most cases less expensive than other multicolor printing methods.	Ranges from low-cost hand presses to very expensive automated units. Drying, destaticizers, feeding devices.	Most plastics. Used in applications like coding pipe.	Multi-color print or decoration.
Valley Printing	Uses embossing rollers to print in depressed areas of a product.	Embosser with inking attachment or special package system.	Used largely with PVC, PE for such areas as floor tiles, upholstery.	Generally two-color maximum.
Labeling	From simple paper labels to multi-color decals and new preprinted plastic sleeve labels.	Equipment runs the gamut from hand dispensers to relatively high-speed machines.	Can be used on all plastics. Used mostly for containers and for price marking.	All sorts of colors and types.

Table 9-6. Guide to Plastic-Decorating Methods. (Courtesy of *Plastics Technology*.)

DONE IN THE MOLD	ECONOMICS	AESTHETICS	PRODUCT DESIGN	CHEMISTRY	MANUFACTURING	COMMENTS
1. Engraved mold	*Unit cost:* low *Labor cost:* low *Investment:* moderate	Limited	Unrestricted	Not critical Good durability	No extra operations	Best for simple lettering and texture.
2. In-mold label	*Unit cost:* high *Labor cost:* high *Investment:* none to moderate	Unlimited	Somewhat restricted	Critical Good durability	Longer molding cycles	Good for thermoplastics and thermosets. Automatic loading equipment becoming available.
3. Inserted nameplates	*Unit cost:* high *Labor cost:* high *Investment:* moderate	Partially limited	Restricted	Not critical Good durability	Longer molding cycles	Allows three-dimensional as well as special effects.
4. Two-shot molding	*Unit cost:* high *Labor cost:* high *Investment:* moderate to high	Limited	Somewhat restricted	Not critical Good durability	Two molding operations	Good where maximum abrasion resistance necessary.

DONE AFTER MOLDING

	ECONOMICS	AESTHETICS	PRODUCT DESIGN	CHEMISTRY	MANUFACTURING	COMMENTS
1. Applique	*Unit cost:* high *Labor cost:* high *Investment:* moderate to high	Somewhat limited	Unrestricted	Not critical	Hand operation	Allows unusual effects.
2. Electrostatic	*Unit cost:* low to moderate *Labor cost:* low *Investment:* moderate to high	Limited	Somewhat restricted	Good durability Critical		Dry process, no tool contact with product.
3. Flexographic	*Unit cost:* low *Labor cost:* low *Investment:* moderate to high	Somewhat limited	Restricted Moderate durability	Moderate to good durability Critical	Automates well	Wet process, tool contacts product. Sometimes requires top coat.
4. Hand painting	*Unit cost:* high *Labor cost:* high *Investment:* low	Somewhat limited	Unrestricted	Critical Good durability	Hand operation	Wet process, tool contacts product.
5. Heat transfer	*Unit cost:* low to moderate *Labor cost:* low to moderate *Investment:* low to moderate	Unlimited	Somewhat restricted	Critical Good durability	Requires little floor space	Dry process, tool contacts product. Multicolor graphics. *(continued)*

Table 9-6. Continued.

DONE IN THE MOLD	ECONOMICS	AESTHETICS	PRODUCT DESIGN	CHEMISTRY	MANUFACTURING	COMMENTS
6. Hot stamping	*Unit cost:* low *Labor cost:* low to moderate *Investment:* low to moderate	Limited	Somewhat restricted	Critical Good durability	Requires little floor space	Dry process, tool contacts product. Produces bright metallics.
7. Labeling	*Unit cost:* low to moderate *Labor cost:* low to moderate *Investment:* low to high	Unlimited	Somewhat restricted	Less critical Moderate to good durability	Adaptable to many situations	Dry process, no tool contact with product at times. Multicolor graphics.
8. Metallizing	*Unit cost:* moderate to high *Labor cost:* moderate to high *Investment:* high	Limited	Somewhat restricted	Critical Good durability	Requires special technological know-how	Wet and dry process, no tool contact with product. Produces bright metallics.
9. Nameplates	*Unit cost:* high *Labor cost:* moderate to high *Investment:* low to moderate	Unlimited	Somewhat restricted	Less critical Good durability	Adaptable to many situations	Dry process, tool contacts product. Multicolor graphics.
10. Offset	*Unit cost:* low	Unlimited	Restricted	Critical	Automates well	Wet process, tool contacts product.

Labor cost: moderate *Investment:* high	Unrestricted			Moderate to good durability		Multicolor graphics.
11. Offset intaglio *Unit cost:* low *Labor cost:* moderate *Investment:* moderate	Limited	Critical	Moderate to good durability	Requires little floor space		Wet process, tool contacts product.
12. Silk screen *Unit cost:* moderate *Labor cost:* moderate *Investment:* moderate	Somewhat restricted	Somewhat limited	Critical	Good durability	Flexible operation	New process.
13. Spray *Unit cost:* moderate *Labor cost:* moderate *Investment:* moderate to high	Unrestricted	Limited	Critical	Good durability	Requires much floor space	Wet process, tool contacts product.
14. Wood-graining *Unit cost:* high *Labor cost:* high *Investment:* moderate to high	Specialized	Specialized	Critical	Good durability	Mostly hand operated	Wet process, no tool contact with product.
						Wet process, tool contacts products.

length of time and the depth of the plastic soak. The solvent could cause immediate or delayed damage. If a part contains "excessive" internal strains, the solvent could release the strains and cause cracking, surface defects, and so on. (To evaluate a plastic's reaction with a solvent, after a part is processed, it is immersed in a solvent to determine whether strain patterns exist. The reaction with the solvent can be correlated with processing vs. part performance.)

The solvent action described above does not mean that adhesives are harmful; they have been used successfully for over a century. But no matter what action is taken in joining (Table 9–7) or assembling (Table 9–8), the processor should determine whether there are limitations to its use.

MACHINING

Each type of plastic has unique properties and machining characteristics, which are far different from those of the metallic or nonmetallic materials familiar to many processors. TPs are relatively resilient compared to metals, and require special cutting procedures. Even within a family of plastics (PE, PC, PPS, etc.), the cutting characteristic will change, depending on the fillers and reinforcements.

Elastic recovery occurs in plastics both during and after machining; so provisions must be made in the tool geometry for sufficient clearance to allow for it. This is so because of the expansion of compressed material due to elastic recovery (Chapter 1), which causes increased friction between the recovered cut surface and the cutting surface of the tool. In addition to generating heat, this abrasion affects tool wear. Elastic recovery also explains why, without proper precautions, drilled or tapped holes in plastics often are tapered or become smaller than the diameter of the drills that were used to make them.

As the heat conductivity of plastics is very slow, essentially all the cutting heat generated will be absorbed by the cutting tool. The small amount of heat conducted into the plastic cannot be transferred to the core of the shape; so it causes the heat of the surface area to rise significantly. This heat must be kept to a minimum or be removed by a coolant to ensure a proper cut.

For many commodity TP resins the softening, deformation, and degradation heats are relatively low. Gumming, discoloration, poor tolerance control, and poor finish are apt to occur if frictional heat is generated and allowed to build up. Engineered TP resins (such as nylon and TFE-fluoroplastic) have relatively high melting or softening points. Thus, they have less tendency to become gummed, melted, or crazed in machining than do plastics with lower melting points. Heat buildup is more critical in the plas-

Table 9–7. A Reference Chart to Help in Selecting the Proper Method of Fastening Thermoplastic Materials (33).

THERMOPLASTICS	MECHANICAL FASTENERS	ADHESIVES	SPIN AND VIBRATION WELDING	THERMAL WELDING	ULTRASONIC WELDING	INDUCTION WELDING	REMARKS
ABS	G	G	G	G	G	G	BODY TYPE ADH. RECOMMENDED
ACETAL	E	P	G	G	G	G	SURFACE TREATMENT FOR ADHESIVES
ACRYLIC	G	G	F-G	G	G	G	BODY TYPE ADH. RECOMMENDED
NYLON	G	P	G	G	G	G	
POLYCARBONATE	G	G	G	G	G	G	
POLYESTER TP	G	F	G	G	G	G	
POLYETHYLENE	P	NR	G	G	G-P	G	SURFACE TREATMENT FOR ADHESIVES
POLYPROPLYENE	P	P	E	G	G-P	G	SURFACE TREATMENT FOR ADHESIVES
POLYSTYRENE	F	G	E	G	E-P	G	IMPACT GRADES DIFFICULT TO BOND
POLYSULFONE	G	G	G	E	E	G	
POLYURETHANE TP	NR	G	NR	NR	NR	G	
PPO MODIFIED	G	G	E	G	G	G	
PVC RIGID	F	G	F	G	F	G	

E-EXCELLENT, G-GOOD, F-FAIR, P-POOR, NR-NOT RECOMMENDED

Table 9-8. A Reference Chart to Help in Selecting the Proper Method of Fastening Thermoset Plastic Materials (33).

THERMOSETS	MECHANICAL FASTENERS	ADHESIVES	SPIN AND VIBRATION WELDING	THERMAL WELDING	ULTRASONIC WELDING	INDUCTION WELDING	REMARKS
ALKYDS	G	G	NR	NR	NR	NR	
DAP	G	G	NR	NR	NR	NR	
EPOXIES	G	E	NR	NR	NR	NR	
MELAMINE	F	G	NR	NR	NR	NR	MATERIAL NOTCH SENSITIVE
PHENOLICS	G	E	NR	NR	NR	NR	
POLYESTER	G	E	NR	NR	NR	NR	
POLYURETHANE	G	E	NR	NR	NR	NR	
SILICONES	F	G	NR	NR	NR	NR	
UREAS	F	G	NR	NR	NR	NR	MATERIAL NOTCH SENSITIVE

E-EXCELLENT, G-GOOD, F-FAIR, P-POOR, NR-NOT RECOMMENDED

tics with lower melting points (Tables 1–6 and 2–1). Thermoset resins generally have the fewest problems of any plastics during machining.

Cutting Guidelines

The properties of plastics must be considered in specifying the best speeds, feeds, depths of cuts, tool materials, tool geometries, and cutting fluids. Machining data are available from machinery handbooks as well as plastic material suppliers and cutting machinery suppliers. Note that some plastics may be cut at higher speeds with no loss of reasonable tool life; but higher speeds usually result in thermal problems, especially with the commodity resins.

Guidelines for tool geometry start by reducing frictional drag and heat. It is desirable to have honed or polished surfaces on the tool where it comes in contact with the work. The geometries of the tools should be such that they generate continuous-type chips. In general, large rake angles will serve this purpose because of the force directions resulting from these angles. Care must be exercised to keep rake angles from being so large that brittle fracture of workpieces result and chips become discontinuous.

Drill geometry should be made to differ form that used for metals by means of wide polished flutes combined with low helix angles, to help eliminate the packing of chips, which causes overheating. Also the normally 118-degree point angle is generally modified to 70 to 120 degrees.

Round saws should be hollow-ground, with burrs from sharpening removed by stoning, and hands and jig saws should have enough set to give adequate clearance to the back of the blade. This set should be greater than is usual for cutting steel. It is always better to relieve the feed pressure near the end of a cut to avoid chipping.

The proper rate of feed is important, and, because most sawing operations are hand-fed, experience is required to determine the best rate. Attempts to force the feed will result in heating of the blade, gumming of the plastic, loading of the saw teeth, and an excessively rough cut. Chrome plating of the blade reduces friction and tends to give better cuts. Above all, the saw—whether band or circular—must be kept sharp. Circular saws are usually from $\frac{1}{32}$ to $\frac{1}{8}$ in. thick. The width of bandsaws is usually $\frac{3}{16}$ to $\frac{1}{2}$ in.

Both TP and TS resins can be sawed by use of cutoff machines with abrasive wheels. This equipment is used to cut rods, pipes, L-beams, and so on. With appropriate wheels properly used, clean cuts can be made. If necessary, water is used to prevent overheating.

Chapter 10

TESTING/QUALITY CONTROL

INTRODUCTION

Processors should keep quality under control and demand consistent materials that can be used with a minimum of uncertainty. Plant quality control (QC) is as important to the end result as selecting the best processing conditions with the correct grade of plastic, in terms of both properties and appearance. After the correct plastic has been chosen, its bleeding, reprocessing, and storage stages of operation need to be frequently or continuously updated. The processor should set up specific measurements of quality to prevent substandard products from reaching the end user (1–2, 290–316).

The properties of plastics are directly dependent on temperature, time, and environmental conditions, and these conditions can be related to raw material, processing, and part performance (Chapter 1). The most important testing is that done on the finished part. In turn, tests done on materials and during processing all must be related to part performance.

Unfortunately there is no single set of rules designating which tests are to be conducted in order to manufacture a part repeatedly with zero defects. The tests depend on the required performance. For example, if a part is to operate where any type of failure could be catastrophic to life, then extensive and usually expensive testing is necessary. This chapter has been prepared to make processors aware of some of the different tests that are available. How deeply one gets involved depends on the performance requirements. If all that is required is to weigh the part, that is all that one does.

Testing and QC are the most discussed but often the least understood facets of business and manufacturing. Many companies spend a high percentage of each sales dollar on QC. Usually it involves the inspection of components and parts as they complete different phases of processing. Parts that are within specifications proceed, while those that are out of spec are

326

either repaired or scrapped. The workers who made the out-of-spec parts are notified that they produced defective parts, and that they should correct "their" mistakes.

The approach just outlined is after-the-fact QC; all defects caught in this manner are already present in the part being processed. This type of QC will usually catch defects, and it is necessary, but it does little to correct basic problems in production. One of the problems with add-on QC of this type is that it constitutes one of the least cost-effective ways of obtaining a high-quality part. Quality must be built into a product from the beginning (as illustrated in Fig. 1-1); it cannot be "inspected" into the process. The closest any add-on, after-the-fact quality control can come to improving the quality built into a part is to point out processing defects to the departments or persons responsible for them. The object instead should be to control quality before a part becomes defective.

There are many different approaches to setting up QC. For example, mechanical properties can be considered the most important of all properties, and there are many factors that determine the mechanical behavior of plastics. As reviewed throughout this book, the factors that influence properties include the resin composition (fillers, molecular weight distribution, morphology, etc.), the processing method and machine controls, the capability of auxiliary equipment, and part performance requirements.

Considering what the critical areas of a process are, one can understand that sometimes a test or measurement of resin viscosity is all that is needed.

Industry specifications and standards are continually updated to aid processors in controlling quality and to meet safety requirements, and they are very useful to anyone who must choose tests and QC procedures. For example, ASTM and UL tests are among the most important tests. Organizations involved in specifications and standards preparation include the following:

ASTM: American Society for Testing and Materials (see reference 290)
UL: Underwriters' Laboratories (see reference 291)
ACS: American Chemical Society
ANSI: American National Standards Institute
ASCE: American Society of Chemical Engineers
ASM: American Society of Metals
ASME: American Society of Mechanical Engineers
AWS: American Welding Society
BMI: Battele Memorial Institute
BSI: British Standards Institute
CSA: Canadian Standards Association
DIN: Deutsches Institut, fur Normung, West Germany

EIA: Electronic Industry Association
IEC: International Electrotechnical Commission
IEEE: Institute of Electrical and Electronic Engineers
IFI: Industrial Fasteners Institute
ISA: Instrument Society of America
ISO: International Organization for Standardization
JIS: Japanese Industrial Standards
NADC: Naval Air Development Center
NBS: National Bureau of Standards
NACE: National Association of Corrosion Engineers
NEMA: National Electrical Manufacturers' Association
NFPA: National Fire Protection Association
NPFC: Naval Publications and Forms Center
SAE: Society of Automotive Engineers
SPE: Society of Plastics Engineers
SPI: Society of Plastics Industry
TAPPE: Technical Association of Pulp and Paper Industry

In this chapter only a few tests will be reviewed, to help the processor understand how to evaluate tests. Individuals required to conduct tests should review the specification or standard being followed to ensure that procedures and test equipment are being used. It is very important to make sure that the most up-to-date procedures are being observed.

MECHANICAL PROPERTIES

The most important mechanical property of a plastic is its tensile stress–strain curve (Figs. 10–1 and 7–4). This curve is obtained by "stretching" a sample in a testing machine and measuring its extension and the load required to reach this extension. Plastics show viscoelastic behavior (as reviewed in Chapter 1) that is highly sensitive to temperature and, in some materials, to relative humidity variations; so it is important to use samples of standard shapes, preconditioned at constant and standard temperature and relative humidity before testing. Requirements are explained in the ASTM specifications.

The stress–strain curve provides information about the modulus of elasticity, which is related to the plastic's stiffness and rigidity. This curve also provides information about the yield point, tensile strength, and elongation at break. The curve helps to define toughness (the area under the curve), which is the energy per unit volume required to cause the sample to fail. Tests can be conditioned in different environments such as low and high heat (Figs. 10–2 and 10–3), outdoor weathering, "salt" air, chemicals (Ta-

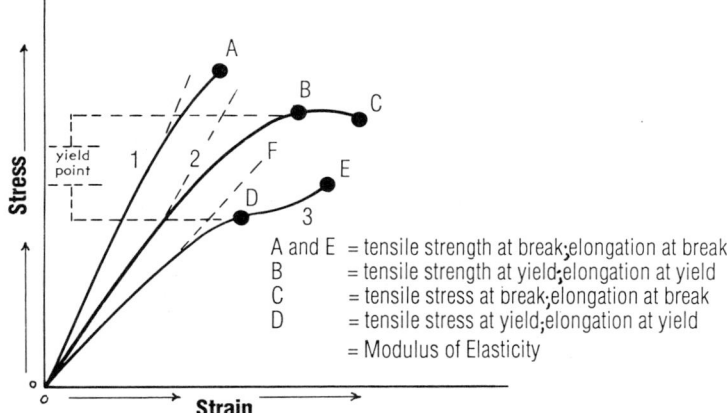

Fig. 10-1. Typical basic tensile stress–strain curves, per ASTM D638.

ble 10-1), water, and so on (Fig. 10-4). Thus, the curve reveals much about a plastic's behavior.

Details on all types of tests (physical, chemical, optical, insulation, etc.) are in the ASTM test methods and standards (Figs. 10-5 and 10-6) (290). These procedures explain the reason for a test, how it is conducted, and how to interpret the results, and they sometimes provide information on long-term test results, and so on. With all the available tests, confusion could exist in deciding which test(s) should be conducted. After a "problem" or potential problem is determined, examine and study the applicable test methods and conduct preliminary tests.

As an example, there could be confusion concerning the relationships between relative humidity index, heat distortion temperature (HDT), and warp temperature. These tests provide specific information, and there is generally no correlation of their results. Clarifications of these tests are given in their respective test-method descriptions.

The relative thermal index is an indicator of a material's long-term resistance to degradation of its electrical, tensile, and impact properties at operating heats. This index is based on tests made to determine the half-life of these properties at elevated heat (i.e., the heat at which the plastic will retain at least one-half of its original properties for the projected life of an electrical appliance or product, up to 100,000 hours), such as summarized in Fig. 10-2.

The heat distortion temperature is the deflection heat under load, which is usually listed on property data sheets. The ASTM D648 test consists of a bar (usually $\frac{1}{2} \times \frac{1}{2} \times 5$ in.) supported at both ends (4 in. apart), with

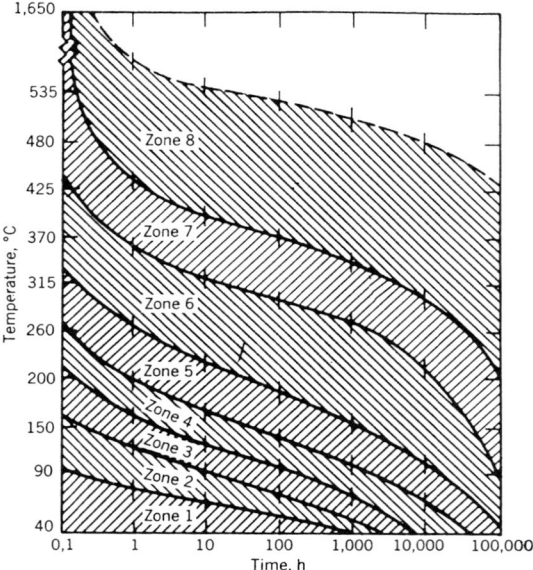

Heat-resistance properties of resins, retaining 50% of properties obtainable at RT with resin exposure and testing at elevated temperature. Zone 1: acrylic; cellulose esters; crystallizable block copolymers; LDPE; PS; vinyl polymers; SAN; SBR; urea–formaldehyde. Zone 2: acetal; ABS; chlorinated polyether; ethyl cellulose; ethylene–vinyl acetate copolymer; furan; ionomer; phenoxy; polyamides; PC; HDPE; PET; PP; PVC; urethane. Zone 3: polychlorotrifluoroethylene; vinylidene fluoride. Zone 4: alkyd; fluorinated ethylene–propylene; melamine–formaldehyde; phenol–furfural; polysulfone. Zone 5: acrylic (thermoset); diallyl phthalate; epoxy; phenol–formaldehyde; polyester; polytetrafluoroethylene. Zone 6: parylene; polybenzimidazole; polyphenylene; silicone. Zone 7: polyamide-imide; polyimide. Zone 8: plastics now being developed using rigid linear macromolecules rather than crystallization and cross-linking.

Fig. 10–2. Heat-resistance properties of different plastics.

a specific load at midspan, which is heated until the load causes a deflection of 0.010 in. This test only provides a reference point and is not related to any specific end-item performance (Fig. 10–6). It can be used as a guide to processability. If the plastic had residual stress or strain due to a processing procedure, the HDT reading would be lower than expected because the heat would first stress-relieve the plastic (through an annealing action); the 0.010 in. deflection would occur at a lower temperature.

The warp temperature, also called mold stress relief, is normally evaluated on the finished part. After exposure to a heat–time period, the part is visually examined for warpage or distortion.

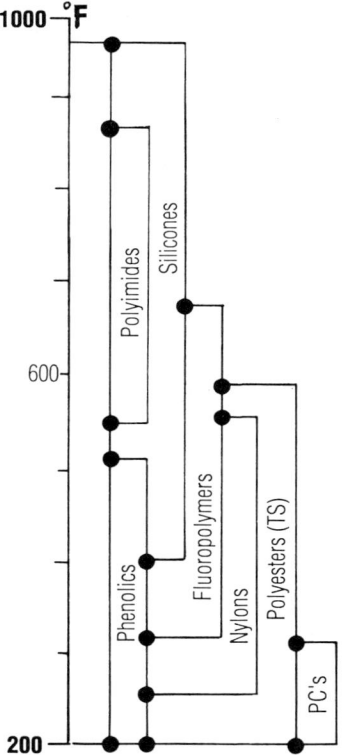

Fig. 10–3. Classifying plastics by range of continuous heat.

PLASTICS FLOW/RHEOLOGY

As reviewed in Chapter 1 and succeeding chapters, the flow (rheology) characteristics of a plastic can provide useful information about its processability and final product performance. This section will review a few of the more practical and common procedures, which are very useful. However, very significant and more meaningful tests can be conducted to determine the real flow/rheological behaviors, and they are used as needed to meet very tight requirements, including processing cost reductions. They are more sophisticated and expensive to run than the tests described here. (For information on these techniques see references 1, 2, 8, 85, and 316.)

Table 10-1. Effects of Elevated Temperature and Chemical Agents on Stability of Plastics (a Rating of "1" Equals Greatest Stability).

ENVIRONMENT	AROMATIC SOLVENTS		ALIPHATIC SOLVENTS		CHLORINATED SOLVENTS		WEAK BASES AND SALTS		STRONG BASES		STRONG ACIDS		STRONG OXIDANTS		ESTERS AND KETONES		24-H WATER ABSORPTION
TEMPERATURE F	77	200	77	200	77	200	77	200	77	200	77	200	77	200	77	200	% CHANGE BY WEIGHT
PLASTIC MATERIAL																	
Acetals	1-4	2-4	1	2	1-2	4	1-3	2-5	1-5	2-5	5	5	5	5	1	2-3	0.22–0.25
Acrylics	5	5	2	3	5	5	1	3	2	5	4	4-5	5	5	5	5	0.2–0.4
Acrylonitrile-Butadiene-Styrenes (ABS)	4	5	2	3-5	3-5	5	1	2-4	1	2-4	1-4	5	1-5	5	3-5	5	0.1–0.4
Aramids (aromatic polyamide)	1	1	1	1	1	1	2	3	4	5	3	4	2	5	1	2	0.6
Cellulose Acetates (CA)	2	3	2	3	3	4	2	3	3	5	3	5	3	5	5	5	2–7
Cellulose Acetate Butyrates (CAB)	4	5	1	3	3	4	2	4	3	5	3	5	3	5	5	5	0.9–2.0
Cellulose Acetate Propionates (CAP)	4	5	1	3	3	4	1	2	3	5	3	5	3	5	5	5	1.3–2.8
Diallyl Phthalates (DAP, filled)	1-2	2-4	2	3	2	4	2	3	2	4	1-2	2-3	2	4	3-4	4-5	0.2–0.7
Epoxies	1	2	1	2	1-2	3-4	1	1-2	1	2	2-3	3-4	4	4-5	2	3-4	0.01–0.10

332

Material																
Ethylene Copolymers (EVA) (Ethylene-Vinyl Acetates)	5	5	5	5	5	1	1	2	5	1	5	1	5	5	2	5	0.05–0.13
Ethylene/Tetrafluoro-ethylent Copolymers (ETFE)	1	1	1	1	1	1	1	1	1	1	1	1	1	1	1	1	<0.03
Fluorinated Ethylene Propylenes (FEP)	1	1	1	1	1	1	1	1	1	1	1	1	1	1	1	1	<0.01
Perfluoroalkoxies (PFA)	1	1	1	1	1	1	1	1	1	1	1	1	1	1	1	1	<0.03
Polychlorotrifluoro-ethylenes (CTFE)	1	1	1	3	4	1	1	1	1	1	1	1	1	1	1	1	0.01–0.10
Polytetrafluoroethylenes (TFE)	1	1	1	1	1	1	1	1	1	1	1	1	1	1	1	1	0
Furans	1	1	1	1	1	2	2	2	2	1	1	5	5	1	1	1	0.01–0.20
Ionomers	2	4	4	4	4	1	4	1	4	2	4	1	5	5	1	4	0.1–1.4
Melamines (filled)	1	1	1	1	1	2	3	2	3	2	1	2	3	1	1	2	0.01–1.30
Nitriles (high barrier alloys of ABS or SAN)	1	4	2–4	1–4	2–5	1	2–4	1	2–4	1	2–5	3–5	5	5	1–5	5	0.2–0.5
Nylons	1	1	1	1	2	1	2	2	3	1	2	5	5	5	1	1	0.2–1.9
Phenolics (filled)	1	1	1	1	1	2	3	3	5	2	1	4	5	5	2	2	0.1–2.0
Polyallomers	2	4	4	4	5	1	1	1	1	1	5	1	4	1	3	3	<0.01

FLUOROCARBONS

(continued)

333

Table 10-1. Continued.

PLASTIC MATERIAL	AROMATIC SOLVENTS		ALIPHATIC SOLVENTS		CHLORINATED SOLVENTS		WEAK BASES AND SALTS		STRONG BASES		STRONG ACIDS		STRONG OXIDANTS		ESTERS AND KETONES		24-H WATER ABSORPTION % CHANGE BY WEIGHT
TEMPERATURE F	77	200	77	200	77	200	77	200	77	200	77	200	77	200	77	200	
Polyamide-imides	1	1	1	1	2	3	1	1	3	4	2	3	2	3	1	1	0.22–0.28
Polyarylsulfones (PAS)	4	5	2	3	4	5	1	2	2	2	1	1	2	4	3	4	1.2–1.8
Polybutylenes (PB)	3	5	1	5	4	5	1	2	1	3	1	3	1	4	1	3	<0.01–0.3
Polycarbonates (PC)	5	5	1	1	5	5	1	5	5	5	1	1	1	1	5	5	0.15–0.35
Polyesters (thermoplastic)	2	5	1	3–5	3	5	1	3–4	2	5	3	4–5	2	3–5	2	3–4	0.06–0.09
Polyesters (thermoset-glass fiber filled)	1–3	3–5	2	3	2	4	2	3	3	5	2	3	2	4	3–4	4–5	0.01–2.50
Polyethylenes (LDPE—low-density to high-density)	4	5	4	5	4	5	1	1	1	1	1–2	1–2	1–3	3–5	2	3	0.00–0.01
Polyethylenes (UHMWPE—ultra high molecular weight)	3	4	3	4	3	4	1	1	1	1	1	1	1	1	3	4	<0.01

334

Material															
Polyimides	1	1	1	1	1	2	3	4	5	3	4	2	5	1	1	0.3–0.4
Polyphenylene Oxides (PPO) (modified)	5	2	3	4	5	1	1	1	1	1	2	1	2	2	3	0.06–0.07
Polyphenylene Sulfides (PPS)	1	1	1	1	2	1	1	1	1	1	1	1	2	1	1	<0.05
Polyphenylsulfones	4	4	1	5	5	1	1	1	1	1	1	1	1	3	4	0.5
Polypropylenes (PP)	4	2	4	2–3	4–5	1	1	1	1	1	2–3	2–3	4–5	2	4	0.01–0.03
Polystyrenes (PS)	4	5	5	5	5	1	5	1	5	4	5	4	5	4	5	0.03–0.60
Polysulfones	4	4	1	5	5	1	1	1	1	1	1	1	1	3	4	0.2–0.3
Polyurethanes (PUR)	4	4	4	4	5	2–3	3–4	2–3	3–4	2–3	3–4	4	4	4	5	0.02–1.50
Polyvinyl Chlorides (PVC)	4	5	5	5	5	1	5	1	5	1	5	2	5	4	5	0.04–1.00
Polyvinyl Chlorides—Chlorinated (CPVC)	4	4	2	5	5	1	2	1	2	2	2	2	3	4	5	0.04–0.45
Polyvinyidene Fluorides (PVDF)	1	1	1	1	1	1	1	1	2	1	2	1	2	3	5	0.04
Silicones	4	4	3	4	5	1	2	4	5	3	4	4	5	2	4	0.1–0.2
Styrene Acrylonitriles (SAN)	5	5	4	3	5	1	3	1	3	1	3	3	4	4	5	0.20–0.35
Ureas (filled)	3	3	3	1	3	2	3	2	3	4	5	2	3	1	2	0.4–0.8
Vinyl Esters (glass fiber filled)	1–2	3	2–4	2	4	1	3	1	3	1	2	3	3–4	4–5	0.01–2.50

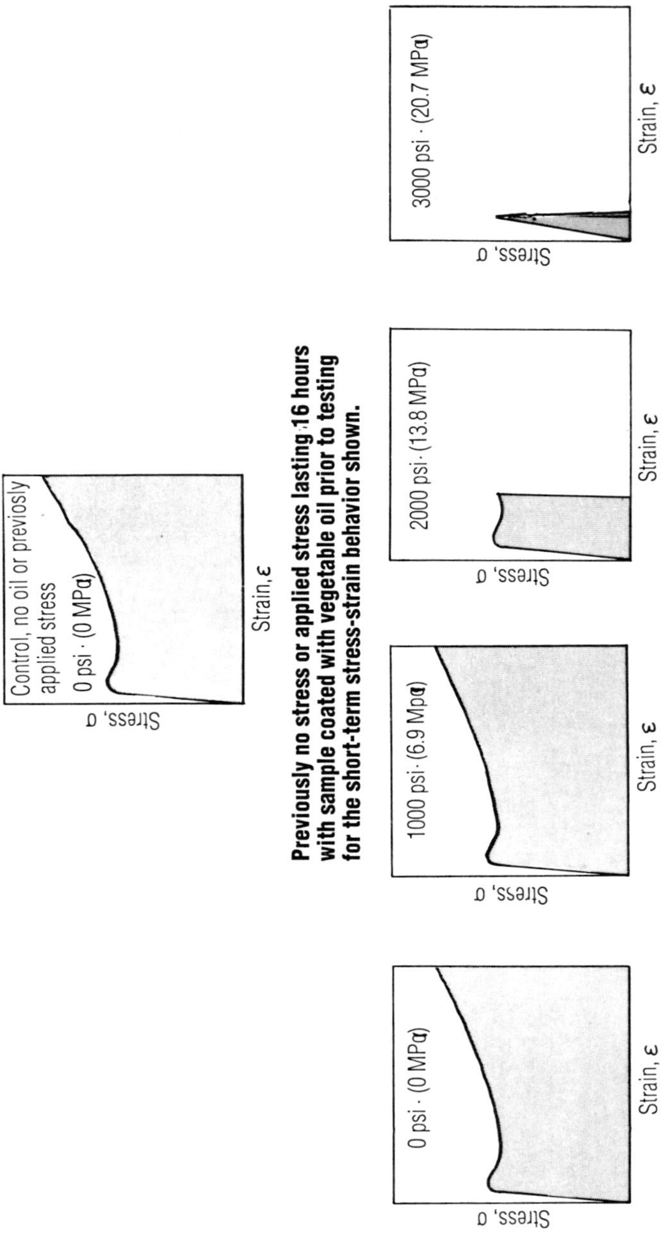

Control, no oil or previosly applied stress

0 psi · (0 MPa)

Stress, σ

Strain, ε

Previously no stress or applied stress lasting 16 hours with sample coated with vegetable oil prior to testing for the short-term stress-strain behavior shown.

3000 psi · (20.7 MPa)

Stress, σ

Strain, ε

2000 psi · (13.8 MPa)

Stress, σ

Strain, ε

1000 psi · (6.9 Mpa)

Stress, σ

Strain, ε

0 psi · (0 MPa)

Stress, σ

Strain, ε

Fig. 10–4. Example of the influence of prior stress and environment on ductility of a specific plastic.

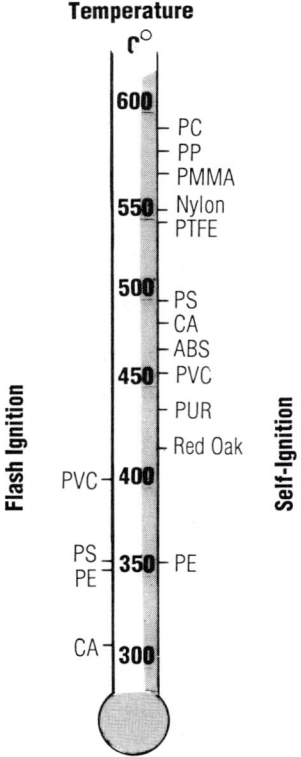

Fig. 10–5. Basic guide to flash-ignition and self-ignition temperatures, per ASTM D1929.

Melt Index Test

The melt indexer (extrusion plastometer) is the most widely used rheological device for examining and studying plastics in many different fabricating processes. It is not a true viscometer, in the sense that a reliable value of the viscosity cannot be calculated from the flow index, which is normally measured. However, it does measure isothermal resistance to flow using an apparatus and test method that are standard throughout the world. Standards used include ASTM D1238 (U.S.A.), BS 2782-105C (U.K.), DIN 53735 (West Germany), JIS K7210 (Japan), ISO R1133/R292 (international), and others.

In this instrument (Fig. 10–7) the polymer is contained in a barrel equipped with a thermometer and surrounded by an electrical heater and an insulating jacket. A weight drives a plunger that forces the melt through

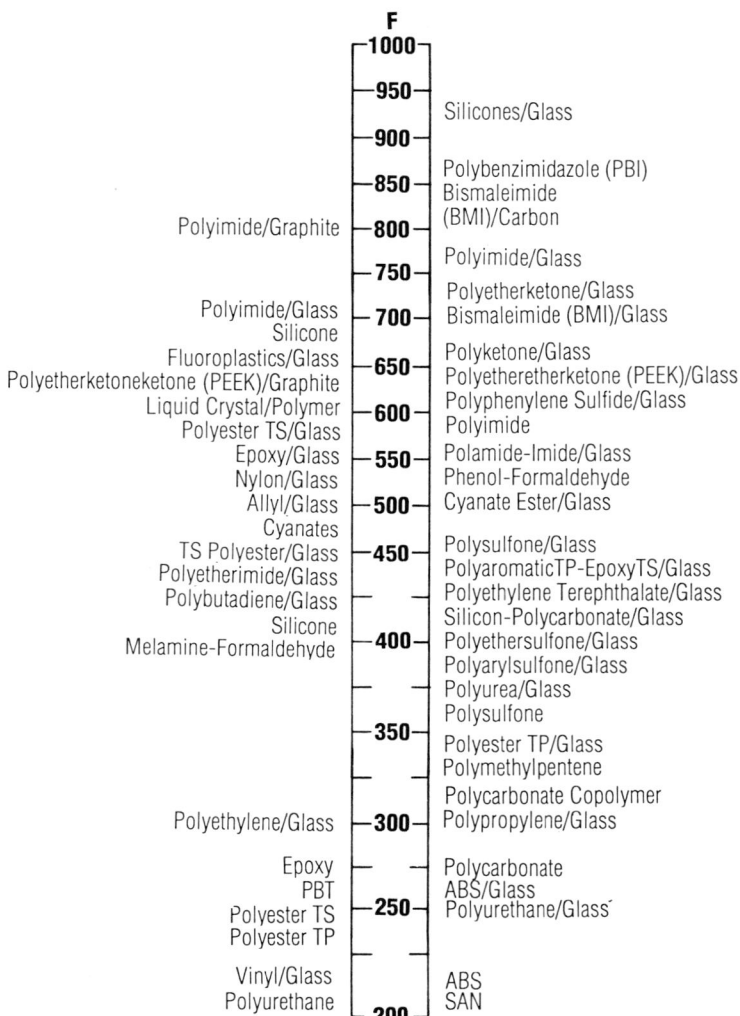

Note: Plastics with reinforcements contain 30 percent, by weight. Higher temperatures obtained by using reinforcements such as carbon, graphite, aramid, etc. (see Chapter 7). TP = thermoplastic and TS = thermoset

Fig. 10–6. Guide to heat resistance based on heat distortion temperature test method, per ASTM D648 at 264 psi.

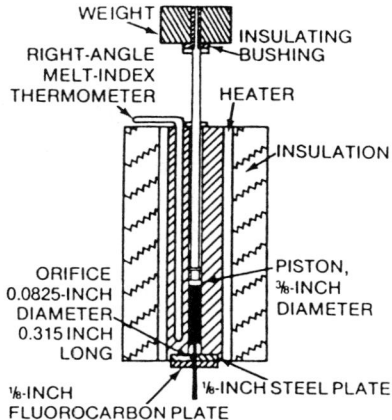

Fig. 10-7. Melt index test, per ASTM D1238.

the die opening, using a standard opening of 2.095 mm (0.0824 in.) and a length of 8 mm (0.315 in.). The standard procedure involves the determination of the amount of polymer extruded in 10 min. The flow rate (expressed in g/10 min.) is reported. As the flow rate increases, the viscosity decreases. Depending on the flow behavior, changes are made to standard conditions (die opening size, temperature, etc.) in order to obtain certain repeatable and meaningful data applicable to a specific processing operation.

The MI (melt indexer) is easy to operate and relatively low-cost; thus, it is widely used for quality control and for distinguishing between members of a single family of polymers. Specifically, this MI makes a single-point test that provides information on resistance to flow only at a single shear rate. Because variations in branching or molecular weight distribution (MWD) can alter the shape of the viscosity curve, the MI may give a false ranking of plastics in terms of their shear rate resistance to flow. To overcome this problem, extrusion rates are sometimes measured for two loads, or other modifications are made. Different companies produce MIs, as listed in different magazines and the literature of test equipment suppliers (1, 2, 316).

Summarizing, the MI is an indicator of the average molecular weight (MW) of a plastic, and is also a rough indicator of processability. Low MW material have high MIs and are easy to process. High MW materials have low MIs and are more difficult to process, as they have more resistance to flow, but they are processable. End-use physical properties improve as the MI decreases (Fig. 10-8). Because processability simultaneously decreases,

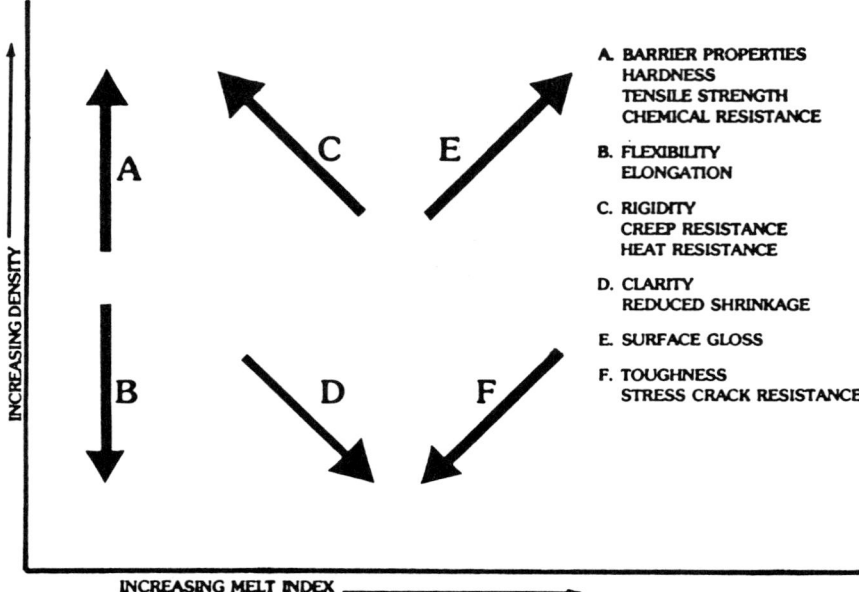

Fig. 10-8. The effects of density and melt index (MI) changes on the properties of polyethylene, with properties increasing in the direction of the arrows.

MI selection for a given application is a compromise between properties and processability. Table 10-2 lists typical MI ranges for the more common plastics processes. Materials with other MIs are processable, but they usually require more sophisticated start-up procedures and process controls.

Spiral Flow Test

This test is used primarily with compression and injection molding of TP and TS resins. Some processors cannot directly relate classic capillary or oscillatory rheology data (MI, etc.) to the real world of molding. They are more comfortable with "flow length" data, especially for comparing various resins for ease of fill. Because flow tests are run in an actual machine (compression or IM), the results are direct, tangible, and easy to interpret. Typically the flow length mold makes a part in the form of a graduated spiral (Fig. 10-9). Different patterns of the spiral and the cross-sectional flow channel are used, based on the characteristics of melts and on experience. The length of flow is measured and can be related to part flow in an actual production mold. It can be used as a QC tool for material.

Table 10-2. Typical Melt Index Ranges for Common Polymer Processes.

PROCESS	MI RANGE
Injection molding	5-100
Rotational molding	5-20
Film extrusion	0.5-6
Blow molding	0.1-1
Profile extrusion	0.1-1

Test method: ASTM D1238.

With TS molding compounds, the flow can be defined as a measure of melt viscosity, gelation rate, and subsequent polymerization, or cure (Chapters 1, 6, and 7). Flow occurs within a dynamic chemical reaction initiated by the application of the above three factors, with the rate of reaction accelerating in proportion to the increase in heat (Fig. 7-5). A valid rule of

Fig. 10-9. Spiral flow test mold and plastic molded spiral test specimen, using an injection molding machine.

thumb is that the rate of reaction (cure) will approximately double with each 50°F (10°C) of increased temperature.

Other Tests

In addition to the spiral flow test for TSs, other tests also are used, such as the ASTM cup closing test, disc flow, and Brabender flow. The ASTM cup closing test, D731, consists of a flash-type compression mold for a cup of specific configuration. The test measures the time required to close the mold to a specific flash thickness, expressed in "X seconds closing time"; the longer the time, the stiffer the flow. This test is quite valid for measuring the flow of compounds intended for compression molding but is inadequate for transfer or screw injection molding.

There are two disc flow tests. "Disc Flow I" is a very simple test in which a measured amount of room-temperature loose compound is compressed between two heated die plates at a specific pressure and temperature. The resultant molded disc is measured for thickness to determine the flow; the thicker the disc, the stiffer the flow. The "Disc Flow II" procedure is similar to "I" except that the mold cavity has five concentric rings. The diameter of the molded disc is measured; the larger the disc, the softer the flow.

As with the cup closing test, the two disc flow tests are best suited for measuring the flow of compounds that will be compression-molded. They are used for QC of material and can detect color contamination or changes, as well as material contamination, and so on.

The Brabender plasti-corder flow test is used with screw injection molding of TS resins to graphically and accurately measure the flow as it relates to IM. Concern about whether a compound would have a sufficient "flow life" during its residence time in the barrel led to the use of this test. It is basically a torque rheometer designed to measure and record torque (in meter-grams) vs. the mixing time. It produces a flow curve that is a measure of the compound's processability.

The device is made up of a heated mixing head containing a pair of sigma-shaped blades that are driven by a dynamometer run at a controlled speed. This test provides very accurate and useful data for the processor. The Brabender test also is used with many other materials, principally TPs, for controlling materials used in many different processes. For decades it has been used to evaluate PVC for extruders, calendering, and so on.

VOLUME-CHANGE TEST

The volume-change indicator for molding plastic is a simple mechanical device that monitors the change in volume of a compression-molded TS compound (Fig. 10–10). The TS is shaped during compression in a thin-walled tube, as both force and balance mandrels move through the die. It is essential that this changing cavity configuration not increase in volume during this phase, or the physical properties of the molded part will be impaired. This possibility can be eliminated by a monitor that tracks the relative positions of the two mandrels.

The monitor has two concentric discs, each of which is independently driven by one of the rams of the mold. The outside disc is rotated by movement of the force ram, and the inside disc is rotated by movement of the balance ram. If the volume between the rams changes during the displacement, the index of one of the discs turns to the red or green portion of the other disc, indicating expansion or contraction, respectively, of the molding compound. (Color coding helps to provide an easy identification.) A green indication is desired, as that denotes decreasing volume and increasing plastification of the compound. A red indication is undesirable, as expansion means that the molding conditions are not properly regulated.

THERMOANALYTICAL TESTS

Thermoanalytical (TA) methods characterize a system in terms of the temperature dependency of its thermodynamic properties and the physiochemical reaction kinetics of TPs and TSs. The techniques reviewed here only include differential scanning calorimetry (DSC), thermogravimetric analysis (TGA), and thermomechanical analysis (TMA). Others also are available, which are useful in the processing plant (1, 2).

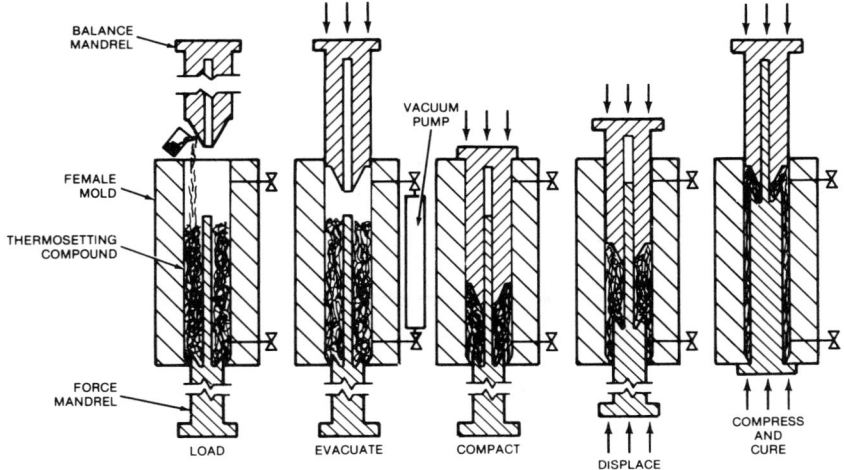

Fig. 10–10. Schematic for the volume-change indicator for thermoset resins.

Differential Scanning Calorimetry (DSC)

Differential scanning calorimetry directly measures the heat flow to a sample as a function of temperature. A sample of the material weighing 5 to 10 g is placed on a sample pan and heated in a time- and temperature-controlled manner. The temperature usually is increased linearly at a predetermined rate. DSC is used to determine specific heats (Fig. 10–11), glass transition temperatures (Fig. 10–12), melting points (Fig. 10–13) and melting profiles, percent crystallinity, degree of cure, purity, thermal properties of heat-seal packaging and hot-melt adhesives, effectiveness of plasticizers, effects of additives and fillers (Fig. 10–14), and thermal history.

DSC also is used to determine the percentage of crystallization (Fig. 10–13). A significant consideration in using polyolefins is their susceptibility to crystallization. The molder needs to know how rapidly material crystallizes as it is cooled. A comparison of materials from different lots will indicate whether they will crystallize in the same manner under the same molding conditions. (Polyolefins are provided in both nucleated and non-nucleated grades. A nucleating agent is added to a material to increase the material's rate of crystallization, a factor bearing on the performance of parts molded from that material.)

DSC is a very useful technique for monitoring the level of antioxidant in, for example, polyolefins such as polypropylene. One of the materials most susceptible to oxidation, polypropylene experiences some brittleness and cracking, with the amount depending partly on the end-use of the

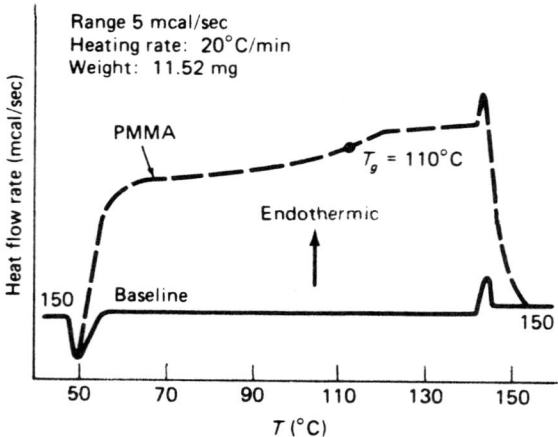

Fig. 10–11. DSC used to determine heat capacity of PMMA near the glass transition temperature.

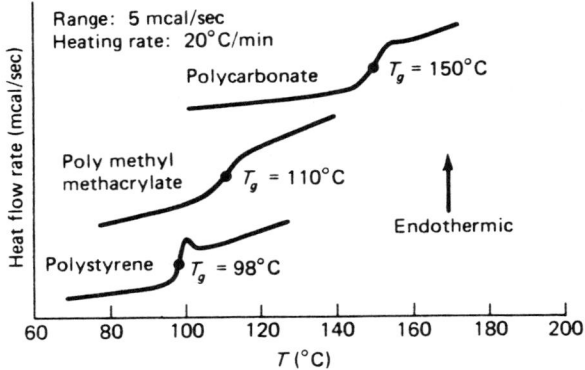

Fig. 10–12. DSC identifies the glass transition temperature for amorphous plastics PC, PMMA, and PS, indicating a minimum temperature for processing the plastics.

molded part. Antioxidants are added to extend the service life and to protect the material during the molding operation, but they are sacrificially oxidized to protect the polymer during molding; and once the antioxidants have been depleted, the material is again vulnerable to oxidation. The end-user of the part needs the antioxidant protection, however, and will not be well served if the antioxidant is used up during the molding operation. Therefore, the molder needs to ensure that sufficient antioxidant is in the raw material before processing and that enough antioxidant remains in the material after molding to meet the customer's needs.

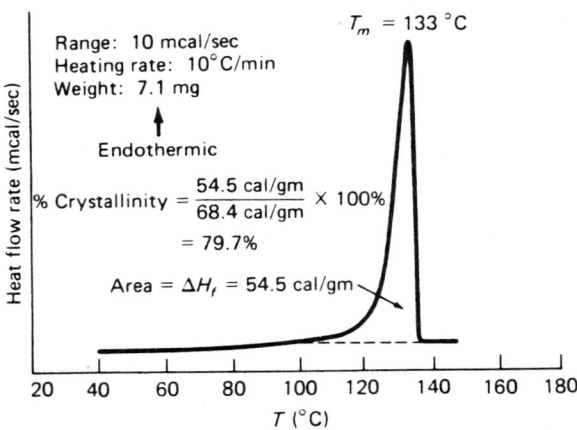

Fig. 10–13. DSC determines the melting point and percent crystallinity of HDPE.

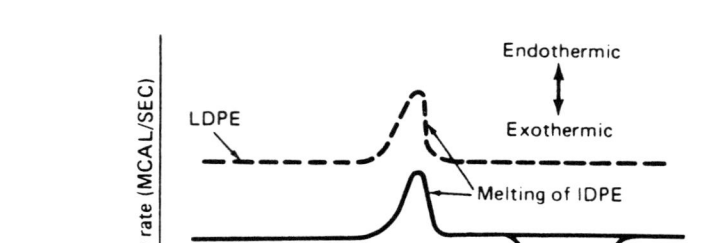

*Differential scanning calor*imetry is used for determining
the *effects of additives and fillers* from a process and
quality-control point of view. The above graph
characterizes *LDPE foam.*

Fig. 10-14. DSC relates to the effects of additives and fillers that can be used in quality control for LDPE foam.

Thermogravimetric Analysis (TGA)

This method measures the weight of a substance heated at a controlled rate as a function of time or temperature. To perform the test, a sample is hung from a balance and heated in the small furnace on the TGA unit according to a predetermined temperature program. As all materials ultimately decompose on heating, and the decomposition temperature is a characteristic property of each material, TGA is an excellent technique for the characterization and quality control of materials (Figs. 10-15 and 10-16).

Properties measured include thermal decomposition temperatures, relative thermal stability, chemical composition, and the effectiveness of flame retardants. TGA also is commonly used to determine the filler content of many thermoplastics.

A typical application of TGA is its use in compositional analysis. For example, a particular polyethylene part contained carbon black and a mineral filler. The electrical properties were important in the use of this product and could be affected by the carbon black content. TGA was used to determine the carbon black content and mineral-filler content for various lots, which were considered either acceptable or unacceptable. The samples were heated in nitrogen to volatilize the PE, leaving carbon black and a mineral-filler residue. The carbon content was then determined by switching to an

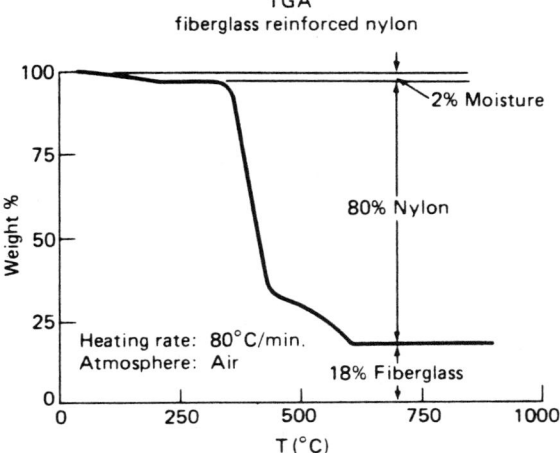

Fig. 10-15. TGA determines the amount of glass fiber reinforcement in nylon.

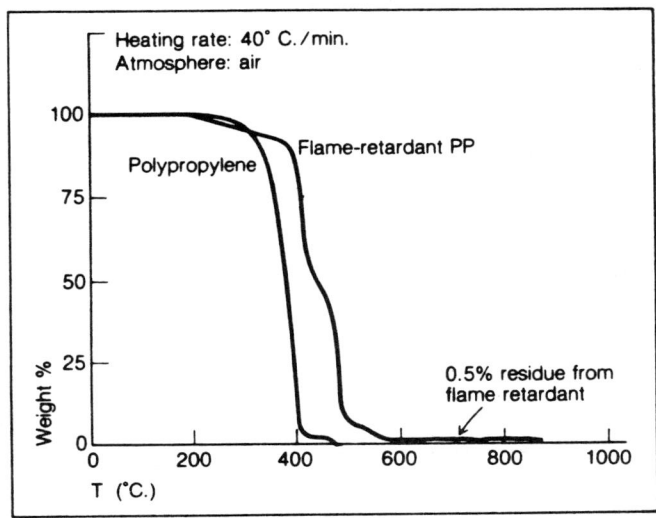

Fig. 10-16. Characterizing the flame retardant in polypropylene using TGA.

air environment to burn off the carbon black. The weight loss was a direct measure of the carbon black content.

Thermomechanical Analysis (TMA)

This system measures dimensional changes as a function of temperature. The dimensional behavior of a material can be determined precisely and rapidly with small samples in any form—powder, pellet, film, fiber, or molded part. The parameters measured by thermomechanical analysis are the coefficient of linear thermal expansion, the glass-transition temperature (Fig. 10–17), softening characteristics, and the degree of cure. Other applications of TMA include the taking of compliance and modulus measurements and the determination of deflection temperature under load.

Tensile–elongation properties and the melt index can be determined by using small samples such as those cut directly from a part. Part uniformity can be determined by using samples taken from several areas of the molded part. Samples also can be taken from an area where failure has occurred or continues to occur. This permits comparisons of material properties in a failed area with properties measured either at an unfailed section or from a sample of new material. Samples also may be taken from within a material blend to ensure that a uniform blend is being supplied. The results of such testing can be used either for evaluation of part failure or in the acceptance testing of incoming materials or parts.

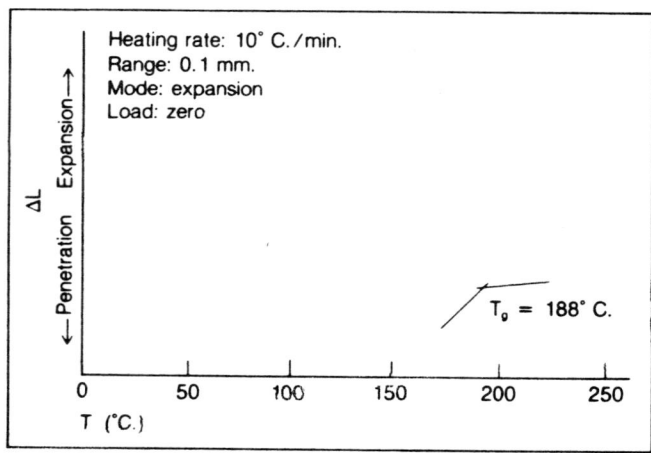

Fig. 10–17. TMA determines the coefficient of expansion and the glass transition temperature of epoxy–graphite composite.

In basic mechanical testing, the mechanical characteristics that can be tested include expansion, penetration, extension, flexure, and compressive compliance. Photoelastic-stress analysis allows the stress distribution to be visually displayed, and strain gauging allows the stress distribution to be approximated. Residual stress, also known as molded-in stress, can be measured by a variety of techniques (1).

NONDESTRUCTIVE TESTING

In the familiar form of testing known as destructive testing, the original configuration of a specimen is changed, distorted, or even destroyed for the sake of obtaining such information as the amount of force that the specimen can withstand before it exceeds its elastic limit and permanently distorts (usually called the yield strength) or the amount of force needed to break it (the tensile strength). The data collected in this instance are quantitative and could be used to design structural parts that would withstand a certain oscillating load or heavy traffic usage. However, one could not use the tested specimen in the part. One would have to use another specimen and hope that it would behave exactly the same as the one that was tested (1).

Nondestructive testing (NDT), on the other hand, examines a specimen without impairing its ultimate usefulness. It does not distort the test specimen's configuration, but provides a different type of data. NDT allows suppositions about the shape, severity, extent, configuration, distribution, and location of such internal and subsurface defects as voids and pores, shrinkage, cracks, and the like.

Most materials contain some flaws, which may or may not be cause for concern. Flaws that grow under operating stresses can lead to structural or component failure, whereas other flaws may present no safety or operating hazards. Nondestructive evaluation provides a means for detecting, locating, and characterizing flaws in all types of materials, while the component or structure is in service, if necessary, and often before the flaw is large enough to be detected by more conventional means. The following is a brief guide to nondestructive evaluation methods (1).

Radiography

Radiography is the most frequently used nondestructive test method. X rays and gamma rays passing through a structure are absorbed distinctively by flaws or inconsistencies in the material, so that cracks, voids, porosity, dimensional changes, and inclusions can be viewed on the resulting radiograph.

Ultrasonics

In ultrasonic testing, the sound waves from a high-frequency ultrasonic transducer are beamed into a material. Discontinuities in the material interrupt the sound beam and reflect energy back to the transducer, providing data that can be used to detect and characterize the flaws.

When an electromagnetic field is introduced into an electrical conductor, eddy currents flow in the material; and any variations in material conductivity due to cracks, voids, or thickness changes can alter the path of the eddy current. Probes are used to detect the current movement and thus describe the flaws.

When flaws or cracks grow, minute amounts of elastic energy are released and propagate in the material as an acoustic wave. Sensors placed on the surface of the material can detect these acoustic waves, providing information about the location and rate of flaw growth. These principles form the basis for the acoustic emission test method.

Although commercially available for the past 25 years or so, ultrasonic detectors never really caught on as a diagnostic or maintenance tool. The biggest problem with ultrasonic detectors was their inability to produce measurements as accurately or as consistently as could many competing devices used for nondestructive testing. The advent of microprocessing is dramatically improving the ability of ultrasonics to detect the wall thickness of metals and plastics, to determine particle dispersion in suspensions, and to detect potential leakage and faulty parts.

Liquid Penetrants

The liquid penetrant method is used to identify surface flaws and cracks. Special low-viscosity fluids containing a dye will penetrate into a flaw or crack when placed on the surface of a part. When the surface is washed, the residual penetrants contained in the part reveal the presence of flaws.

Acoustics

In acoustical holography, computer reconstruction provides the means for storing and integrating several holographic images. A reconstructed stored image is a three-dimensional picture that can be electronically rotated and viewed in any image plane. The image provides full characterization and detail of buried flaws.

Photoelastic Stress Analysis

Photoelastic stress analysis helps the processor to determine why a part broke and how to prevent similar failures in the future. Parts ranging in size from structural composites to tiny thermoplastic heat valves can all be tested easily. The test method also is a valuable tool for predicting where prototype parts may fail (1).

Manufacturers of plastic products want to be sure that their parts will withstand service stresses, especially since they are now faced with increasingly rigorous safety requirements, strict liabilities, and extended product warranties. Mechanical failure due to thermal or mechanical stress is a strong possibility if any of the three aspects of manufacturing—design, processing conditions, or assembly techniques—is mishandled. Poorly designed features such as corners, ribs, or holes are common causes of failure. So are improper processing conditions, including a poor mold design or an inconsistent mold temperature.

Photoelastic analysis, one of several related testing techniques, is easy to use and usually a more economical and positive method than computer analysis. From the information it provides, the test can lead to better-designed, lower-cost products. Traditionally used to test the integrity of metal parts, photoelastic analysis is now being used to physically test thermoplastics as well as thermosets. For transparent plastics, the analysis can be made directly on the plastic. For nontransparent plastics, a transparent coating is used. Actual parts and representative models can be tested by a simple procedure. The former may be stressed under actual use conditions, whereas models are tested under simulated conditions.

Although theoretical analytical methods such as finite element analysis offer one a chance to solve complex stress problems, there are many causes of strain in parts that cannot be reliably tested by these expensive computer-oriented techniques. For instance, strains due to the assembly of components and those caused during processing are extremely difficult problems to analyze without physically testing the part.

Photoelastic analysis is more than just another pretty experimental stress test. When examined under a polariscope, the colorful interference pattern can be used to survey stress distribution and the degree of strain. This analysis ultimately helps one to pinpoint which manufacturing function—design, processing conditions, or assembly techniques—led to part failure or might do so in the future. Interference patterns for coatings and models are analyzed in the same way. The photoelastic color sequence shows stress distribution in the part. In order of increasing stress, the sequence is black, gray, yellow, red, blue-green, yellow, red, and green. Black and gray areas show low strains, whereas a continued repetition of red and green color bands indicates extremely high concentrations of stress. An area of uniform color is under a uniform stress.

The degree of strain is indicated by a fringe order, which is simply a collection of black bands appearing in close proximity to each other between colors in the stress pattern. As the stress configuration increases, so do the number of black bands in a fringe order.

Chapter 11

SUMMARY

INTRODUCTION

Plastics provide the processor with materials that are useful, meet product requirements, produce simple to complex shapes, and are economically beneficial (Fig. 11–1). They can be made to have a long life, to resist corrosive environments, to be degradable, to be recyclable, and to meet prac-

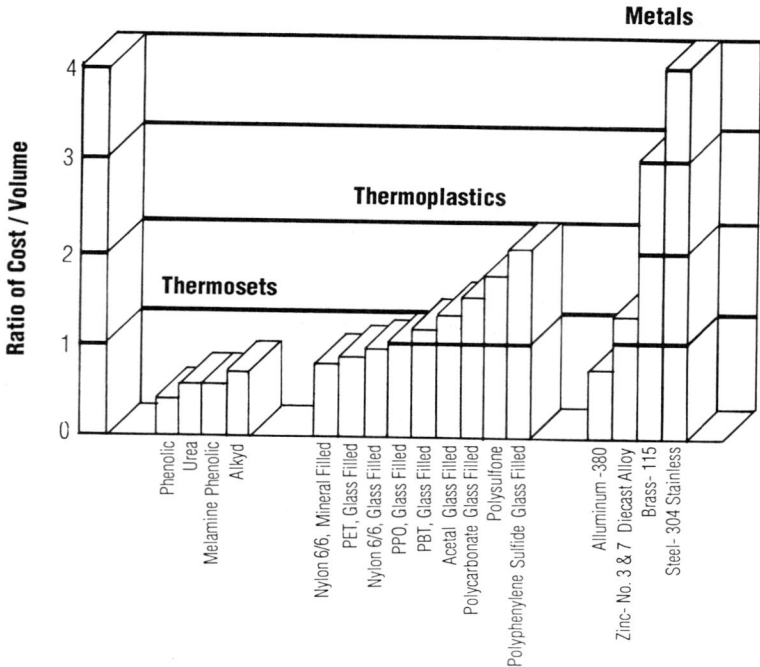

Fig. 11–1. General cost comparison, based on volume, for general classifications of materials.

tically any performance requirements. They also permit the fabrication of parts whose manufacture is difficult or impossible with other materials (steel, aluminum, glass, wood, etc.).

However, plastics processors must continually update their procedures and/or acquire additional knowledge on how to process plastics. New developments in this field are unlimited. This book has emphasized that it is not difficult to process plastics, and has reviewed the many fabricating processes used to produce many different sizes and shapes of thermoplastic and thermoset commodity and engineering resins, which are used either unreinforced or reinforced (in composites). As explained, process selection depends basically on product performance requirements, shape, required dimensional tolerances, plastics processing characteristics, production volume, and cost (1–3, 317–326).

Some plastics can be used with many different processes, but certain others require a specific process. Generally, process selection takes place before material selection, as a range of materials may be available for different processes; or only one method of processing may be available. The "one process" situation could be very unprofitable or restrictive to product performance. Figures 11–2 through 11–7 provide summary guides to selection procedures for processes, products, and materials (3). It is important to realize that the fabrication process can markedly influence product performance.

PLANT CONTROL

It is seldom the case that a processing plant has only one processing machine; and if it has more, it is not what happens on the individual machine that determines profitability, but it is the average performance of all machines. With many machines, it can become very difficult to keep track of all the details that go into the plant's overall operation—hundreds to even thousands of such details. It also becomes increasingly difficult for processors, quality control people, maintenance people, and others always to be present when needed—and it may become very hard for personnel to make decisions as needed.

Modern central control and management systems are changing this situation, however. These systems have been called supervisory control, distributed control, CAD/CAM/CAE, and—the latest—CIM (computer integrated manufacturing; Fig. 2–16). All these designations refer to a system that can monitor all operating parameters for every machine, every piece of materials handling equipment, and all other equipment in the plant. The system receives inputs on all parameters and can issue instructions to each machine to ensure efficient and profitable operation.

354

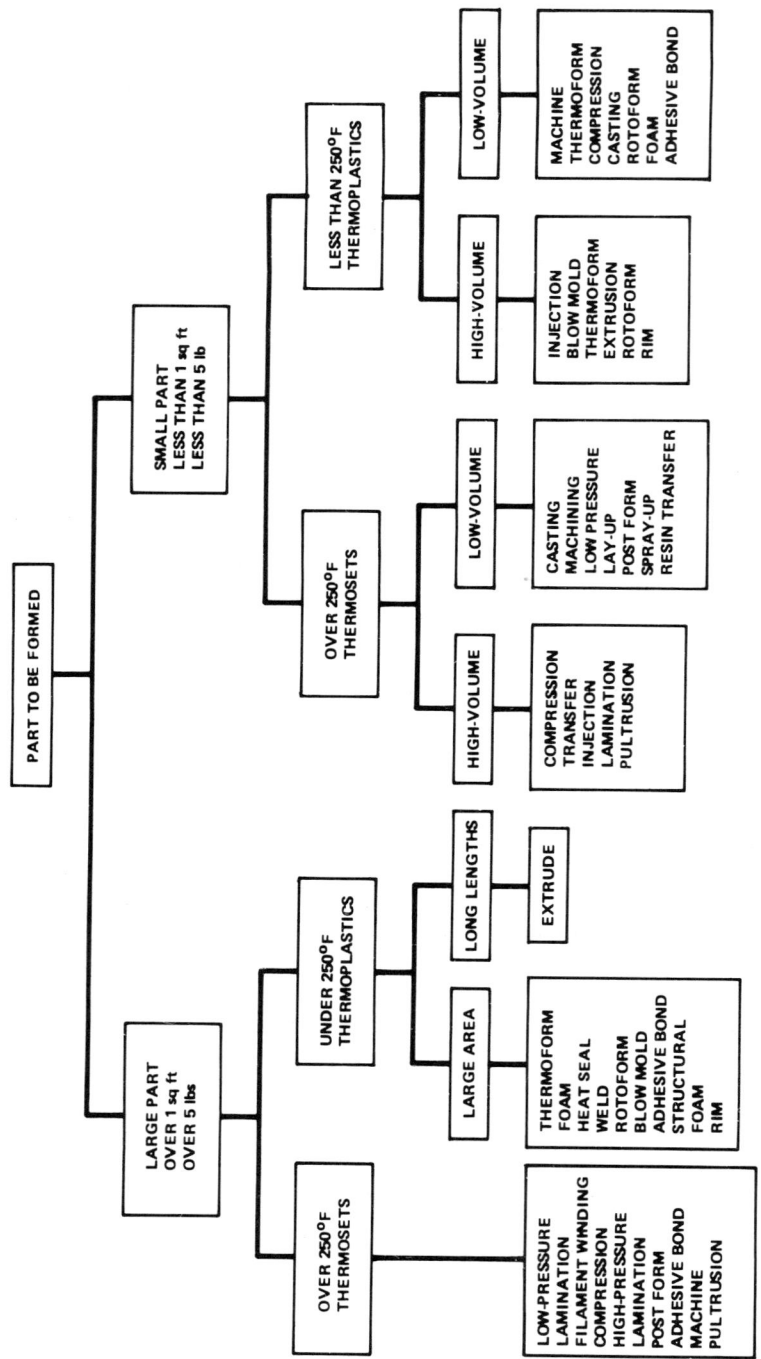

Fig. 11-2. Guide to process selection.

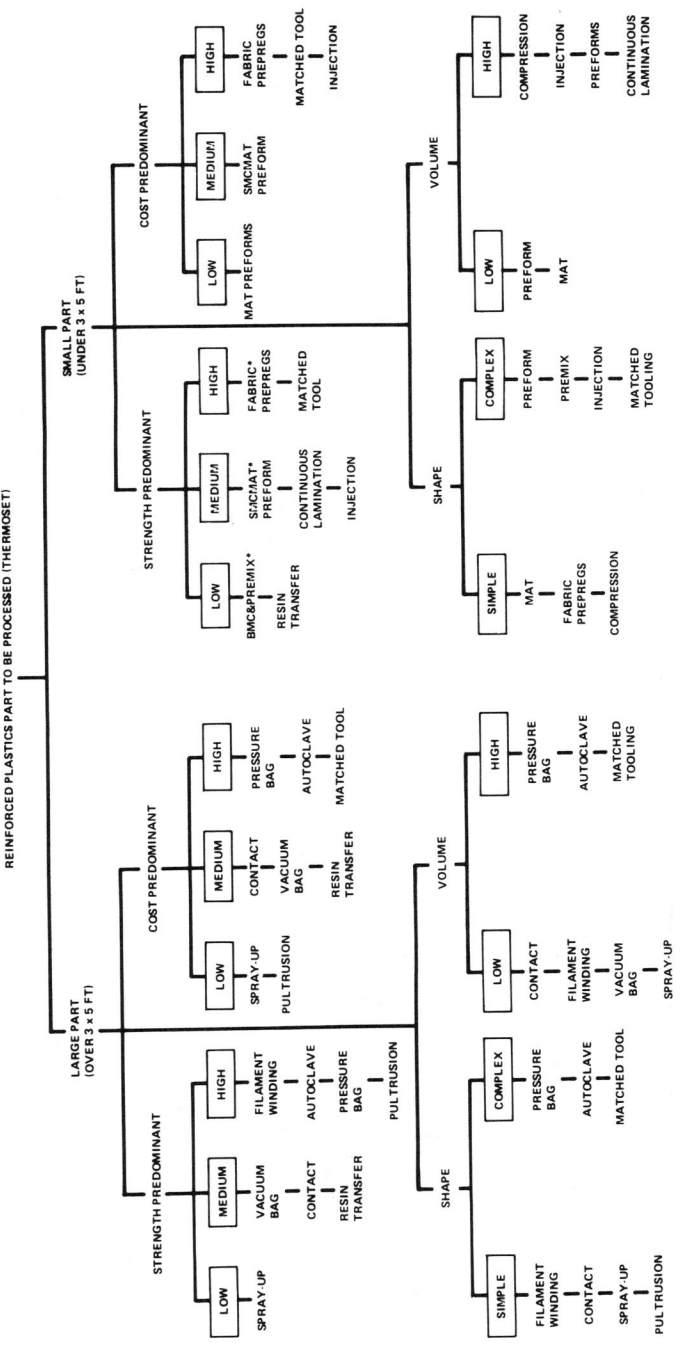

Fig. 11-3. Detailed guide to reinforced thermoset plastics process selection.

*BMC, PREMIX PREFORMS AND PREPREGS - ALL MATCHED TOOL COMPRESSION MOLDING

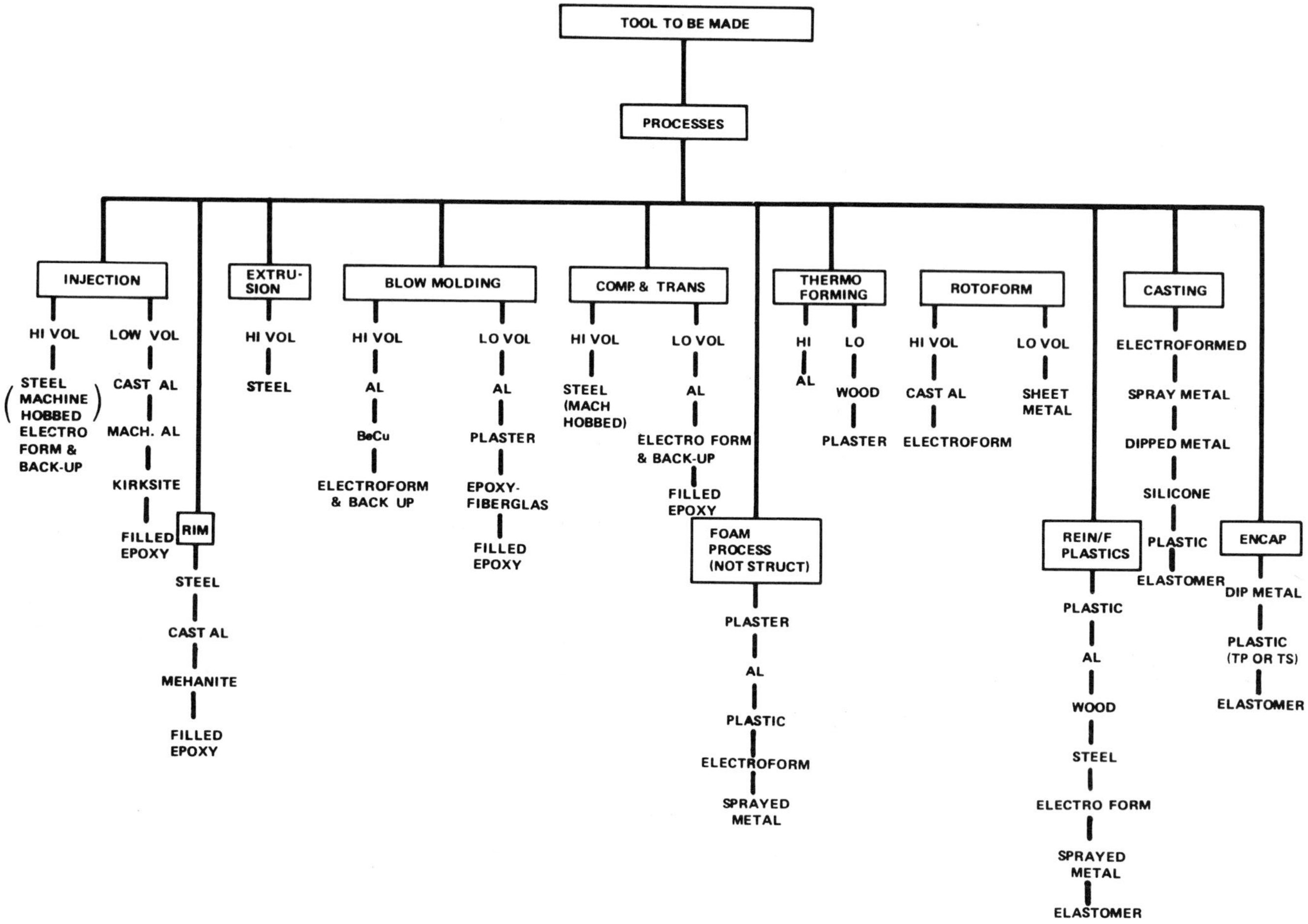

Fig. 11-4. Guide to tooling selection.

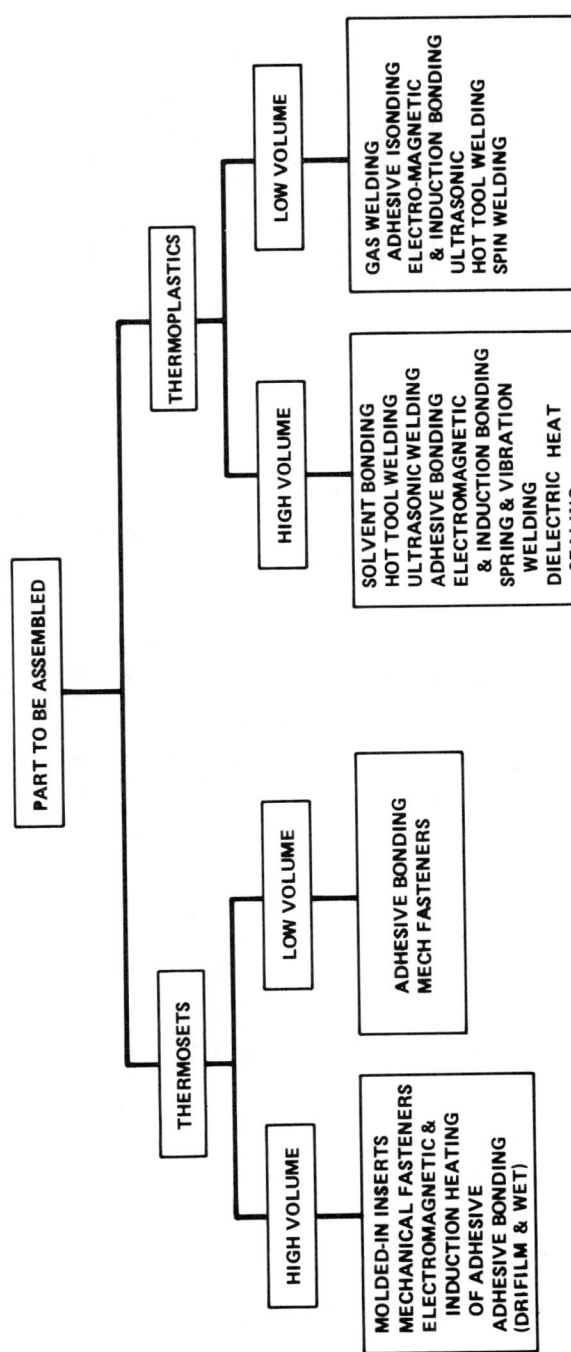

Fig. 11-5. Guide to part assembly selection.

357

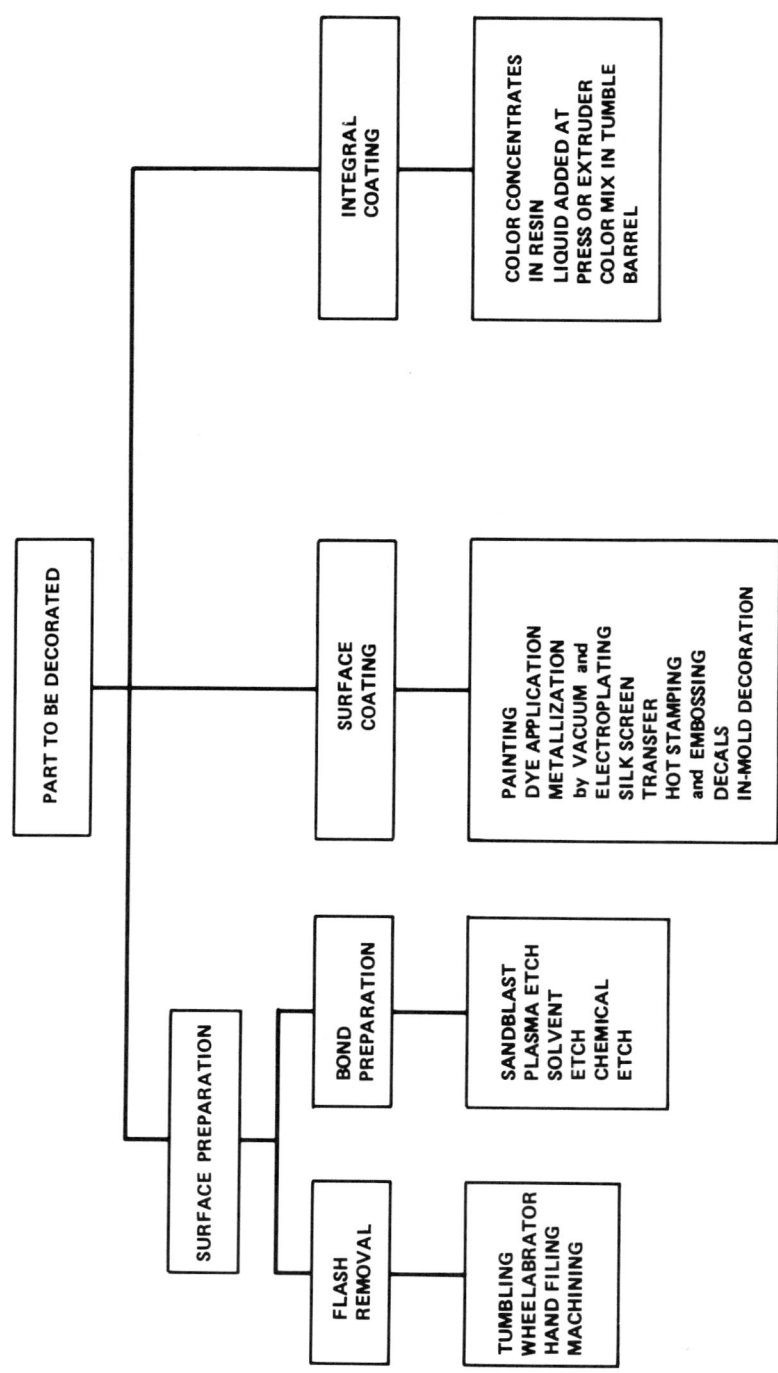

Fig. 11-6. Guide to part decorating selection.

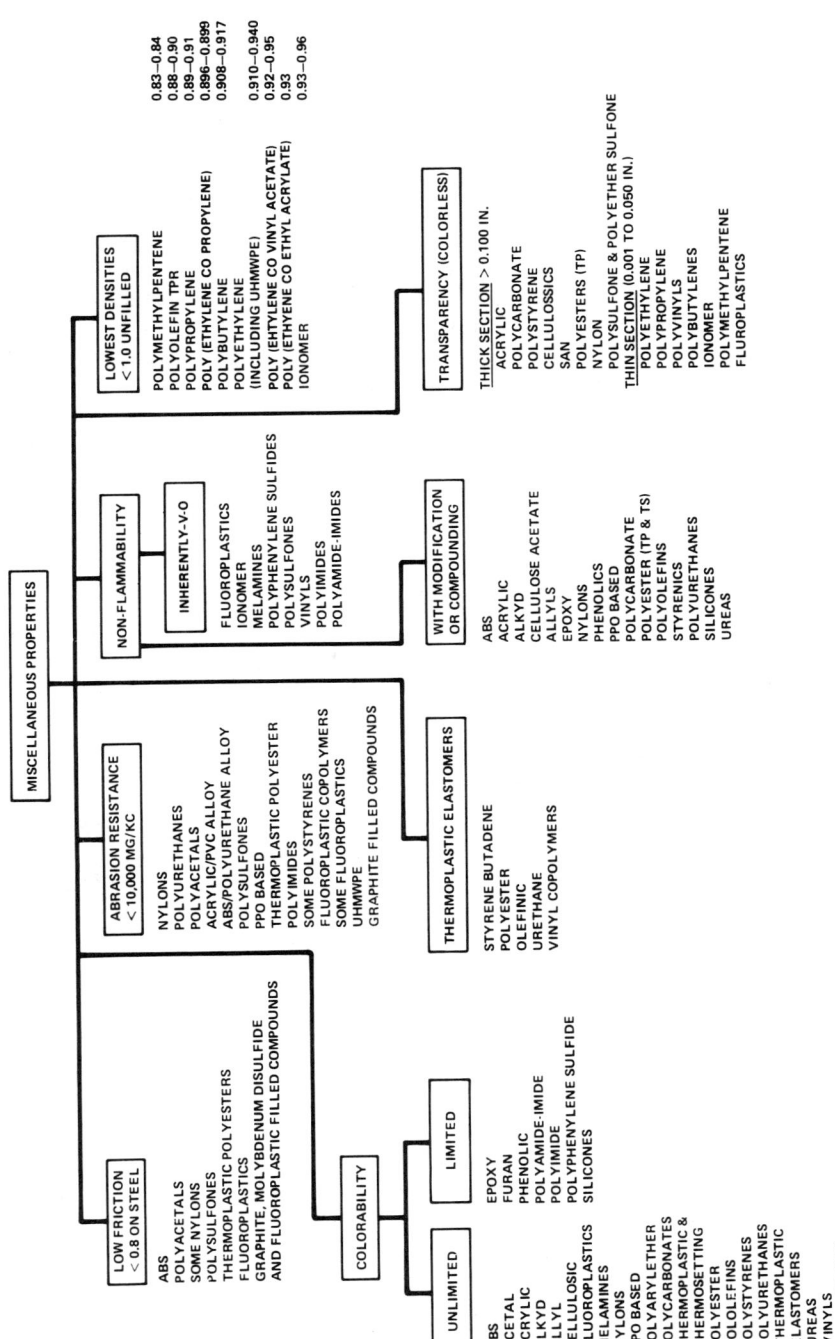

Fig. 11-7. Guide to material selection.

359

For these systems to operate efficiently, talented people are needed to completely integrate the system. These people must be available and must know what is required for all plant operations, including the processing machines. Based on these requirements, a system control is obtained. In turn, these individuals must establish start-up procedures for all plant equipment, making it correspond to the system control. They build limits into the system control, and interface them with control instructions that are best suited to keeping the machines manufacturing parts that meet performance requirements at the lowest processing cost. Thus, it is essential that people working with these control systems be properly trained, to understand how best to operate the processing equipment.

Controls of any type are useful only if it has been determined that they can help the process, and they are properly used. This book was prepared to help people understand how to maximize processing machine efficiency.

PREVENTIVE MAINTENANCE

It is important to set up preventive maintenance procedures on all equipment in the plant. Equipment is built to operate if proper maintenance is used. Processors should make regular machine checkups and maintenance a habit. A thoroughly implemented machine maintenance program will reduce downtime and operating costs as well as rejects of poor parts. Periodic machine checkups and regular maintenance should become a habit.

The machine operator or attendant will not always be able to perform all necessary checking and maintenance steps. For example, to maintain such equipment as a drive mechanism or special instruments, the person may need the help of an electrician or an instrument service specialist, who may not be present or available. The goal is to not require this type of service. With a proper maintenance schedule, it may not be necessary or may be needed very infrequently.

Examples of good preventive maintenance activities include: (1) establish the frequency of lubrication and what types of oil or grease must be used; (2) check for oil leaks and have a procedure to correct/eliminate them; (3) check heaters, thermocouples, pressure transducers, and so on; (4) set up schedules and procedure to clean machines and molds/dies (barrels, screws, sliding mechanisms, clamps, etc.); (5) check control circuits (electrical, hydraulic, mechanical, etc.); (6) schedule checks of conditions wherever questions of alignment, level, parallelism (mold parts, mold press, die system, etc.), and other similar situations exist; (7) set up a schedule to check safety devices on all equipment; and (8) schedule sessions to repeat instructions on safety equipment procedures to all personnel. Figure 11–8 shows where accidents usually occur (324, 325).

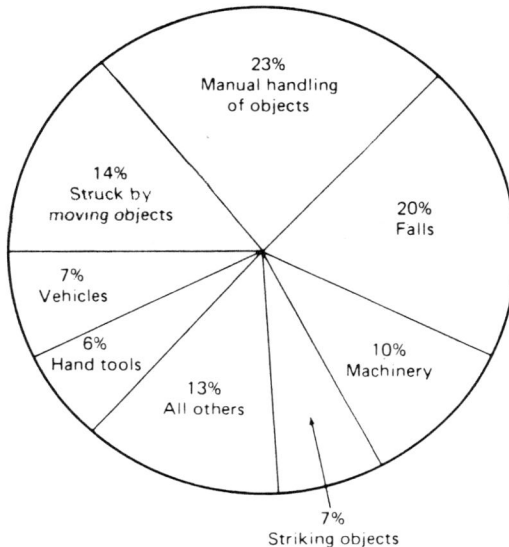

Fig. 11-8. Where accidents occur in all types of manufacturing plants, including plastics plants.

EQUIPMENT IMPROVEMENTS

New equipment always offers potential or significant improvements in processing capabilities. Designers always plan for equipment to aid in meeting goals of zero defects, as well as reduce production costs. Figures 11-9 and 11-10 show that there has been a significant improvement in single-screw extruder outputs (318).

Although it is one of the oldest types of machinery for plastics processing, the single-screw extruder has not yet reached the end of its development—as can be seen in improvements over the past few years, based on the continuing industrial demand for improvements. Output, a major focus years ago, remains paramount; but qualities such as wear and versatility are of equal importance. Improvements in all equipment are ongoing, with additional new developments occurring in injection molding machines (Chapter 2).

There are many "old" machines in operation, particularly in the United States; so there is ample room to improve and simplify plastics processing. In the United States it is estimated that about 80,000 injection molding machines, 12,000 extruders, and 6,000 blow molders are operating. For each type of machine, about 30 percent are under five years old, at least 35 percent are five to ten years old, and the rest are over ten years old. Present

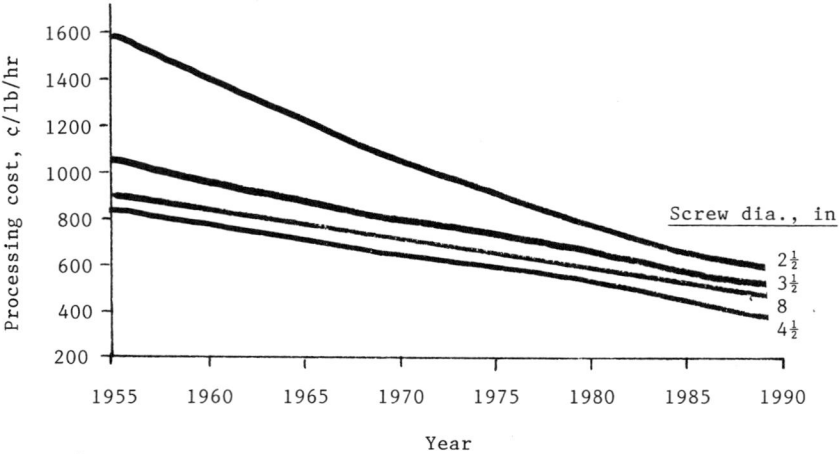

Fig. 11–9. Illustration of steadily dropping processing costs for extrusion, based on pound-per-hour output for different-size screws.

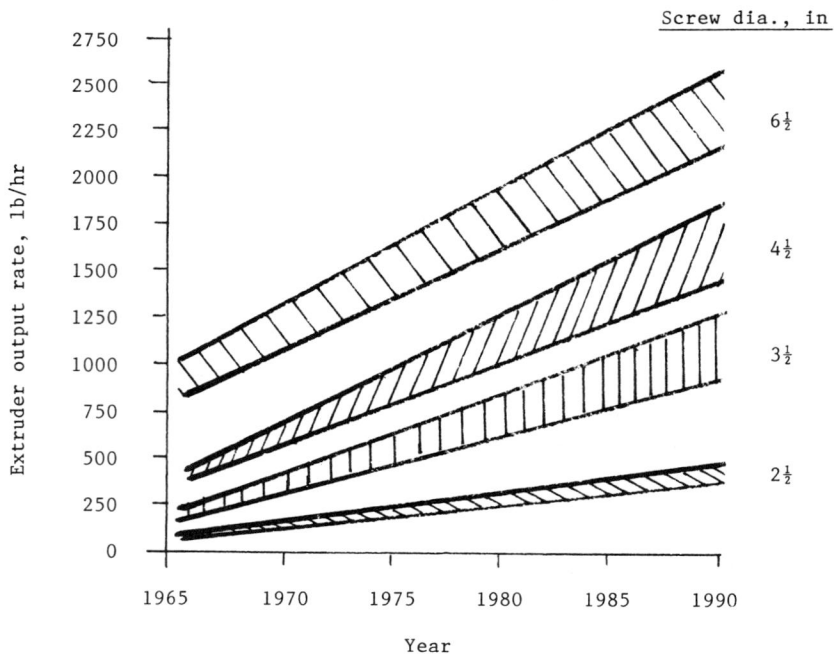

Fig. 11–10. Illustration of increasing outputs for various-size extruders processing the more commonly used plastics.

annual equipment purchases in the United States are about 5,000 injection molding machines, 1,500 extruders, and 1,000 blow molding machines. Also about 10 percent of U.S. machines use robots, compared to about 90 percent using robots in Japan.

TECHNICAL COST MODELING

The adoption of any technology for producing manufactured products is characterized by a wide range of processing (Fig. 11–11), materials, and economic consequences. Although considerable talent can be brought to bear on the processing and engineering aspects, economic questions remain. Cost problems are particularly acute when the technology that will be employed is not fully understood, as much of cost analysis is based on historical data, past experience, and individual accounting practices.

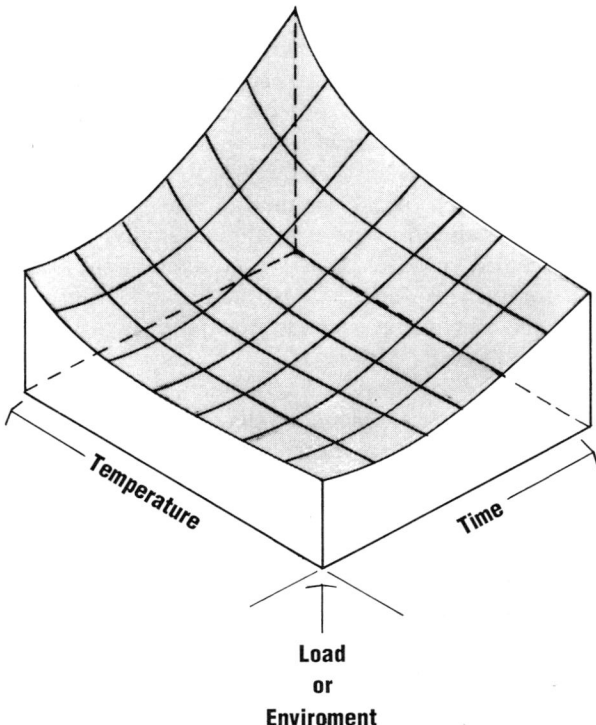

Fig. 11–11. Basics of processing with plastics; three-dimensional illustration of the contour plots.

Historically, technologies have been introduced on the shop floor incrementally, with their economic consequences measured directly. Although incorporating technical changes in the plant to test their viability may have been appropriate in the past, it is economically infeasible to explore today's wide range of alternatives in this fashion. Technical cost modeling (TCM) has been developed as a method for analyzing the economics of alternative manufacturing processes without the prohibitive economic burden of trial-and-error innovation and process optimization (2).

TCM is an extension of conventional process modeling, with particular emphasis on capturing the cost implications of process variables and economic parameters. By coordinating cost estimates with processing knowledge, critical assumptions (processing rates, energy used, materials consumed, etc.) can be made to interact in a consistent, logical, and accurate framework of economic analysis, producing cost estimates under a wide range of conditions.

For example, TCM can be used to determine the plastic process that is best for production without extensive expenditures of capital and time. Not only can TCM be used to establish direct comparisons between processes, but it can also determine the ultimate performance of a particular process, as well as identifying the limiting process steps and/or parameters. (TCM models have been produced by J. V. Busch and F. R. Field III; see reference 2.)

TCM uses an approach to cost estimating in which each of the elements that contribute to total cost is estimated individually. These individual estimates are derived from basic principles and the manufacturing process. It reduces the complex problem of cost analysis to a series of simpler estimating problems, and brings processing expertise, rather than intuition, to bear on solving these problems.

In dividing cost into its contributing elements, the first distinction that can be made is that some cost elements depend upon the number of products produced annually, whereas others do not. For example, the cost contribution of the plastic is the same regardless of the number of items produced unless the material price is discounted because of very high volume. On the other hand, the per-piece cost of tooling will vary with changes in production volume. These two types of cost elements, which are called variable and fixed costs, respectively, create a natural division of the elements of manufacturing part cost.

Variable cost elements are those elements of piece cost whose values are dependent of the number of pieces produced. For most plastics fabrication processes the principal variable cost elements are material, direct labor, and energy costs.

Fixed costs are those elements of piece cost that are a function of the

annual production volume. Fixed costs are called fixed becuase they typically represent one-time capital investments (building, silo, processing machine, etc.) or annual expenses unaffected by the number of parts produced
(building rent, engineering support, administrative personnel, etc.). Typically, these costs are distributed over the total number of parts produced
in a given period. For plastics processes the principal elements are main
machine cost, auxiliary equipment cost, tooling cost, building cost, overhead labor cost, maintenance cost, and cost of capital.

To demonstrate the use of such a comparative cost analysis, production
of a panel was analyzed according to different processes (Fig. 11-12). In
these case studies, the following conditions existed: (1) panels measured 24
× 36 in. (61 × 91 cm) with the wall thickness dictated by process and part
requirements, so that the weights of the panels differed; (2) production was
at a level of 40,000/yr; (3) plastics for all panels were of the same type,
except that different grades had to used based on process requirement, so
that costs changed; (4) each panel received one coat of paint, except that
the structural foam also had a primer coating; and (5) costs were allocated

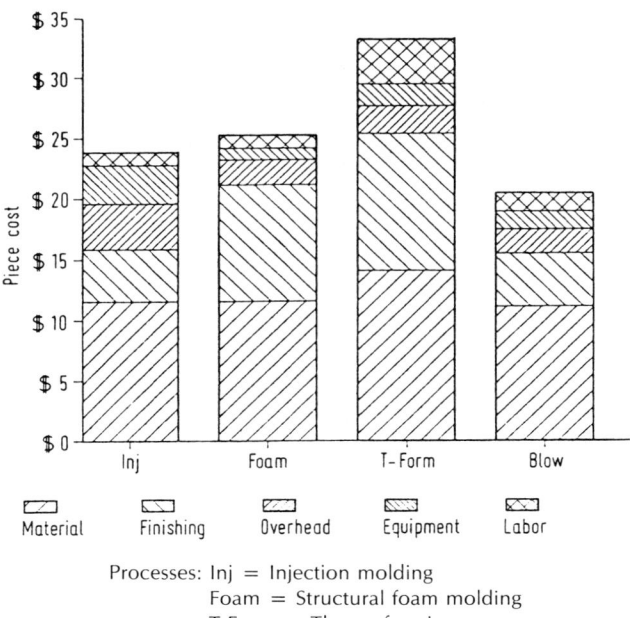

Processes: Inj = Injection molding
           Foam = Structural foam molding
           T-Form = Thermoforming
           Blow = Blow molding

Fig. 11-12. Cost comparison of panel production using a technical cost modeling program,
showing blow molding with the lowest piece cost.

as needed to those processes that required trimming and other secondary operations.

TCM can keep cost data current, based on cost changes from day to day, region to region, and so on. Of course, the means of keeping these data updated requires that those costs be obtained on a "regular" basis and incorporated into the TCM.

PLASTICS GROWTH

Plastics are among the nation's and the world's most widely used materials, having surpassed steel on a volume basis in 1983 (Fig. 11–13). By the end of this century, plastics will surpass steel on a weight basis (Fig. 11–14). Plastics materials and products cover the entire spectrum of the world's economy, so that fortunes are not tied to any particular business segment. Plastics manufacturers are in a position to benefit in a wide variety of markets: packaging, building and construction, electronics/electrical items, furniture, apparel, appliances, agriculture, housewares, luggage, transportation (automotive, aircraft, boat, etc.), medicine and health care, recreation, and so on (188).

Effective exploitation of processing opportunities is the key to success. In turn, success hinges on other factors, such as proper product design, proper selection of materials, and use of the best available processing equipment (Fig. 11–15). As new equipment usually is designed to be more productive and produce better-quality products, one should stay abreast of new

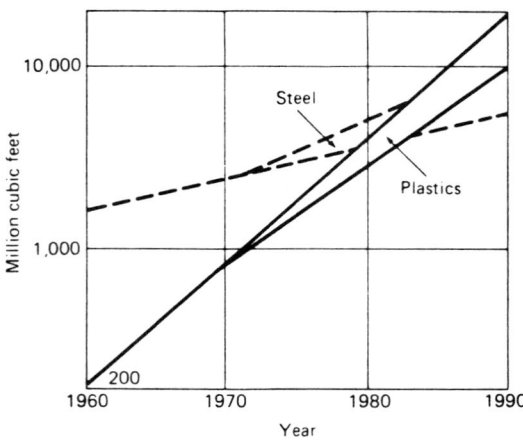

Fig. 11–13. World consumption of plastics by volume.

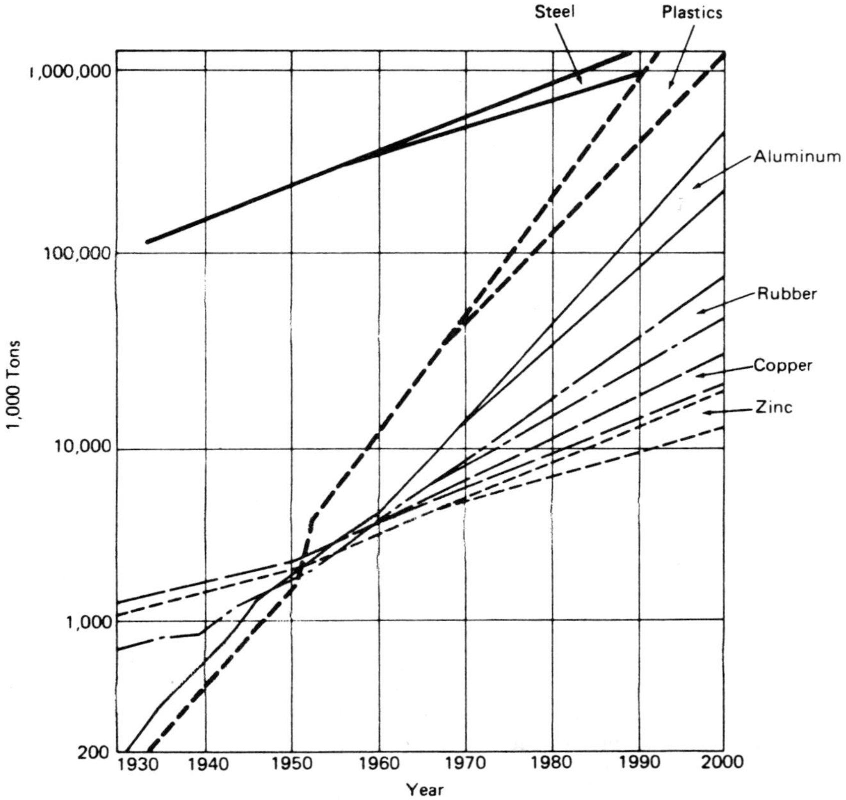
Steel
Plastics
1,000,000
100,000
10,000
1,000
200
1,000 Tons
Aluminum
Rubber
Copper
Zinc
1930
1940
1950
1960
1970
1980
1990
2000
Year

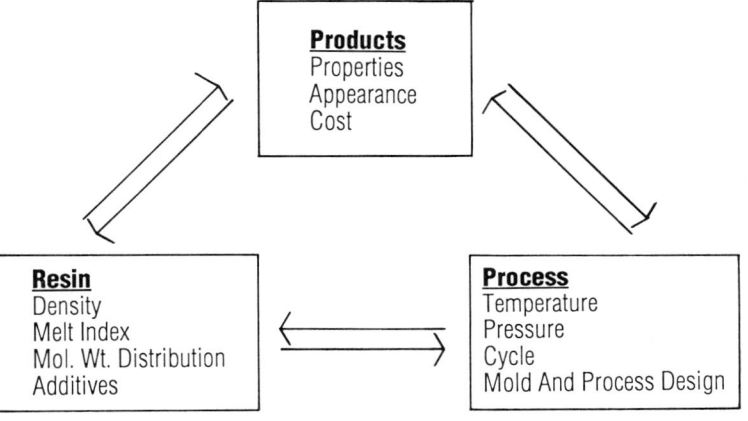

Fig. 11-14. World consumption of plastics by weight.

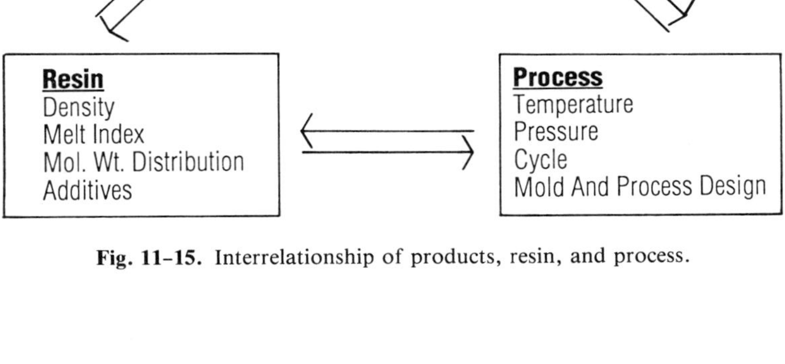
Products
Properties
Appearance
Cost
Resin
Density
Melt Index
Mol. Wt. Distribution
Additives
Process
Temperature
Pressure
Cycle
Mold And Process Design

Fig. 11-15. Interrelationship of products, resin, and process.

Become aware that for any gain there could be a loss
not originally included in the design performance.

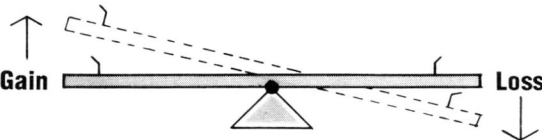

When you gain "something" there will be a loss .....does
that loss influence product performance (for any material:
plastic, wood, steel, glass, etc.).

Fig. 11–16. With any gain there is some type of loss, which could be either an advantage or a disadvantage.

equipment developments and evaluate them logically. Figure 11–16 can serve as a reminder that there are gains and losses to be balanced in any operation. With processing equipment, there is an extremely vast area for improving profitability by ensuring that the best available equipment is being used to meet specific performance requirements.

REFERENCES

1. Rosato, D. V. and Rosato, D. V., *Injection Molding Handbook,* Van Nostrand Reinhold, 1986.
2. Rosato, D. V. and Rosato, D. V., *Blow Molding Handbook,* Van Nostrand Reinhold, 1989.
3. Schwartz, S. S. and Goodman, S., *Plastics Materials and Processes,* Van Nostrand Reinhold, 1982.
4. *Engineering Plastics,* Vol. 2, ASM International, 1988.
5. Chanda, M. and Roy, S. K., *Plastics Technology Handbook,* Marcel Dekker, 1987.
6. Dym, J. B., *Injection Molds and Molding,* Van Nostrand Reinhold, 1979.
7. Du Bois, J. H. and Pribble, W. I., *Plastics Mold Engineering Handbook,* Van Nostrand Reinhold, 1978.
8. Rauwendaal, C., *Polymer Extrusion,* Hanser, 1986.
9. Rao, N. S., *Designing Machines and Dies for Polymer Processing with Computer,* Hanser, 1981.
10. *Melt Temperature Profile Revealed,* PT, pp. 33–37, Jul. 1988.
11. Frados, J., *Plastics Engineering Handbook,* Van Nostrand Reinhold, 1947.
12. Du Bois, J. H. and John, F. W., *Plastics,* Van Nostrand Reinhold, 1974.
13. Harper, C. A., *Handbook of Plastics and Elastomers,* McGraw-Hill, 1975.
14. Bruins, P. F., *Polyblends and Composites,* Wiley, 1970.
15. Szycher, M. and Robinson, W. J., *Synthetic Biomedical Polymers,* Technomic, 1980.
16. Brady, G. S. and Clauser, H. R., *Materials Handbook,* McGraw-Hill, 1985.
17. Boyer, H. E., *Selection of Materials for Service Environments; Source Book,* ASM, 1987.
18. *International Plastics Selector,* Cordura Publ., San Diego, CA, Annually.
19. *Plastics Technology Data Bank,* Bill Publ., Annually.
20. *Plastiserve-Plastics Information Systems,* Data Service Inc., Annually.
21. Plastics Technical Evaluation Center (Plastec), Picatinny Arsenal, NJ.
22. Throne, J. L., *Thermoforming,* Hanser, 1987.
23. *Structural Plastics Selection Manual,* No. 66, ASCE, 1984.
24. Dorgham, M. A. and Rosato, D. V., *Designing with Plastics and Advanced Composites,* Interscience Enterprises, Geneva, Switzerland, 1986.
25. Griffith, A. A., *The Phenomena of Rupture & Flow in Solids,* Phil. Trans. Royal Society of London, Ser. A, 221, 163–198, 1921; and *The Theory of Rupture,* Proceedings of First International Congress—Applied Mechanics, p. 55, 1924.
26. Nielsen, L. E., *Mechanical Properties of Polymers and Composites,* Vols. 1 and 2, Marcel Dekker, 1974.
27. Middleman, S. S., *Fundamentals of Polymer Processing,* McGraw-Hill, 1977.
28. Kusy, P. F., *Plastics Materials Selection Guide,* No. 760663, SAE, Sept. 1976.
29. Kampf, G., *Characterization of Plastics by Physical Methods; Experimental Techniques and Practical Applications,* Hanser, 1986.

30. Rosato, D. V., *Materials Selection,* pp. 357–379, in Mark-Bikales-Overberger-Menges: *Encyclopedia of Polymer Science and Engineering,* Vol. 9, Wiley, 1987.
31. Timoshenko, S. and Young, D. H., *Elements of Strength of Materials,* Van Nostrand Reinhold, 1962.
32. Rosato, D. V., *Thermosets,* pp. 350–391, in Mark-Bikales-Overberger-Menges: *Encyclopedia of Polymer Science and Engineering,* Vol. 14, Wiley, 1988.
33. Beck, R. D., *Plastics Product Design,* Van Nostrand Reinhold, 1980.
34. Dym, J. B., *Product Design with Plastics,* Industrial Press, 1983.
35. Levy, S. and DuBois, J. H., *Plastics Product Design Engineering Handbook,* Van Nostrand Reinhold, 1977.
36. Ehrenstein, G. W. and Erhard, G., *Designing with Plastics,* Hanser, 1987.
37. MacDermott, C., *Selecting Thermoplastics for Engineering Applications,* Marcel Dekker, 1984.
38. Ezrin, M., *Plastics Analysis Guide,* Hanser, 1984.
39. Benjamin, B. S., *Structural Design with Plastics,* Van Nostrand Reinhold, 1982.
40. Harper, C. A., *Handbook of Materials and Processes for Electronics,* McGraw-Hill, 1970.
41. Miller, E., *Plastics Products Design Handbook,* Part A, Materials and Components, Marcel Dekker, 1981.
42. Miller, E., *Plastics Products Design Handbook,* Part B, Processes and Design for Processes, Marcel Dekker, 1983.
43. Bruins, P. F., *Polyblends and Composites,* Wiley, 1970.
44. Diamond, W. J., *Practical Experiment Designs,* Hanser, 1988.
45. Sarwell, J. A., *Electronic Packaging,* ASM, 1986.
46. Young, J. E. and Shane, R. S., *Materials and Processes: Optimize Design,* Marcel Dekker, 1985.
47. Bower, H. E., *Selection of Materials for Component Design—Source Book,* ASM, 1986.
48. Wessels, B. M., *Advances in Electronic Materials,* ASM, 1986.
49. Rosato, D. V., *Advanced Engineering Design—Short Course,* National Design Engineering Conference, ASME, 1983.
50. Griffith, A. A., *Phenomena of Rupture and Flow of Solids,* Phil. Trans. Royal Society, 1921.
51. Bauer, E., *Engineering Design for Plastics,* Van Nostrand Reinhold, 1964.
52. Rosato, D. V. and Schwartz, R. T., *Environmental Effects on Polymeric Materials,* Vols. 1 and 2, Wiley, 1968.
53. Schnabel, W., *Polymer Degradation: Principles and Applications,* Hanser, 1981.
54. Burns, R., *Polyester Molding Compounds,* Marcel Dekker, 1982.
55. Turi, E. A., *Thermal Characterization of Polymer Materials,* Technomic, 1981.
56. Iwen, E. D., *Degradation and Stabilization of PVC,* Elsevier, 1984.
57. Titow, W. V., *PVC Technology,* 4th Edition, Elsevier, 1984.
58. *Engineering with Rigid PVC: Processability and Applications,* Marcel Dekker, 1984.
59. Chapoy, L., *Recent Advances in Liquid Crystal Polymers,* Elsevier, 1985.
60. Gacher, R., *Plastics Additives,* Hanser, 1987.
61. Mascia, L., *The Role of Additives in Plastics,* Wiley, 1974.
62. Matthews, G. A. R., *Polymer Mixing Technology,* Elsevier, 1982.
63. Pearson, J. and Richardson, S., *Computational Analysis of Polymers,* Elsevier, 1983.
64. Walker, B. M., *Handbook of Thermoplastic Elastomers,* Van Nostrand Reinhold, 1979.
65. Culbertson, B. and Pittman, C., *New Monomers and Polymers,* Plenum, 1984.
66. Ceresa, J. A., *Block Graft Copolymerization,* Wiley, 1973.
67. Carraher, C. and Moore, J., *Modifications of Polymers,* Vol. 21, Plenum, 1984.
68. Allen, N. S., *Degradation and Stabilization of Polyolefins,* Elsevier, 1983.

69. Bark, L. S. and Allen, N. S., *Analysis of Polymer Systems*, Elsevier, 1982.
70. Hall, I. H., *Structure of Crystalline Polymers*, Elsevier, 1984.
71. Lange, A. and Ezrin, M., *Plastics Analysis Guide*, Hanser, 1983.
72. VanVlack, L. H., *Materials for Engineering: Concept and Applications*, Addison-Wesley, 1982.
73. Young, J. E. and Shane, R. S., *Materials and Processes: Optimize Design*, Marcel Dekker, 1985.
74. DuBois, J. H., *Plastics*, Van Nostrand Reinhold, 1967.
75. *Guide to Materials Engineering Data and Information*, ASM, 1986.
76. Goodman, S. H., *Handbook of Thermoset Plastics*, Noyes Publ., 1986.
77. *SAE Handbook*, Vol. 1: *Materials*, SAE, 1986.
78. Oertel, G., *Polyurethane Handbook*, Hanser, 1984.
79. Schroder, E., *Guide to Polymer Characterization*, Hanser, 1988.
80. Mitchell, J. Jr., *Applied Polymer Analysis and Characterization*, Hanser, 1987.
81. Bralla, J. G., *Handbook of Product Design for Manufacturing: A Practical Guide to Low-Cost Production*, McGraw-Hill, 1986.
82. Bernhardt, E. C. and Bertacchi, G., *TMC Plastics Molding and Cost Optimization*, SPE ANTEC, Chicago, May 1983.
83. Stoeckhert, K., *Mold Making Handbook*, Hanser, 1983.
84. *Injection Molding Operations*, Husky IM Systems Ltd., Bolton, Canada LOP 1AO, 1980.
85. Sors, L., et al., *Plastics Molds and Dies*, Van Nostrand Reinhold, 1981.
86. Rubin, I. I., *Injection Molding Theory and Practices*, Wiley, 1972.
87. Pye, R. C. W., *Injection Mold Design*, The Plastics and Rubber Inst., England, 1980.
88. Stoeckert, K., *Injection Molds: 102 Proven Designs*, Hanser, 1983.
89. Johannaber, F., *Injection Molding Machines*, Hanser, 1984.
90. Suh, N. P. and Sung, H., *Science and Technology of Polymer Processing*, MIT Press, 1979.
91. Ceresa, R. J., *Block and Graft Copolymerization*, Wiley, 1973.
92. McKelvey, J. M., *Polymer Processing*, Wiley, 1962.
93. Lahti, G. P., Calculation of Pressure Drops and Outputs, *SPE Journal*, Jul. 1963.
94. Brydson, J. A., *Flow Properties of Polymer Melts*, Van Nostrand Reinhold, 1970.
95. Throne, J. L., *Plastics Processing Engineering*, Marcel Dekker, 1979.
96. Mark-Bikales-Overberger-Menges, *Encyclopedia of Polymer Science and Engineering*, Vols. 1–15, 2nd Edition, Wiley, 1987.
97. Gutfinger, C., et al., *Polymer Engineering Sciences*, 15, 515, 1975.
98. Kamal, M. R. and Kenig, S., *Polymer Engineering Sciences*, 12, 294, 303, 1972.
99. Manzione, L. T., *Cyclic Time in Injection Molding*, SPE IMD Newsletter, No. 7, 1984.
100. Colby, P., Screw and Barrel Technology, *Spirex Bulletin*, Spirex Corp., Youngstown, OH 44512 USA, 1988.
101. *Plasticizing Performance Test Procedure for Screw Machines*, SPI Injection Molding Div., Washington, DC, 1968.
102. Glanvill, A. B., *The Plastics Engineer's Data Book*, Machinery Publ., 1971.
103. Barr, R. A., *High Performance Screw Is Custom Designed*, SPE IMD Newsletter, No. 18, 1988.
104. Chung, C. I., *A New Look at the Mechanism of Melting*, Plastics Engineering, Vol. 32, No. 6, Jun. 1976.
105. Chung, C. I. and Barr, R., *A Novel Energy Efficient Extruder Screw Design*, SPE ANTEC, 168, May 1983.
106. Nichols R. and Kheradi, G., *Melting in Counter-Rotating Tangential Twin Screw Extruder*, SPE ANTEC, 134, May 1983.
107. Olmsted, B. A., *Solving the Shot to Shot Variation Dilemma*, PM&E, p. 38, Dec. 1987.

108. Olmsted, B. A., *Influence of Melt Quality,* SPE IMD Newsletter, No. 5, 1987.
109. Snyder, J. A., *Mold Design Software Makes Part Shrinkage Predictable,* MP, pp. 10–11, Jan. 1989.
110. Bernhardt, E. C., *Computer Aided Engineering for Injection Molding,* Hanser, 1983.
111. Bernhardt, E. C., *Computer Integrated Injection Molding,* SPE IMD Newsletter, No. 16, 1987.
112. Bernhardt, E. C. and Bertacchi, G., *New Tool for Mold Design: Computerized Shrinkage Analysis,* PT, pp. 81–85, Jan. 1986.
113. Wenskus, J. J., *Statistical Process Control for Injection Molding,* Eastman Kodak Co., Rochester, NY, Document EK No. 248026V, Sept. 7, 1988.
114. Wenskus, J. J., *SPC for IM,* SPE IMD Newsletter, No. 13, 1987.
115. Richardson, P. A., *Introduction to Extrusion,* SPE, 1974.
116. Fisher, E. C., *Extrusion of Plastics,* The Plastics and Rubber Inst., 1976.
117. Levy, S., *Plastics Extrusion Technology Handbook,* Industrial Press, 1981.
118. Janssen, L. P., *Twin Screw Extrusion,* Elsevier, 1978.
119. Griff, A. A., *Plastics Extrusion Technology,* Van Nostrand Reinhold, 1962.
120. *Grooved Feed Extrusion,* Davis Standard Keystone Review, Sept. 1988.
121. Rosato, D. V., *Extrusion: Technology, Markets, Economics Seminar,* University of Lowell, 1986.
122. Schenkel, G., *Plastics Extrusion Technology and Theory,* Elsevier, 1966.
123. Esposita, R. A., *Add-On Mixers Boast Extruder Performance,* Plastics Engineering, Oct. 1986.
124. Schott, N. R., *Plastics Extrusion Processing Technology Seminar,* University of Lowell, 1988.
125. Cramm, R. H. and Sibbach, W. R., *Coextrusion Coating and Film Fabrication,* TAPPI, 1983.
126. Martelli, F. G., *Twin Screw Extruders: A Basic Understanding,* Van Nostrand Reinhold, 1983.
127. Michaeli, W., *Extrusion Dies; Design and Engineering Computations,* Hanser, 1984.
128. Bikales, M., *Extrusion and Other Plastics Operation,* Wiley, 1971.
129. Fisher, E. G. and Whitfield, J. L., *Extrusion of Plastics,* Wiley, 1976.
130. Tadmor, Z. and Klein, J., *Engineering Principles of Plasticating Extrusion,* Van Nostrand Reinhold, 1970.
131. Burkhardt, U., et al., *Twin-Screw Extruders,* SPE ANTEC, May 1977.
132. Burkhardt, U., et al., *Twin-Screw Extruders,* PW, Nov. 1977.
133. Menges, G., et al., *Extrusion,* SPE ANTEC, May 1972.
134. Duska, J. J. and Kruder, G. A., *Extrusion Developments,* SPE ANTEC, May 1976.
135. Nichols, R. J. and Kruder, G. A., *Extrusion,* SPE ANTEC, May 1974.
136. McKelvey, J. M., *Extrusion,* May 1978.
137. Maddock, B. H., *Screw Designs,* SPE Journal, No. 15, p. 383, 1959.
138. Martelli, F., *Calculating Extruder Energy Efficiency,* Plastics Compounding, Apr./May 1980.
139. Nichols, R. J., *Counter-Rotating, Tangential Twin Screw Extruders,* SPE ANTEC, May 1982.
140. Martelli, F., Twin-Screws, *European Plastics News,* pp. 24–25, Feb. 1984.
141. Shirato, T., et al., *Counter vs. Corotating Twin Screws: Where Each Fits,* PT Productivity, Series Four, pp. 5–10, 1987.
142. Laake, H. J. and Rabiger, N., *Temperature Development in a Rubber Processing Pin Barrel Extruder,* Kunststoffe, 78, pp. 833–837, Sept. 1988.
143. Kruder, G. A., Venting: *Critical for Quality Extrusion,* PM&E, p. 22, May 1987.

144. McKelvey, J. M., *New Data Show How Gear Pumps Boost Extrusion Productivity,* PT, Sept. 182.
145. Naitove, M. H., *Time to Get Smart about Gear Pumps,* PT, Feb. 1985.
146. Malloy, A. M., *Closed Loop Inlet Pressure Control of Melt Pump Extrusion,* SPE ANTEC, May 1984.
147. Kramer, W. A., *Gear Pump Characteristics and Applications,* SPE ANTEC, May 1985.
148. Bak, D. J., *Gear Pump Enhances Extruder,* Design News, Nov. 1982.
149. *Gear Pump Technical Bulletin TSD-4,* Harrel Inc., CT, 1988.
150. Rice, W. T., *Conservation of Energy and Raw Materials by Utilization of Gear Pumps with Extruders,* ASME Pub. 80-PET 2, Feb. 1980.
151. McKelvey, J. M. and Rice, W. T., *Retrofitting Plasticating Extruders with Gear Pumps,* Chemical Engr., Jan. 1983.
152. Bernhardt, E. C., *Processing of Thermoplastics,* Van Nostrand Reinhold, 1953.
153. Levy, S., *High-Tech Extrusion: Conditioning the Melt,* PM&E, pp. 31–32, Jan. 1989.
154. Kruder, G., *The Basics of Die Design,* PM&E, p. 22, Jan. 1989.
155. Crochet, M. J., et al., *Numerical Simulation of Non-Newtonian Flow,* Elsevier, 1984.
156. Michaeli, W., *Extrusion Dies,* Macmillan, 1984.
157. Murray, T. A., *Here's Your Guide to Die Extrusion,* PT, pp. 99–105, Feb. 1978.
158. Metzger, A. P. and Matlack, J. D., *Comparative Swelling Behavior of Various Thermoplastic Polymers,* SPE ANTEC, 1967.
159. Cloeren, P. P., *Feedback Coextrusion Systems,* TAPPI Laminations and Coatings Conference Proceedings, pp. 501–512, 1987.
160. Drennan, W. C., *Tie-Layer Resins: More Than Ties to "Bind,"* Plastics Packaging, pp. 32–35, Sept./Oct. 1988.
161. Smith, D. J., *Blown Stretch Film Processing Parameters,* TAPPI Laminations and Coatings Conference Proceedings, 163–165, 1987.
162. *Petrothene Polyolefins—A Processing Guide,* USI Chem. National Dist., 1986.
163. Hensen, F. and Braun, S., *Manufacturing of Film Tapes,* Industrial and Production Engr., Mar. 1978.
164. Wiebaden, W. S., *Improving the Properties of Plastic Film by Orientation,* Kunststoffe, pp. 52–54, Oct. 1988.
165. Wright, W. D., *Blown Film Cooling Joins the March toward More Efficient Processing,* PE, pp. 30–33, Sept. 1981.
166. Hoechst Celanese Corp. *Technical Literature.*
167. Halter, H. H., *What a Blown Film Control System Should Do,* PT Productivity, Series Four, pp. 11–18, 1987.
168. *Blown Film's a Cinch with New Air Ring,* PT, pp. 33–37, Sept. 1988.
169. *Bubble Configurations Affect Film Cooling Dynamics,* PW, p. 39, Oct. 1982.
170. Paderborn, J. W., *Developments in Blown Film Extrusion,* Kunststoffe, pp. 54–57, Oct. 1988.
171. Virginski, W. B., *High Quality Sheet Production Keyed to Roll-Stand Expertise,* PM&E, pp. 25–26, Jul. 1988.
172. Troisdorf, N. B., *Profile Extrusion Lines,* Kunststoffe, pp. 59–61, Oct. 1988.
173. Levy, S., *Processes and Equipment for In-line Post Extrusion Forming,* PM&E, pp. 26–29, Aug. 1978.
174. Fow, L. and Bush, F., *Precision Profile Extrusion,* SPE Extrusion Newsletter, pp. 3–4, Feb. 1984.
175. Hanau, G. B., *Vacuum Web Coatings of Plastics,* Kunststoffe, pp. 6–7, Sept. 1988.
176. Wortberg, J., *Justifying the Cost of Automation in Blown Film Extrusion,* PT Productivity, Series Three, pp. 5–9, 1986.

177. Harris, P. W., *Use of Web Guiding Can Boost Output, Profitability,* Paper Film and Foil Converter, pp. 40–41, Jan. 1989.

178. Levy, S., *On-line Gauging of Extrusion Dimensions,* PM&E, pp. 29–31, Dec. 1978.

179. Menges, G., et al., *Systems of Process Control in Extrusion,* Kunststoffe, pp. 45–48, Oct. 1988.

180. *Transverse Gauging Control Debuts for Blown Film, Windmoeller and Hoelscher,* PM&E, p. 9, Jan. 1989.

181. Infante, R., *Utilization of Pressure Transducers for Improved Control of Extrusion Process,* SPE ANTEC, Apr. 1988.

182. Murray, C. J., *Closed-Loop Feedback Controls Web Tension,* Design News, pp. 82–83, Dec. 1988.

183. Shapiro, S. I., *On-line Discrimination of Multilayer Structures,* TAPPI Laminations and Coatings Conference Proceedings, pp. 519–522, 1987.

184. Wood, R., *Film Winding,* PM&E, p. 26, Dec. 1988.

185. Fitzgerald, K. R., *Troubleshooting in Extrusion Begins with Good Management,* Plastics World, Yearly Directory, 1988.

186. Rosato, D. V., *Choosing the Blow Molding Machine to Meet Product Demands,* Plastics Today, Mar. 1989.

187. Rosato, D. V., *What to Consider in Picking an Injection Molding Machine,* Plastics Today, Jan. 1989.

188. Rosato, D. V., *Markets for Plastics,* Van Nostrand Reinhold, 1969.

189. Stoeckhert, K., *Economic Evaluation of Interior Cooling of Blow Molded Products,* Ind. and Prod. Engr., Feb. 1981.

190. Jummrich E., *Improving Efficiency of CO_2 Interior Cooling in the Extrusion Blow Molding,* Ind. and Prod. Engr., Feb. 1981.

191. Galli, E., *Postcooling: Effective but Is It Economical,* Plastic Packaging, pp. 36–39, Sept./Oct. 1988.

192. Lodge, C., *Troubleshooting Problems in Container Blow Molding,* PW Directory, pp. 405–408, 1988.

193. *Polyolefin Blow Molding—An Operating Manual,* USI Chemicals, 1986.

194. *PET Bottles Injection/Stretch Blow Molded in Single Stage,* PM&E, p. 10, Jan. 1989.

195. Lohrbacher, V. and Hoven-Nievelstein, W. B., *Extrusion Blow Molding,* Kunststoffe, pp. 61–64, Oct. 1988.

196. *Upgraded Line of Blow Molders Offers Faster Cycles; Coextrusion Capabilities,* MP, p. 126, May 1987.

197. *Husky Introduces its First Injection Stretch System,* PW, pp. 58–63, Jan. 1989.

198. Kandt, A., *From Design to Production-Vehicle Fuel Tanks,* European Plastics News, pp. 31–34, Mar. 1986.

199. Oliversen, G., *Blow Molding Steps Up to Distributed Controls,* PE, pp. 31–34, Nov. 1986.

200. Lounsbury, D. C., *What Processors Should Know about Extruder Maintenance,* PT, Dec. 1981.

201. Galli, E., *Designed for Blow Molding,* Plastics Design, pp. 21–28, Jan./Feb. 1989.

202. Bonn, W. D., *Blow Molding of Engineering Plastics,* Kunststoffe, pp. 3–6, Sept. 1988.

203. Smoluk, G. R., *Large Parts Blow Molding Take on a New Measure of Cost Efficiency,* MP, pp. 61–64, Oct. 1967.

204. *Tooling Techniques for Blow Molding Technical Parts,* PM&E, pp. 40–43, Nov. 1988.

205. Miller, B., *K'86 Spotlights Big Gains in Blow Molding,* PW, pp. 45–49, Feb. 1987.

206. Bonn, R. H., *Development of Blow Molding from Beginnings to the Present Day,* Ind. and Prod. Engr., Jan. 1980.

207. *Blow Molding—New Materials and New Markets Forge on Engineering Change,* MP, pp. 91–96, Jan. 1989.
208. Irwin, C., *Blow Molding, Encyclopedia of Polymer Science and Engineering,* Vol. 2, pp. 447–478, Wiley, 1985.
209. Kruder, G. A., *Matching the Extruder to Operating Requirements,* PM&E, p. 26, Sept. 1988.
210. *Blow Molding Report—Expertise to Satisfy Market Demands,* European Plastics News, pp. 17–25, Jun. 1987.
211. Sneller, J. A., *Blow Molding Controls Begin to Make CIM More Likely Prospect,* MP, pp. 57–62, Aug. 1988.
212. Smoluk, G. R., *Hot Competition among Builders; Better Days for Blow Molders,* MP, pp. 50–56, Aug. 1988.
213. Murphy, C., *Radiant Heater Panels Improve Thermoforming,* MP, pp. 78–82, May 1987.
214. *Paper/Plastic Can Provides Good Barrier at Low Cost,* Packaging Technology, pp. 146–147, Nov. 1986.
215. Levy, S., *Processes and Equipment for In-line Post-Extrusion Forming,* PM&E, pp. 26–29, Aug. 1978.
216. Malpass, V. E. and Kempthorn, J. T., *Setting Conditions for Polyolefin Thermoforming,* PE, pp. 53–57, Aug. 1986.
217. Titus, J. B., *Solid Phase Forming/Cold Forming of Plastics,* Plastec Report R-42, Plastec, Picatinny Arsenal, Dover, NJ, 1976.
218. *In-Line Thermoforming of Performance Packaging,* European Plastics News, pp. 14–16, Feb. 1986.
219. Smoluk, G. R., *Quiet Thermoforming Evolution Begins to Make Its Impact,* MP, pp. 50–56, Dec. 1988.
220. Singleton, R. W., *Electric Infrared: Textile Applications in the 1980's,* Proceedings, AATCC National Technical Conference, p. 201, 1980.
221. Hylton, D. C. and Cheng, C. Y., *Thermoforming Polypropylene on Conventional Equipment,* PE, pp. 55–57, Apr. 1988.
222. Galli, E., *Thermoforming CPET Containers,* PM&E, pp. 23–26, Apr. 1986.
223. Lodge, C., *Partnering Heats Up Cambell's Package Design With CPET,* PW, pp. 55–57, Dec. 1988.
224. Brinkman, R. D., *Coated vs. Uncoated Webs in Thermoforming/Fill/Seal,* Paper, Film and Foil Converter, pp. 78–82, Oct. 1984.
225. Bigg, D. M., et al., *High Strength Stampable Thermoplastic Composites,* PE, pp. 51–54, Mar. 1988.
226. Keim, H. J., *Thermoforming of Thermoplastic Blanks,* Kunststoffe, pp. 81–82, Oct. 1988.
227. Malpass, V. E. and Dean, A. F., *Estimating Thermoforming Behavior of Mineral Filled Polypropylenes,* PE, pp. 27–31, Jan. 1989.
228. *Focused IR Bonds Shapes TP's,* Tech. News, PT, pp. 32–41, Jan. 1989.
229. McConnell, W. K., Jr., *Industrial Thermoforming Symposium and Workshop,* Arlington, TX, SPE, Mar. 12–14, 1985.
230. Wright, R. E., *Thermoset Molding Manual,* Rogers Corp., Manchester, CT, 1984.
231. Hull, J. L., Molding Plastics correspondence and literature, 1989.
232. Lubin, G., *Handbook of Composites,* Van Nostrand Reinhold, 1982.
233. Gaylord, M. W., *Reinforced Plastics—Theory and Practice,* Cahners, 1974.
234. White, R. B., *Premix Molding,* Van Nostrand Reinhold, 1964.
235. Stuttgard, H. C., *Water Hydraulics in Plastics Processing,* Kunststoff, pp. 18–20, 1988.

236. Milewski, J. V. and Rosato, D. V., *History of Reinforced Plastics,* National ACS Meeting, Houston, TX, Mar. 23–28, 1980.
237. Taggert, D. G. and Pipes, R. B., *Processing Induced Fiber Orientation in Transfer and Injection Molding,* Univ. of Delaware, 1979.
238. Fitts, B. B., *The Influence of Processing on the Behavior of Glass Fiber Reinforced Plastics,* Rogers Corp., 1982.
239. Ishida, H., *Interfaces in Polymer, Ceramic and Metal Matrix Composites,* Elsevier, 1988.
240. Sanders, B., *Short Fiber Reinforced Composite Materials,* ASTM, STP 772, 1982.
241. *Composites Annual Conference Preprint Books,* SPI, Issued Yearly since 1944.
242. Rosato, D. V., *Nonwoven Fibers in Reinforced Plastics,* Industrial and Engineering Chemistry, Vol. 54, No. 8, pp. 30–37, Aug. 1962.
243. *Plastics for Aerospace Vehicles,* MIL-HDBK-17 & 17A, Supt. of Documents, U.S. Government Printing Office, 1981.
244. *Structural Sandwich Composites,* MIL-HDBK-23, Supt. of Documents, U.S. Government Printing Office, 1974.
245. Faupel, J. H., *Engineering Design: A Synthesis of Stress,* Wiley, 1981.
246. Pritchard, G., *Developments in Reinforced Plastics,* Elsevier, 1985.
247. McCormac, J. C., *Design of Reinforced Concrete,* Harper & Row, 1978.
248. Green, W. A. and Micunovic, M. V., *Mechanical Behavior of Composites and Laminates,* Elsevier, 1987.
249. Marshall, I. H. and Demuts, E., *Supportability of Composite Airframes,* Elsevier, 1988.
250. *Fiberglas Plus Design: A Comparison of Materials and Processes for Fiber Glass Composites,* Owens-Corning Fiberglas Corp., July 1985.
251. *Manufacturing Handbook and Buyers' Guide,* PT Annual, 1988/1989.
252. Jellinek, K., and Bollig, F. J., *The Future of Thermoset Molding Materials,* European Plastic News, pp. 26–30, Mar. 1985.
253. Marker, L. and Ford, B., *Rheology and Molding Characteristics of Glass Fiber Reinforced Sheet Molding Compounds,* SPI Composites Inst., Tech. Conf., 1977.
254. Miller, B., *Composite Structures: Next Wave in Detroit,* PW, pp. 33–34, Nov. 1986.
255. Rosato, D. V., *Filament Winding,* Wiley, 1964.
256. Rosato, D. V., *Reinforced Plastics Seminar,* Univ. of Lowell, 1986.
257. Redwitz, W. B., *State and Direction of Molding Process-Analysis of Thermoset Plastics,* Industrial and Production Engineering, pp. 117–121, Feb. 1984.
258. *Proceedings of the American Society for Composites, Third Tech. Conference,* Univ. of Washington and Boeing Co. Sponsors, Technomic, Sept. 1988.
259. *Resin Transfer Molding,* Owens-Corning Fiberglas Corp., 1981, 1988.
260. Hull, J. L., *Injection Molding Thermosets Seminar,* Hull Corp., 1986.
261. Han, C. D., et al., *Development of a Mathematical Model for the Pultrusion Process,* Polymer Engr. and Science, Wiley, Vol. 26, No. 6, 1986.
262. Nachtrab, W. C., *Design and Construction of Tooling for Sheet Molding Compound,* PM&E, pp. 43–47, Aug. 1986.
263. LaVerne, L., *Composite Cutting Comes of Age,* Advanced Composites, pp. 43–47, Sept./Oct. 1986.
264. Rosato, D. V., *Designing with Plastics Lectures,* Rhode Island School of Design, 1989.
265. *Materials Reference Guide, Machine Design,* Annual Reference Issue, Apr. 1989.
266. *Phenolic RTM News Release,* PM&E, p. 27, Jan. 1989.
267. Braun, H. J. and Stuttgart, P. E., *PUR-RIM and RRIM Technology: Advances Made and Profitability,* Kunststoffe, pp. 76–80, Oct. 1988.
268. Schlotterbeck, D. G., et al., *Polyurea/Amide Elastomers—A New RIM Generation Debuts,* PE, pp. 37–40, Jan. 1989.
269. *Corvette Advances Bumper Technology News,* Design News, p. 24, Jan. 1989.

270. Hasloch, R. K., *Processing of EPS Moldings and Blocks,* Kunststoffe, pp. 72–76, Nov. 1988.

271. Pirmasens, G. M., *Conveying, Mixing and Metering,* Kunststoffe, pp. 6–14, Oct. 1988.

272. Smoluk, G. R., *Now Meter-Mix-Dispense Is More Sophisticated,* MP, pp. 57–65, Jan. 1989.

273. Miller, B., *Styrene Emissions from Open Molds,* PW, p. 39, Jan. 1989.

274. Rogers, J. K., *Recycling Grows Up: Where It's Headed,* PT, pp. 50–56, Dec. 1988.

275. *Laser Ablation Promotes Adhesion in SMC Surfaces,* Search-Bulletin, GM Research Lab., Warren, MI, Vol. 23, No. 5, Nov./Dec. 1988.

276. *In-Mold Sulfonation, New Blends Expand Options For Auto Fuel Tanks,* PT News, pp. 25–29, Jan. 1989.

277. *Drying TP Polyurethanes and Other Hygroscopic Polymers,* DSG Report No. 20, D. S. Gilmore Lab., Upjohn, 1981.

278. Paulson, J., *Effective Means For Reducing Formation of Fines and Streamers in Air Conveying Systems,* SPE RETEC, Houston, TX, Feb. 1978.

279. Hoppe, H., *Pellet Cleaning by Air Separation,* Plastics Compounding, Jun. 1983.

280. Bozzelli, J. W., et al., *Moisture Content Analyzer Provides Critical Processing Control,* MP, pp. 80–82, Nov. 1988.

281. Kirkland, C., *Are Gravimetric Blenders Worth The Cost?,* PW, pp. 74–77, Nov. 1988.

282. *Gravimetrics Are In,* PT News, pp. 23–29, Sept. 1988.

283. *New Blender Weighs in to Heavy Gravimetric Competition,* PM&E, p. 9, Dec. 1988.

284. Galli, E., *In-Mold Coating,* PM&E, pp. 27–30, Oct. 1988.

285. Darmstadt, N. V., *Advances In The Electrostatic Flocking of Plastics Moldings,* Kunststoffe, pp. 89–92, Oct. 1988.

286. Hanau, G. B., *Vacuum Coating of Plastics,* Kunststoffe, pp. 6–7, Sept. 1988.

287. Heck, W., et al., *Painting of Plastics in the Outer Skin of a Car Body,* Kunststoffe, pp. 84–89, Oct. 1988.

288. Dominik, M., *Coating of Web-like Carrier Materials,* Kunststoffe, pp. 82–84, Oct. 1988.

289. McCarthy, L. R., *Here are Tips on Solving Common Moisture Problems,* PW Directory, pp. 313–317, 1989.

290. *Plastics Annual Standards,* Vol. 8.01 Test Methods, Vol. 8.02 Materials, Vol. 8.04, etc., ASTM, Philadelphia, PA, Annuals.

291. *Industry Standards,* Underwriters' Laboratories (UL), Northbrook, IL, USA.

292. Blair, J. A., *Codes and Standards Update,* Plastics Today, p. 20, Jan. 7, 1989.

293. Tobin, W. J., *Quality Control for Plastics,* T/C Press, Los Angeles, CA, 1986.

294. Parker, S. P., *Dictionary of Scientific and Technical Terms,* McGraw-Hill, 1989.

295. Blake, A., *Handbook of Mechanics; Materials and Structures,* Wiley, 1985.

296. Orberge, G. and Jones, F., *Machinery Handbook,* Industrial Press, 1981.

297. Baumeister and Marks, *Standard Handbook for Mechanical Engineers,* McGraw-Hill, 1978.

298. Williams, J. G., *Stress Analysis of Polymers,* Wiley, 1973.

299. Hsu, T. H., *Stress and Strain Handbook,* Gulf Publ., 1986.

300. Boyer, H. E., *Atlas of Stress–Strain Curves,* ASM, 1986.

301. Belding, W. E., *Handbook of Engineering Mathematics,* ASM, 1983.

302. Barton, L. O., *Mechanism Analysis: Simplified Graphics and Analytical Techniques,* Marcel Dekker, 1984.

303. Baumeister, T., *Mark's Standard Handbook of Mechanical Engineers,* McGraw-Hill, 1978.

304. Adams, R. D. and Wake, W. C., *Structural Adhesive Joints in Engineering,* Elsevier, 1984.

305. Boyer, H. E., *Atlas of Fatigue Curves,* ASM, 1986.

306. Altshuler, T. L., *Fatigue: Life Predictions for Materials Selection and Design,* ASM Software, 1988.
307. Hertzberg, R. W. and Manson, J. A., *Fatigue of Engineering Plastics,* Academic Press, 1980.
308. Hertzberg, R. W., *Deformation and Fracture Mechanics of Engineering Materials,* Wiley, 1976.
309. *Failure Analysis and Prevention,* ASM, 1986.
310. Schnabel, W., *Polymer Degradation; Principles and Applications,* Hanser, 1981.
311. Ferry, J. D., *Viscoelastic Properties of Polymers,* Wiley, 1970.
312. Smoluk, G. R., *Rheology—Study of Flow,* SPE Journal, Vol. 27, pp. 20–30, Dec. 1971.
313. Addleman, R. and Barrie, I. T., *How to Assess the Flow of Molten Polymers in Real Situations,* European Plastics News, Feb. 1988.
314. Heier, W. C., *Volume Change Indicator for Molding Plastic,* Report LAR-12280, Langley Research Center, Hampton, VA, 1976.
315. Rosato, D. V., *Testing and Quality Control Seminar, Rheology Without the Mathematics,* Univ. of Lowell, 1986.
316. Dealy, J. M., *Rheometrics for Molten Plastics,* Van Nostrand Reinhold, 1981.
317. Plastics World, *Monthly Publ.,* Newton, MA.
318. Plastics Technology, *Monthly Publ.,* New York, NY.
319. Plastics Engineering, *Monthly Publ.,* Brookfield Center, CT.
320. Plastics Machinery & Equipment, *Monthly Publ.,* Denver, CO.
321. Plastics Design & Processing, *Monthly Publ.,* Libertyville, IL.
322. Kunststoffe, *Monthly Publ.,* Munich, West Germany.
323. European Plastics News, *Monthly. Publ.,* London, England.
324. American National Standards Institute, New York, NY.
325. *Accident Prevention Manual for Industrial Operations,* National Safety Council, Chicago, IL, Annual.
326. *Facts and Figures of the U.S. Plastics Industry,* SPI, Annual.

APPENDIX

METRIC CONVERSION CHART (318)

To Convert . . . U.S. System	To . . . Metric System	Multiply by . . .	To Convert . . . Metric System	To . . . U.S. System	Multiply by . . .
		Density			
lb/in³	kg/m³	27 680	kg/m³	lb/in³	0.000 036
lb/ft³	g/cm³	0.0160	g/cm³	lb/ft³	62.43
lb/ft³	kg/m³	16.0185	kg/m³	lb/ft³	0.0624
lb/in³	g/cm³	27.68	g/cm³	lb/in³	0.036 13
		Temperature			
in/(in • °F)	m/(m • °C)	1.8	m/(m • °C)	in/(in • °F)	0.556
°F	°C	(°F −32)/(1.8)	°C	°F	1.8°C +32
°F	K	(°F +459.67)/(1.8)	K	°F	1.8K −459.67
		Pressure			
psi	kPa	6.8948	kPa	psi	0.145
psi	MPa	0.006 89	MPa	psi	145
psi	GPa	0.000 0068 9	GPa	psi	145 038
psi	bar	0.0689	bar	psi	14.51
		Energy and Power			
ft • lbf	J	1.3558	J	ft • lbf	0.7376
in • lbf	J	0.113	J	in • lbf	8.850
ft • lbf/inch	J/m	53.4	J/m	ft • lbf/inch	0.0187
ft • lbf/inch	J/cm	0.534	J/cm	ft • lbf/inch	1.87
ft • lbf/in²	kJ/m²	2.103	kJ/m²	ft • lbf/in²	0.4755
kW	metric horsepower	1.3596	metric horsepower	kW	0.7355
U.S. horsepower	kW	0.7457	kW	U.S. horsepower	1.3419
Btu″	J	1055.1	J	Btu″	0.000 95
Btu″	W • h	0.2931	W • h	Btu″	3.412
Btu″ • in/ (h • ft² • °F)	W/(m • K)	0.1442	W/(m • K)	Btu″ • in/ (h • ft² • °F)	6.933
Btu″/lb	kJ/kg	2.326	kJ/kg	BTU″/lb	0.4299
Btu″/(lb • °F)	J/(kg • °C)	4187	J/(kg • °C)	Btu″/(lb °F)	0.000 239
V/mil	MV/m	0.0394	MV/m	V/mil	25.4
		Output			
lb/min	g/s	7.560	g/s	lb/min	0.1323
lb/h	kg/h	0.4536	kg/h	lb/h	2.2046
		Velocity			
in/min	cm/s	0.0423	cm/s	in/min	23.6220
ft/s	m/s	0.3048	m/s	ft/s	3.2808
		Viscosity			
poise	Pa • s	0.1	Pa • s	poise	10

METRIC CONVERSION CHART (318)

To Convert . . . U.S. System	To . . . Metric System	Multiply by . . .	To Convert . . . Metric System	To . . . U.S. System	Multiply by . . .
Length					
mil	millimeter	0.0254	millimeter	mil	39.37
inch	millimeter	25.4	millimeter	inch	0.0394
inch	centimeter	2.54	centimeter	inch	0.3937
foot	centimeter	30.48	centimeter	foot	0.0328
foot	meter	0.3048	meter	foot	3.2808
yard	meter	0.9144	meter	yard	1.0936
Area					
inch^2	millimeter^2	645.16	millimeter^2	inch^2	0.0016
inch^2	centimeter^2	6.4516	centimeter^2	inch^2	0.155
foot^2	centimeter^2	929.03	centimeter^2	foot^2	0.0011
foot^2	meter^2	0.0929	meter^2	foot^2	10.7639
yard^2	meter^2	0.8361	meter^2	yard^2	1.1960
Volume, Capacity					
inch^3	centimeter^3	16.3871	centimeter^3	inch^3	0.061
fluid ounce	centimeter^3	29.5735^3	centimeter^3	fluid ounce	0.0338
quart (liquid)	decimeter^3 (liter)	0.9464	decimeter^3 (liter)	quart (liquid)	1.0567
gallon (U.S.)	decimeter^3 (liter)	3.7854	decimeter^3 (liter)	gallon (U.S.)	0.2642
gallon (U.S.)	meter^3	0.0038	meter^3	gallon (U.S.)	264.17
foot^3	decimeter^3	28.3169	decimeter^3	foot^3	0.0353
foot^3	meter^3	0.0283	meter^3	foot^3	35.3147
yard^3	meter^3	0.7646	meter^3	yard^3	1.3079
in^3/lb	m^3/kg	0.000 036	m^3/kg	in^3/lb	27 680
ft^3/lb	m^3/kg	0.0624	m^3/kg	ft^3/lb	16.018
Mass					
ounce (avdp.)	gram	28.3495	gram	ounce (avdp.)	0.03527
pound	gram	453.5924	gram	pound	0.0022
pound	kilogram	0.4536	kilogram	pound	2.2046
pound	metric ton	0.000 45	metric ton	pound	2204.6
U.S. ton (short)	metric ton	0.9072	metric ton	U.S. ton (short)	1.1023
Force					
lbf	N	4.448	N	lbf	0.225

Acknowledgement: Data checked by, and according to style confirmed by National Bureau of Standards' Metric Information Office.

Standard Metric Symbols				Metric Prefixes[b]		
				Numerical Value	Term	Symbol
A	ampere	kg	kilogram	10	deka	da
bar	bar	L	liter	10^2	hecto	h
cd	candela	m	meter	10^3	kilo	k
C	celsius"	N	newton	10^6	mega	M
g	gram	Pa	pascal	10^9	giga	G
h	hour	S	siemens	10^{12}	tera	T
Hz	hertz	s	second	10^{-1}	deci	d
J	joule	t	metric ton	10^{-2}	centi	c
K	kelvin	V	volt	10^{-3}	milli	m
"Formerly called		W	watt	10^{-6}	micro	μ
Centigrade				10^{-9}	nano	n
				10^{-12}	pico	p

[b]These prefixes may be used with all metric units

MATHEMATICAL SYMBOLS AND ABBREVIATIONS

$+$	plus (addition)	a', a''	a-prime, a-second
$-$	minus (subtraction)	a_1, a_2	a-sub one, a-sub two
$\pm \quad \mp$	plus or minus, (minus or plus)		
x	times, by (multiplication)	$(\,),[\,],\{\,\}$	parentheses, brackets, braces
$\div, /$	divided by		
:	is to (ratio)	$\angle \perp$	angle, perpendicular to
: :	equals, as, so is	a^2, a^3	a-square, a-cube
$\therefore$	therefore	a^{-1}, a^{-2}	$\dfrac{1}{a}, \dfrac{1}{a^2}$
$=$	equals		
$\sim \approx$	approximately equals	$\sin^{-1}a$	the angle whose sine is
$>$	greater than	π	pi $= 3.141593+$
$<$	less than	μ	microns $= .001$ millimeter
$\geqq$	greater than or equals	$m\mu$	micromillimeter$=.000001$ mm.
$\leqq$	less than or equals	Σ	summation of
$\neq$	not equal to	ε, e	base of hyperbolic, natural or
$\doteq$	approaches		Napierian logs $= 2.71828+$
$\propto$	varies as	Δ	difference
∞	infinity	g	acceleration due to gravity
$\parallel$	parallel to		(32.16 feet/sec. per sec.)
$\sqrt{}, \sqrt[3]{}$	square root, cube root	E	coefficient of elasticity
$\square$	square	v	velocity
$\bigcirc$	circle	f	coefficient of friction
$\circ$	degrees (arc or thermometer)	P	pressure of load
$'$	minutes or feet	HP	horsepower
$''$	seconds or inches	RPM	revolutions per minute

GREEK ALPHABET

A, α	Alpha	H, η	Eta	N, ν	Nu	T, τ	Tau
B, β	Beta	Θ, ϑ	Theta	Ξ, ξ	Xi	Y, υ	Upsilon
Γ, γ	Gamma	I, ι	Iota	O, o	Omicron	Φ, φ	Phi
Δ, δ	Delta	K, $\varkappa$	Kappa	Π, π	Pi	X, χ	Chi
E, ε	Epsilon	Λ, λ	Lambda	P, ϱ	Rho	Ψ, ψ	Psi
Z, ζ	Zeta	M, μ	Mu	Σ, σ, ς	Sigma	Ω, ω	Omega

INDEX

INDEX